AF598118

METHODS IN MOLECULAR BIOLOGY™

Series Editor
John M. Walker
School of Life Sciences
University of Hertfordshire
Hatfield, Hertfordshire, AL10 9AB, UK

For other titles published in this series, go to
www.springer.com/series/7651

Signal Transduction Immunohistochemistry

Methods and Protocols

Edited by

Alexander E. Kalyuzhny

R&D Systems, Inc., Minneapolis, MN, USA

Editor
Alexander E. Kalyuzhny, Ph.D.
R & D Systems, Inc.
Minneapolis, MN
USA
Alex.Kalyuzhny@rndsystems.com

ISSN 1064-3745 e-ISSN 1940-6029
ISBN 978-1-61779-023-2 e-ISBN 978-1-61779-024-9
DOI 10.1007/978-1-61779-024-9
Springer New York Dordrecht Heidelberg London

Library of Congress Control Number: 2011921260

Printed on acid-free paper

Humana Press is part of Springer Science+Business Media (www.springer.com)

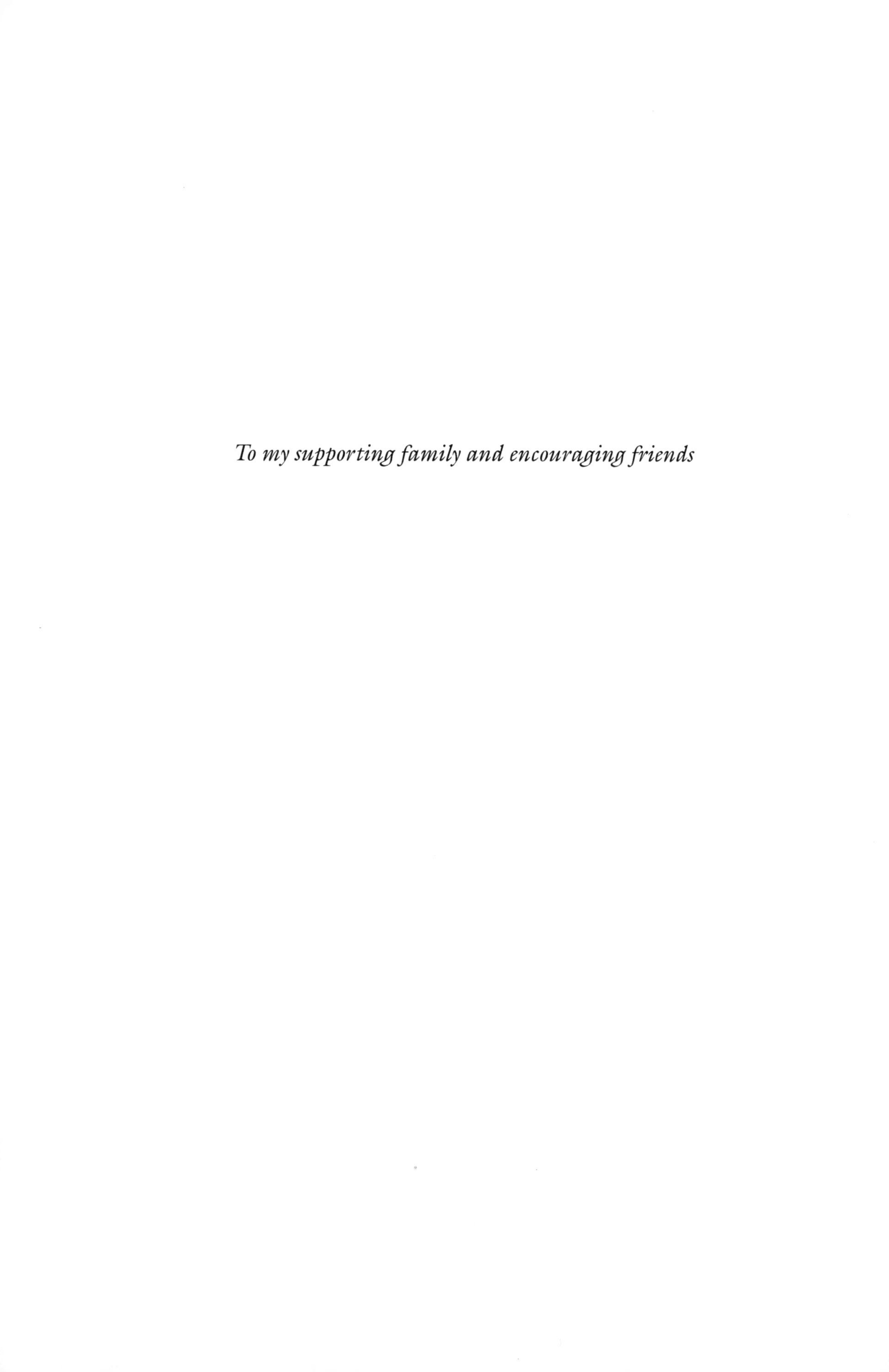

To my supporting family and encouraging friends

Preface

Immunohistochemistry (IHC) is one the most valuable research and diagnostic tools in biomedical research. Unlike detecting constitutively expressed targets, immunohistochemical detection of labile, low abundance, and short-lived signal transduction molecules appears to be a very challenging task. This book represents a set of detailed protocols written by IHC experts addressing the challenges of signal transduction immunohistochemistry (ST-IHC); because it would be fair to say that ST-IHC as a discipline is in its infancy and the chapters in the first part are of a more introductory nature which should help new investigators in their orientation in the field. The second part is dedicated to techniques used for the preservation of antigens and their unmasking. The third part presents protocols in digital imaging and image analysis of stained cells and tissues and high-throughput data collection and data analysis. The fourth part is focused on ST-IHC techniques used in neuroscience as well as cancer and stem cell research. And finally, the fifth part presents novel ST-IHC techniques that can be easily adopted for a wide variety of research tasks. This book can be used as a guide by novices and has a wealth of ideas that can be exploited by experienced researchers who are always on the lookout for new experimental tricks and hints. It can also serve as a troubleshooting guide for researchers in academia and in industry.

I wish to thank all the authors who, in addition to their own research projects, devoted a lot of time working on book chapters. In addition, I wish to thank R&D Systems, Inc., where I run the IHC department, for their support and for giving me the opportunity to gain invaluable IHC experience by validating thousands of antibodies over the years.

Minneapolis, MN *Alexander E. Kalyuzhny*

Contents

Contributors

AMY J. ARCHULETA • *PhosphoSolutions LLC, Aurora, CO, USA*
JAYANT AVVA • *Case Western Reserve University, Cleveland, OH, USA*
CHRISTIANNE BANDEIRA-MELO • *Laboratório de Inflamação, Instituto de Biofísica Carlos Chagas Filho, Universidade Federal do Rio de Janeiro, Rio de Janeiro, RJ, Brazil*
DEBABRATA BANERJEE • *Department of Medicine and Pharmacology, The Cancer Institute of New Jersey, Robert Wood Johnson Medical School, UMDNJ, New Brunswick, NJ, USA*
AFSAR BARLAS • *Developmental Biology Program, Molecular Cytology Core Facility, Memorial Sloan-Kettering Cancer Center, New York, NY, USA*
JURAJ BODO • *Department of Clinical Pathology, Cleveland Clinic, Cleveland, OH, USA*
MATS BORÉN • *Denator AB, Dag Hammarskjöldsv. 10A, Uppsala, Sweden*
PATRICIA T. BOZZA • *Laboratório de Imunofarmacologia, Instituto Oswaldo Cruz, Fundação Oswaldo Cruz, Rio de Janeiro, RJ, Brazil*
MICHAEL D. BROWNING • *PhosphoSolutions LLC, Aurora, CO, USA*
KATHY BRUMBAUGH • *R&D Systems, Inc, Minneapolis, MN, USA*
ROBERTO CAMPOS-GONZALEZ • *R&D Systems, Inc, Minneapolis, MN, USA*
CHARLES CHAVKIN • *Department of Pharmacology and Program for Neurobiology and Behavior, University of Washington, Seattle, WA, USA*
DANIEL CIZNADIJA • *Department of Molecular Biology, Memorial Sloan-Kettering Cancer Center, New York, NY, USA*
MICHAEL B. COHEN • *Department of Pathology, The University of Iowa, Iowa City, IA, USA*
JEFF COOPER • *R&D Systems, Inc, Minneapolis, MN, USA*
SKYLER DILLON • *Arthritis and Immunology Program, Oklahoma Medical Research Foundation, Oklahoma City, OK, USA; Department of Medicine, University of Oklahoma Health Sciences Center, Oklahoma City, OK, USA*
YASER DORRI • *Arthritis and Immunology Program, Oklahoma Medical Research Foundation, Oklahoma City, OK, USA; Department of Medicine, University of Oklahoma Health Sciences Center, Oklahoma City, OK, USA*
ANIL DSOUZA • *Arthritis and Immunology Program, Oklahoma Medical Research Foundation, Oklahoma City, OK, USA; Department of Medicine, University of Oklahoma Health Sciences Center, Oklahoma City, OK, USA*
NING FAN • *Developmental Biology Program, Molecular Cytology Core Facility, Memorial Sloan-Kettering Cancer Center, New York, NY, USA*
NIELS T. FOGED • *Visiopharm A/S, Hoersholm, Denmark*
JILLIAN FRISCH • *R&D Systems, Inc, Minneapolis, MN, USA*

SHO FUJISAWA • *Developmental Biology Program, Molecular Cytology Core Facility, Memorial Sloan-Kettering Cancer Center, New York, NY, USA*
MICHAEL GRAHEK • *R&D Systems, Inc, Minneapolis, MN, USA*
MICHAEL GRUNKIN • *Visiopharm A/S, Hoersholm, Denmark*
JODI HAGEN • *R&D Systems, Inc, Minneapolis, MN, USA*
YINGWEI HE • *Abgent, Inc, San Diego CA, USA*
J. P. HOUCHINS • *R&D Systems, Inc, Minneapolis, MN, USA*
ERIC D. HSI • *Department of Clinical Pathology, Cleveland Clinic, Cleveland, OH, USA*
JAMES W. JACOBBERGER • *Case Western Reserve University, Cleveland, OH, USA*
WADE JOHNSON • *R&D Systems, Inc, Minneapolis, MN, USA*
ALEXANDER E. KALYUZHNY • *R&D Systems, Inc, Minneapolis, MN, USA*
BIJI T. KURIEN • *Arthritis and Immunology Program, Oklahoma Medical Research Foundation, Oklahoma City, OK, USA; Department of Medicine, University of Oklahoma Health Sciences Center, Oklahoma City, OK, USA*
JULIA C. LEMOS • *Department of Pharmacology and Program for Neurobiology and Behavior, University of Washington, Seattle, WA, USA*
WEN-CHIEH LIAO • *R&D Systems, Inc, Minneapolis, MN, USA*
MONG-SHANG LIN • *R&D Systems, Inc, Minneapolis, MN, USA*
VERNON C. MAINO • *BD Biosciences Immunocytometry Systems, San Jose, CA, USA*
KATIA MANOVA • *Developmental Biology Program, Molecular Cytology Core Facility, Memorial Sloan-Kettering Cancer Center, New York, NY, USA*
LEOPOLDO MENDOZA • *Solulink, Inc, San Diego, CA, USA*
PRAVIN J. MISHRA • *Department of Medicine and Pharmacology, The Cancer Institute of New Jersey, Robert Wood Johnson Medical School, UMDNJ, New Brunswick, NJ, USA*
JENNIFER NGUYEN • *Solulink, Inc, San Diego, CA, USA*
KRISTIN M. NIXON • *PhosphoSolutions LLC, Aurora, CO, USA*
JAKOB RAUNDAHL • *Visiopharm A/S, Hoersholm, Denmark*
CLARISSE A. ROTH • *Department of Pharmacology, University of Washington, Seattle, WA, USA*
JORDAN SCHOEPHOERSTER • *R&D Systems, Inc, Minneapolis, MN, USA*
DAVID SCHWARTZ • *Solulink, Inc, San Diego, CA, USA*
R. HAL SCOFIELD • *Arthritis and Immunology Program, Oklahoma Medical Research Foundation, Oklahoma City, OK, USA; Department of Medicine, University of Oklahoma Health Sciences Center and Veterans Affairs Medical Center, Oklahoma City, OK, USA*
JERRY SEDGEWICK • *Sedgewick Initiatives, Saint Paul, MN, USA*
RADINA P. SOEBIYANTO • *Case Western Reserve University, Cleveland, OH, USA*
SREE N. SREENATH • *Case Western Reserve University, Cleveland, OH, USA*
STEVEN STOESZ • *R&D Systems, Inc, Minneapolis, MN, USA*
CRYSTAL A. STUTZKE • *PhosphoSolutions LLC, Aurora, CO, USA*
MARIA A. SUNI • *BD Biosciences Immunocytometry Systems, San Jose, CA, USA*
JOSEPH SWEET • *R&D Systems, Inc, Minneapolis, MN, USA*
SERGEI I. SYRBU • *Immunopathology Laboratory, Department of Pathology, The University of Iowa, Iowa City, IA, USA*

MESRUH TURKEKUL • *Developmental Biology Program, Molecular Cytology Core Facility, Memorial Sloan-Kettering Cancer Center, New York, NY, USA*
WEI WANG • *Abgent, Inc, San Diego, CA, USA*
MICHAEL C. WEIS • *Case Western Reserve University, Cleveland, OH, USA*
PETER F. WELLER • *Department of Medicine, Beth Israel Deaconess Medical Center, Harvard Medical School, Boston, MA, USA*
CHUN WU • *Abgent, Inc, San Diego, CA, USA*

Part I

Antibodies as a Tool: From Concept to Design and Application

Chapter 1

Overview of the Generation, Validation, and Application of Phosphosite-Specific Antibodies

Kathy Brumbaugh, Wade Johnson, Wen-Chieh Liao, Mong-Shang Lin, J.P. Houchins, Jeff Cooper, Steven Stoesz, and Roberto Campos-Gonzalez

Abstract

Protein phosphorylation is a universal key posttranslational modification that affects the activity and other properties of intracellular proteins. Phosphosite-specific antibodies can be produced as polyclonals or monoclonals in different animal species, and each approach offers its own benefits and disadvantages. The validation of phosphosite-specific antibodies requires multiple techniques and tactics to demonstrate their specificity. These antibodies can be used in arrays, flow cytometry, and imaging platforms. The specificity of phosphosite-specific antibodies is key for their use in proteomics and profiling of disease.

Key words: Antibody, Phosphosite-specific, Western blotting, ELISA, Multiplex, Flow cytometry, Immunocytochemistry

1. Introduction

Protein phosphorylation, like many other posttranslational modifications, introduces changes in mass and charge to an acceptor protein. This change alters the conformation of the acceptor protein, as well as its activity, binding properties, and subcellular distribution. Phosphorylation at key amino acids within a protein is considered a hallmark of the change in the protein's activity. Because of the rapid and reversible protein changes induced by phosphorylation, eukaryotic cells have preserved this modification and it has evolved as a tightly controlled regulator of key cellular processes, such as cell division, motility, neurotransmission, and metabolism. In eukaryotic cells, reversible protein

Alexander E. Kalyuzhny (ed.), *Signal Transduction Immunohistochemistry: Methods and Protocols*, Methods in Molecular Biology, vol. 717, DOI 10.1007/978-1-61779-024-9_1,

phosphorylation occurs primarily on serine, threonine, and tyrosine amino acids (1). In addition, dysregulated protein phosphorylation has been closely associated with several diseases, including cancer (2).

The phosphorylation status of a protein is due to the balanced activities between a protein kinase that transfers a phosphate from ATP to its target polypeptide, and a phosphatase that removes it from the polypeptide; thus, many phosphorylations are transient by nature (3). There are ~520 different protein kinases, the "kinome," in the human genome that are responsible for most cellular phosphorylations. Kinases have a degree of specificity and selectivity for their target proteins based on recognition and substrate-binding domains within their amino acid sequence (2). Some kinases, like MEK1, are very selective and may have only two protein substrates, ERK1 and ERK2, while other kinases, such as Akt1, are capable of recognizing and phosphorylating multiple protein substrates. On the other hand, there are approximately 150 phosphatases in the human genome (4). Thus, phosphatases appear not to be as selective as their kinase counterparts in choosing a protein substrate.

Approximately 30 years ago, the preferred and most widely used method to investigate protein phosphorylation was labeling cells and proteins with ^{32}P. Radioactive labeling of phosphoproteins was used to determine if a protein contained phosphate, to elucidate the type of phospho-amino acid, and to identify protein substrates and their corresponding kinases (5). This radioisotope as ^{32}P-ATP was used to label cells and proteins followed by lysis and immunoprecipitation, if required, electrophoretic separation and autoradiography of gels. Once the bands of interest were identified, they were excised and digested with enzymes like trypsin, followed by two-dimensional mapping and sequencing (6). While radio-labeling of proteins with ^{32}P is, without a doubt, one of the most sensitive ways to assess phosphorylation, these associated methods are remarkably cumbersome and stressful. The advent of the first successful antibodies to phospho-tyrosine (pTyr) and subsequent phosphosite-specific antibodies facilitated the study of phosphorylation and rapidly accelerated the study of this posttranslational modification in cellular events (7–10).

The initial pTyr antibodies were rapidly adopted by scientists and used to discover many phosphorylations that had not been seen before, e.g., after stimulation of cells with growth factors or oncogene activation. The same antibodies were used to further isolate and purify these novel phospho-proteins and to develop tools for their study. Among the proteins that were discovered by this immunopurification protocol are Insulin Receptor Substrate-1 (11), Caveolin (12), and pp120 Catenin (13). Because of the combination of the newly developed reagents and Western blotting, it became possible to generate a phospho-protein profile from cells under many different conditions at a pace several-fold faster than with ^{32}P-labeling (14).

One of the major challenges is to obtain highly specific and sensitive antibodies capable of capturing intracellular phosphorylation events of low frequency or abundance, because of the rarity and transient nature of phosphorylations (15). Although with exceptions, on activation of a signaling pathway, only a small fraction (<10%) of a typical protein kinase target becomes phosphorylated with a defined kinetics of few minutes followed by de-phosphorylation by phosphatases or degradation terminating the signaling event (4, 16, 17). These phospho-protein's traits are useful when determining the specificity of an antibody but at the same time some of these properties pose serious limitations, therefore requiring antibodies of high affinity to fit in with methods of superior sensitivity.

According to Phosphosite (http://www.phosphosite.org), perhaps one of the most comprehensive database of phosphorylations today, there are more than 81,000 unique phosphorylations described in mammalian cells. Given a conservative average of three phosphorylations per protein, it is estimated that there are many more new phosphorylations in proteins yet to be discovered. Clearly, the scientific community still does not have phosphosite-specific antibodies to all of the already discovered sites to study and evaluate their individual importance and role in biomedicine.

Although currently there are several other very sensitive techniques, like mass spectrometry, to aid in the study of protein phosphorylation (18), antibodies will still provide a rapid, economical, and adaptable avenue to study these important modifications in years to come. As new phosphorylations are being discovered and validated by mass spectrometry or mutagenesis, phosphosite-specific antibodies will remain ideal for the everyday experiment to study the phosphorylation under a myriad of conditions, cells, and tissues.

There are two major issues in the world of antibodies determining their usefulness, one of them is specificity and the second is the applicability of the antibodies in different platforms or instruments. With this in mind, we have two main goals in writing this short review. First, to summarize the major techniques utilized in characterizing and validating phosphosite-specific antibodies. We believe that these techniques provide a minimum set of tools to determine these important reagents. The second goal is to illustrate some of the exciting uses for phosphosite-specific antibodies in biomedicine as powerful tools in elucidating the biology of normal and diseased cells. We should keep in mind that the strategies and methods described below can also be applied and used when characterizing antibodies to other posttranslational modifications such as acetylation and methylation among others.

2. Materials

2.1. Antibody Generation

There are two main routes to obtain antibody-based affinity reagents for biomedical research, one by the in vivo immunization of individual animals inducing B cells to secrete the antibodies, and the second by phage display of antibody libraries and screening for binders to the desired phospho-epitope. The former is the most widely used method and requires the active introduction of a foreign biological compound, either as a hapten-carrier conjugate or as a protein, to elicit an immune response in an animal, mostly mice, rats, chickens, and guinea pigs whose immune response is well understood. The latter method requires the availability of a library of arranged genes packed and individually displayed on a filamentous phage. The phage library carrying and expressing a large immunoglobulin gene repertoire is exposed to the antigen of interest, followed by cycles of enrichment and selection until a single "phage clone" is obtained. One of the main limitations to the phage display approach is the low affinity of the binders generated, limiting their use as everyday reagents. Since there are just a handful of published reports in generating antibodies to posttranslational modifications of the phage approach, it will not be covered in this review (19).

The serial immunization of animals takes advantage of the natural ability of the immune system to recognize an injected protein "immunogen," whether bacteria, cell debris, or a phosphopeptide, and to rapidly re-arrange the immunoglobulin genes to produce antibodies until the best fit for the immunogen is found. The immune system re-arranges the heavy and light chains of antibodies accordingly to provide the best antibody response to fight the foreign intruder. Scientists have been taking advantage of the immune system in animals such as rabbits, mice, and chickens, to generate superb antibodies for use in biomedical research, for decades.

Protocols for generation of phosphosite-specific antibodies, both polyclonal and monoclonal, adhere very closely to the ones used for a regular antibody and superbly summarized in the classic book by Harlow and Lane (20). Phosphosite-specific antibodies generated in vivo have been available for several years, and basically these reagents have been obtained by three main protocols: first by immunization with a phospho-peptide (8, 21, 22); second by using cells or mixed phospho-protein complexes (23); and third by fortuitous discovery and diligent characterization (24). In their pioneering publication, Sternberger and Sternberger superbly characterized a series of monoclonal antibodies to neurofilament proteins that reacted only when these proteins were phosphorylated (24). The Sternberger's antibodies have since become a standard for the study of neurodegenerative diseases.

Potentially, there is additional strategy to generate phosphosite-specific antibodies by using a large phosphorylated protein as the immunogen that is seldom used. If the immunogen is a large phospho-protein or a mixture of phosphorylated proteins, the host's immune system might make the many different antibodies based on the immunogenicity of the available epitopes. This may require the availability of proteins or peptides lacking the phosphorylation site of interest and used in a subtractive selection to isolate the antibodies of interest. The use of recombinant or natural phospho-proteins as antigens in theory may be advantageous since we can envision these immunogen-eliciting antibodies to a protein closely resembling its natural state.

2.1.1. Antibodies to Phospho-Amino Acids

The forerunners of phosphosite-specific antibodies were the antibodies raised to individual phospho-amino acids. The typical approaches used to generate both polyclonal and monoclonal antibodies to phosphorylated serine, threonine, or tyrosine requires coupling of the amino acid, or an analog, to a carrier protein and using this complex as the immunogen. Alternatively, there are successful and commercially available antibodies to anti-pTyr of the use of a mixture of phosphorylated proteins as immunogens. Many publications and protocols are already available to generate antibodies to phospho-amino acids and we will not elaborate these methods further (7, 9, 25, 26). These antibodies recognize the phosphorylated amino acid regardless of the surrounding sequence, and have been very useful tools in studying cell activation and dissecting signaling mechanisms. Among the main three phospho-amino acids, pTyr is the most immunogenic, perhaps due to its bulkiness and salient exposure within proteins.

2.1.2. Phosphosite-Specific Antibodies

In contrast to the antibodies mentioned above that recognize either pSer, pThr, or pTyr regardless of the context or position within a protein, phosphosite-specific antibodies recognize the phospho-amino acid but within the context of the amino acids surrounding the phosphorylation site. In other words, the binding of any given phosphosite-specific antibody encompasses stretches upstream and downstream of the phosphorylation site. The value of these reagents is that they often discriminate single amino acid differences that are found between the phospho-sites among proteins. This is a very important feature since key auto-phosphorylations are highly conserved, in their amino acid sequence, among species and protein homologues.

There are two major limiting factors to be considered when planning to develop antibodies to unique phosphosites in a protein. First, the nature and characterization of the immunogen needs to be decided according to the goals, and generated in sufficient amounts for immunization and screening. For instance, the selection of a phosphorylation site within the target protein will require an analysis of potential homologies and production feasibility (peptide synthesis or recombinant phospho-protein). Very often, data interpretation is confounded if the targeted phosphorylation site is similar to or conserved among several known or unknown proteins. The second limiting factor is to have a well-characterized and suitable cellular validation strategy in place.

Using short phospho-peptides as immunogens, or surrogates, to generate antibodies to protein phosphorylation sites allows us to "direct" the immune response toward the phosphosite of interest. However, this strategy also comes with some disadvantages. First, peptides are not very immunogenic in their structure and second, the peptide's conformation is probably far from the natural state of a cellular protein (27). The widespread phospho-peptide approach was made possible, thanks to the simplified synthesis of these reagents. In the first published report, Greengard's group at Yale University used a synthetic peptide phosphorylated in vitro with a GMP-dependent protein kinase followed by purification and conjugation to KLH and then using this conjugate as immunogen in rabbits (8). This approach has been simplified by synthesizing phospho-peptides with the phosphoamino acid incorporated during the peptide synthesis (28). Typically, the desired phosphorylated amino acid is placed in the middle of a 10–14 amino acids-long peptide to allow enough exposure of the phosphorylation, and at the same time restricting antibodies to the nonphosphorylated portions of the peptide sequence (8, 21, 22, 28). Homologies with other phosphorylated proteins may limit, or extend, the length of the phospho-peptide to avoid potential cross-reactivities with other unrelated phospho proteins.

Nonetheless, sometimes antibodies cross-reactive to proteins with similar sequences surrounding the phosphorylation site are desired, and are the primary objective. In these cases, the phospho-peptides immunogens are designed with phospho-motifs that are common among protein families, or protein substrates and protein isoforms. Some of these anti-phospho-motif antibodies are commercially available and have been useful for the identification and immunopurification of multiple protein substrates to a particular kinase. For instance, antibodies specific for MAP kinase protein substrates containing the motif PXpS.pTP (29); or those proteins containing the phosphorylation site for ATM and ATR DpS/pTQ (30).

The quality and format of the phospho-peptide becomes a limiting factor during the immunization. It is always advisable to consult your peptide supplier to provide you with quality assurance of the phospho-peptide. Depending on the sequence surrounding the phospho-amino acid, there may be limitations, poor yields, and difficulties in its synthesis. Furthermore, because of their small size, peptides are poor immunogens and need to be either conjugated to a large protein carrier such as keyhole lympet hemocyanine, bovine serum albumin, ovalbumin (20), or synthesized with a multilysine scaffold to increase their molecular weight and antigenicity (31).

Defining a validation strategy in a cellular environment sets the parameters for defining a positive or a negative phosphosite-specific antibody for its intended use. The validation strategy requires that the availability of the cells or tissues expressing the phospho-protein, the ligand needed for cell activation, and the phosphorylation kinetics each be very well defined. All these parameters need to be well understood with the methodology desired for validation, such as Western blotting, flow cytometry, or immunocytochemistry, and bioassay among others.

Both polyclonal and monoclonal approaches have merits and disadvantages. Historically, most polyclonal phosphosite-specific antibodies have been developed in rabbits due to the lower cost to develop and short time needed to obtain them –typically within 60 days. Rabbits respond quite well to phospho-peptides conjugated to carriers and generate a robust immune response, making these rodents the preferred host. Because of the polyclonal nature of the antibodies binding to different areas of the phospho-peptide, these reagents tend to have a higher avidity than monoclonal antibodies. Rabbit sera require affinity purification through cycles of nonphospho and phospho-peptides to remove undesirable immunoglobulins and other proteins. If the polyclonal sera from rabbits immunized with phospho-peptides give a reasonable titer by ELISA and specificity by Western blotting, the IgG within the sera can be purified to enrich for the desired phosphosite-specific antibodies. It is advisable to purify the sera first through a Protein A column to enrich for the IgG fraction followed by affinity chromatography with the corresponding phospho-peptide. Sometimes, it is advisable to perform an extra affinity purification using the nonphosphorylated peptide to remove any IgG directed to the nonphospho-peptide.

The elution steps from the affinity purification are done with strong acids or bases, sometimes damaging the eluted antibodies. In spite of the low yields of phosphosite-specific antibodies, rabbits can produce antibodies of very high titer and specificity for at least a couple of years. Additionally, there is substantial variability in the immune response among rabbits. In other words, it is often not possible to find two rabbits yielding antibody of similar yield and affinity, thus requiring several rabbits to sustain and produce enough amounts of antibodies with similar affinities for large-scale studies. Polyclonal phosphosite-specific antibodies derived from rabbits are expected to have differences in affinity and specificity from batch to batch. Once the positive rabbits have been identified, you can proceed to a regular boosting and bleeding schedule, if polyclonals, or fusion and selection of stable hybridomas (32). Alternatively, the immunoglobulin genes can be obtained from the antibody-producing B cells and the IgG's expressed in a defined cellular background (32a). The decision of whether to pursue a polyclonal or a monoclonal phosphosite-specific antibody will be also determined by its final intended use(s).

Although phosphorylated residues are often found in surface-exposed areas of a protein, these regions are frequently not good inducers of a robust immune response. Here, the polyclonal antibody route offers an advantage over monoclonal antibodies because the polyclonal antibody mixture will contribute to the final avidity and binding towards its intended target.

One of the main advantages of monoclonal antibodies is their consistency, reproducibility, and the fact that they do not require harsh purification steps that could potentially damage the antigen-binding site of the antibody. On the other hand, the higher cost of monoclonal production definitively favors polyclonal antibody development. However, once you have a high-affinity and specific monoclonal antibody, there is no need to make it again. Furthermore, monoclonal antibodies are usually the preferred reagent for instrumentation in clinical research where reproducibility, consistency, and accuracy are absolute requirements.

Monoclonal antibody technology is widely available and supported with published methodologies that have been used and validated with many years of research (20). It involves the collection of immune B cells from the positive immunized animal and fusing these with a stable and compatible myeloma cell to produce an immortalized hybrid expressing the IgGs of the parental B cell. As with any other immunogen, a key for a successful fusion will depend on the screening strategy.

The typical animals used for monoclonal antibody production are mice, rats, and more recently rabbits (20, 32). The immunizations need to be done according to the local safety and animal care guidelines. It is advisable to immunize from two to five animals and to follow their immune response by standard techniques such as ELISA and Western blot. These assays will monitor the progression of the animals' immune response, titer, and will help one to select the best responding individual. The prefusion assays should be the same as the postfusion assays used to identify and select the hybridomas.

Once a monoclonal antibody and its corresponding secreting cell have been identified, it is possible to isolate the corresponding gene to the antibody by routine molecular biology techniques. Expression of an antibody as a recombinant protein may open the possibilities for improving its expression and modifying the affinity towards its targets. The recombinant antibody can be expressed in bacteria or other suitable cells. This has been done for several anti-phosphotyrosine antibodies by RT-PCR using the mRNA from the hybridomas (33–35) and by expressing the antibody to phospho-ERK as a Fab in a phage display library (36).

2.1.3. Hybridoma Selection

When selecting for phosphosite-specific antibodies, it is advisable to do an initial ELISA differential screen with the phosphorylated immunogen versus its nonphosphorylated version on the ELISA plates. This, in theory, will ensure the selection of hybridomas recognizing only the phosphorylated protein of interest. Ideally, if the immunogen was a phospho-peptide, then the ELISA screen should be done with the full-length phosphorylated protein to avoid antibodies that only recognize the phospho-peptide. This initial ELISA will also help one to reduce the number of potential hybridomas for further evaluation. At this early stage, hybridomas tend to be relatively unstable due to competition with multiple clones, adaptation to the selection media, and chromosomal instability, etc. Thus, it is important to gain information about the specificity of these hybridomas within a cellular context as quickly as possible to avoid losing valuable hybridomas.

As mentioned above, having a well-defined experimental system for validation is essential for a successful fusion and antibody characterization. Consequently, incorporating Western blot, flow cytometry, or immunocytochemistry assay screens using cells from controls and stimulated cells, is essential for the rapid selection of the phosphosite-specific clones. These assays should approximate, very closely, the ultimate intended use of the final hybridomas. In some situations, the endogenous level of the desired phospho-protein is below the limits of detection for commonly used methods like flow cytometry, microscopy, and Western blot. In these cases, it is recommended to have a transfected cell line that on ligand activation induces the phosphorylation of the target protein. Sometimes, this "artificial" approach is a viable avenue to help in the identification and characterization of antibodies. While this is not ideal, it offers true negative controls in nontransfected cells or transfected cells with amino acid substitutions on the site of interest.

2.2. Validation Techniques and Tools

Although the techniques are largely similar, validation of antibodies to individual phosphosites within a protein, versus a regular antiprotein (or total antibody), requires several additional steps to confirm their specificity. These extra requirements are due because of the high selectivity of phosphosite-specific antibodies to discriminate a small percentage of phosphorylated protein from a large excess of nonphosphorylated protein. In addition, it is imperative that the desired antibodies are not recognizing any other phosphorylated amino acid within the same protein, or related isoforms. These rigorous requirements for the phosphosite-specific antibodies are needed for these reagents to be used in many different areas of biomedicine and in sensitive instrumentation with consistency and reproducibility.

Some of the additional tests may require the induction of the desired phosphorylation or removal of the phosphate from the target protein. A single method, or application, is rarely robust enough to validate high-affinity phosphosite-specific antibodies. Therefore, when validating a phosphosite-specific antibody, it is highly recommended to validate these with more than one technique representing the target protein in different conformations and cellular environments.

2.2.1. Enzyme-Linked Immunoassay (ELISA)

Often, the first validation test for a phosphosite-specific antibody is an ELISA using the phosphorylated protein, or phospho-peptide, and the nonphospho-protein, and nonphospho-peptide-carrier as positive and negative controls, respectively. Validation by ELISA allows for a rapid and sensitive evaluation of many different samples, namely sera from immunized mice or hybridoma supernatants. An ELISA does not require a large sample volume since it can be diluted manifolds. The ELISA is rapid, inexpensive, and amenable to automation, if needed.

One of the main limitations of the ELISA protocol is that the antibodies selected will not necessarily detect the phosphoprotein in its cellular environment. This is probably due to the "forced" and random adsorption of an immunogen, or short phospho-peptide, onto a plastic surface. This lack of correlation between ELISA and other techniques is more pronounced when using peptides directly adsorbed to the plates without any linker to keep the phosphorylated amino acid exposed. This could be minimized if the full-length phosphorylated and nonphosphorylated proteins are used in the ELISA screen; however, these reagents are not always available. Nonetheless, ELISA has proven a very useful initial step for the rapid elimination of these antibodies that fail to discriminate the phosphorylated protein, or peptide, from the nonphosphorylated versions and reduce the number of potential hybridomas to the most promising ones. Additional ELISA procedures using libraries of phospho-peptides bound to the plates can further refine the specificity of an antibody to the individual amino acid level. This approach is extremely useful when determining potential antibody cross-reactivities with related phospho-proteins. In addition, if the proper antigen and antibody concentration curves are performed, an ELISA can provide insights into the apparent affinity of a phosphosite-specific antibody towards one or more phospho-epitopes.

2.2.2. Western Blotting

Over the past 30 years, Western blotting (a.k.a. immunoblot, electroblot) has become a powerful and useful technique utilized in the validation of total protein and phosphosite-specific antibodies (37, 38). It involves the lysis of cells (from culture dishes, blood, solid tissues, body fluids) containing all the cellular proteins followed by electrophoretic separation via polyacrylamide gel and subsequent electrotransfer of the separated proteins onto a membrane. The membrane will represent a replica of the cellular proteins, separated by the electrophoresis, into individual entities of unique molecular weight. This technique, however, is limited by the resolution of the polyacrylamide gel and the relative abundance of the protein targets. The protein(s) of interest in the membrane is visualized by its decoration with an antibody (purified, sera, hybridoma supernatant, etc.) tagged with a reporter or a secondary antibody coupled to a detection system like chemiluminescence or fluorescence. The ability to separate multiple lysates simultaneously in a single gel allows the researcher to test control cells against those treated with a particular ligand resulting in the phosphorylation of the desired protein target. For instance, growth factors, e.g., PDGF, EGF, NGF, induce a pleiotropic response in cultured cells with multiple phosphorylations emanating from membrane-bound receptor tyrosine kinases and downstream intracellular kinases. The phosphorylations can be visualized globally with anti-pTyr, pSer, or pThr antibodies, or in individual proteins with antibodies to unique phosphorylation sites. There is a clear distinction between the basal phosphorylation in controls and the cells activated by a ligand, or treatment, examined side by side in the same blot. The expectation for a good phosphosite-specific antibody to a unique site would be to decorate the intended target of the correct molecular weight with the right cell activation. Figure 1 illustrates the specificity of a phosphosite-specific antibody towards its target only when cells are treated with a ligand.

The kinetics of the phosphorylation is also an excellent guide when searching for antibody specificity. The validity of these intra-

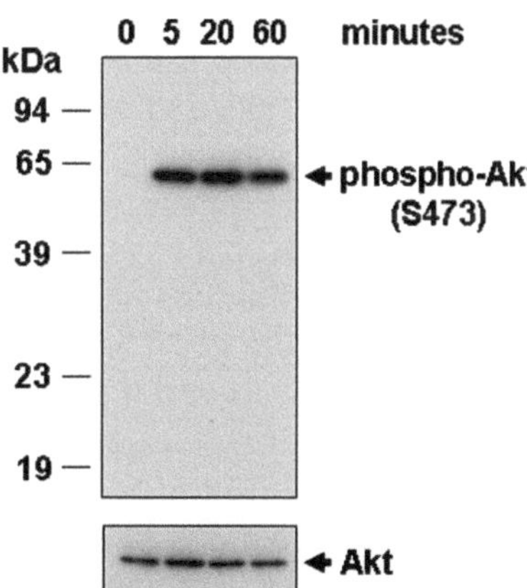

Fig. 1. Western blot detection of Akt phosphorylated at S473. Lysates prepared from human MCF-7 cells treated with 100 ng/mL of IGF-I for the indicated times were immunoblotted with either anti-phospho-Akt (S473) (R&D Systems, Catalog # AF887; *upper panel*) or anti-Akt (pan) (R&D Systems, Catalog # MAB2055; *lower panel*) antibodies.

cellular phosphorylations can be further assessed by incubation of the membrane with phosphatases or competition of the antibodies with ligands such as pTyr, pSer, pThr, analogs like phenylphosphate and purified phospho-proteins. The visualization of the phospho-protein and the correlation with the expected biology and molecular weight are perhaps the major advantages of the Western blotting technique.

However, the results and interpretations are not always straightforward. Because of the transient and labile nature of phosphorylation events, scientists need to be careful when preparing lysates for electrophoresis and Western blotting. There are many different extraction buffers or "cocktails" containing phosphatase and protease inhibitors that prevent, or limit, phosphate removal from proteins. These cocktails may result in different outcomes depending on the cell or tissue type and the concentration of the inhibitor used. One has to be aware of the thermal sensitivity of phosphatases, and lower the temperature between 0 and 4°C while performing extractions and other lysate manipulations. Contrary to most kinases, phosphatases are inactive at low temperatures. In our experience, the immediate lysis of cells and tissues with chaotropic and denaturing agents like SDS or urea result in a very consistent recovery of phosphorylations, perhaps due to the rapid actions of these compounds on phosphatases and proteases.

In general, immunoglobulins are very stable once they are purified, and most IgGs react toward their target at physiological salt concentrations with very rapid kinetics when there is an excess of IgG (39). High-affinity antibodies only require 1 h or less to recognize their target in immunoblots, incubated at room temperature or 37°C. If an antibody needs several hours and lower temperature for binding and reactivity, it may be indicative of an IgG with a low affinity toward its target.

Blocking conditions deserve special attention when working with phosphosite-specific antibodies (40). In our experience, nonfat milk and the appropriate buffer containing Tween-20 give consistent results with the many polyclonal and monoclonal phosphosite-specific antibodies we have evaluated. Bovine serum albumin is not as stringent as a blocker for Western blots, whether the antibody is against the total protein or towards a phosphorylated epitope. The concentration of Tween-20 should be maintained between 0.5 and 0.5% for optimal results. For reasons not yet entirely clear, most anti-pTyr antibodies exhibit high background if nonfat milk is utilized as a blocking buffer; thus, it is recommended to switch to bovine serum albumin or other alternative blocking agents. In conjunction, the use of a high-capacity polyvinylidene difluoride (PVDF, Immobilon) membrane allows for ease of use, resistance to shear, and low background.

The detection of the phospho-protein of interest on the membrane blot requires a detection system to label the antibodies and thus the antigen. After the blocking and incubation with phosphosite-specific antibodies steps, the blot is incubated with a second antibody directed towards the "primary" phosphosite-specific antibody, and this "secondary" antibody is tagged with a detection system. Alternatively, the primary antibody can be directly labeled and used in a single-step protocol. There are several commercially available detection systems for Western blotting and they all offer some benefits and increase in sensitivity, stability, and performance (Amersham, Thermo-Fisher). Incubation with secondary antibodies labeled with alkaline phosphatase followed by the enzyme's substrate provides a colorimetric reaction localizing the complex of primary phosphosite-specific antibodies and the target protein. Although not very sensitive in comparison with alternative chemiluminescence, or radiometric methods, alkaline phosphatase detection is rapid and economical since it does not require further manipulation or equipment. Horseradish peroxidase-conjugated primary or secondary antibodies provide a more sensitive method than alkaline phosphatase. In the presence of H_2O_2 and luminol, the peroxidase modifies the substrate, which emits light that can be captured on film. Antibodies can also be tagged with fluorochromes, like the ones offered by LI-COR in the near-infrared range, to provide very sensitive detection and linearity. However, the laboratory will need access to the appropriate instrumentation to read these results. Other detection methods include using Q-dots that emit fluorescence at specific wavelengths. Q-dots are also amenable to multiplexing by using more than one color (Life Technologies).

There are many different protocols for Western blotting in thousands of publications and from different vendors, and most derived from the original publication by Towbin et al. (14). Once in a while, we encounter some proteins that behave differently under the typical conditions. For instance, Caveolins form oligomers that are resistant to SDS and often appear as a ladder of different molecular weights in a Western blot (41). We strongly recommend the inclusion of a positive control to facilitate the interpretation of the results. High molecular weight proteins (>200 kDa) also pose several problems in their identification. First, they do not resolve well in gradient gels, and often they exhibit a poor electrotransfer. For high molecular weight proteins, we recommend to run 6% gels or 3–8% Tris–acetate gels from Life Technologies, to improve the resolution and transfer of these difficult proteins. Sometimes, the inclusion of 0.05% SDS in the transfer buffer improves the immunodetection of these large proteins.

2.2.3. siRNA and Knockout Mice

Silencing RNA (siRNA) can also be used to manipulate the phosphorylation of a protein by artificially creating a "quasi-negative control" and aiding in determining the specificity of an antibody by alternate means. In this scenario, a short RNA complementary to the mRNA of the protein of interest is artificially introduced into a cell, triggering the degradation of the hybridizing mRNA, and eventually decreasing the total levels of its corresponding protein product. Reducing the total levels of a target by degrading its mRNA is very useful when there is a constitutive phosphorylation of the protein of interest and a lack of inducible phosphorylation protocols. By artificially downregulating the phospho-protein of interest, or the upstream kinase, followed by the induction of the phosphorylation, lysis, and Western blotting, one can determine if the antibody recognizes the phosphorylation of interest. As usual, the proper controls need inclusion to confirm that the total levels of protein have indeed been reduced. Once the siRNA cells have been validated, they can also be used for several other techniques, such as immunocytochemistry and flow cytometry, to further study phosphorylation events. This is important because, depending on the siRNA used, the effect may be visible only on the newly transcribed protein without altering the levels of the phosphorylated protein pool.

siRNA as a tool offers several advantages, including the commercial availability of validated reagents and the expedient nature of the experiments. The downregulation does not need to be 100% complete and the transfections typically involve time periods from 24 to 72 h. Unfortunately, not all cell types are easily manipulated by siRNA and some may require lengthy optimization of the protocols. Furthermore, depending on the nature of the mRNA and protein turnover rate, some targets are not disturbed by siRNA.

Knockout mice lacking the phospho-protein of interest, or ideally the kinase responsible for the phosphorylation, can be very powerful tools to determine the specificity of an antibody. Tissues from these unique mice and normal control animals can be used to compare the reactivity of a phospho-specific signal by either immunohistochemistry, ELISA, Western blotting, or flow cytometric techniques. Alternatively, primary cells or cell lines can be derived from the knockout mice to further explore signaling events and phosphorylation in cell culture conditions.

2.2.4. Phosphatase and Competition Treatments

Phosphatase treatment of cell lysates, fixed cells, tissue sections, and Western blot membranes can be used as a tool to determine if the signal generated by the antibody is phospho. Commercially available enzymes such as alkaline and lambda phosphatases are a reliable source of these tools to remove phosphates. Protocols must be optimized depending on the phosphatase source, optimal pH, and metal ion requirements. Phosphatase treatments can be performed directly on lysates, fixed cells, or Western blot membranes containing surface-bound phospho-proteins. As anticipated, a true phosphosite-specific antibody will show a negative signal in the phosphatase-treated cells compared to the untreated controls. Figure 2 depicts an example of a phosphatase-

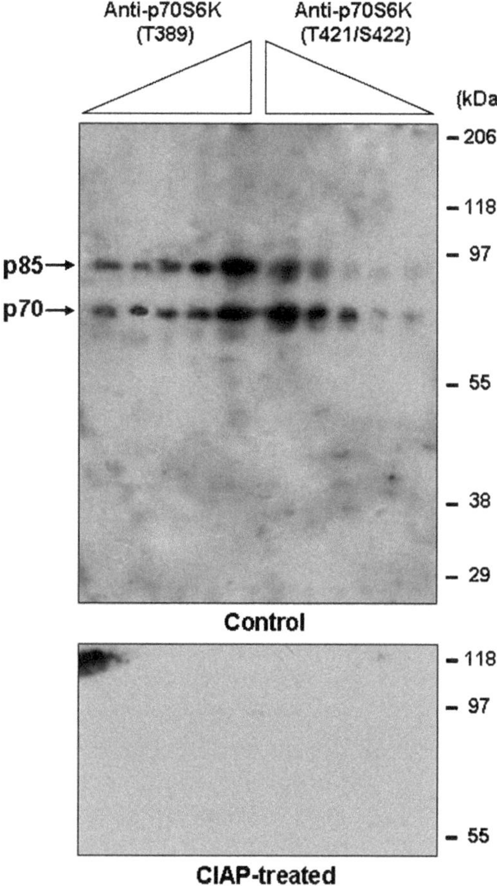

Fig. 2. Phosphatase treatment of membrane shows phosphospecificity. HeLa cell lysates were resolved by SDS-PAGE, transferred to membranes, and the lower membrane was treated with Calf Intestinal Alkaline Phosphatase (CIAP). Following CIAP or control treatment, membranes were blocked and then probed with either anti-human p70 S6Kinase (T389) (R&D Systems Catalog #AF8963) or anti-human p70 S6Kinase (T421/S422) (R&D Systems Catalog #AF8965). These antibodies recognize both p70 and p85 forms of S6Kinase.

treated blot to remove the phosphate on the protein of interest from the activated cell lysates.

Competition experiments are another useful tool to help determine the specificity of an antibody. The antibody is titrated with a wide range of concentrations of phospho, nonphospho peptides, or proteins. These competitions can be done for techniques such as ELISA and Western blot. These experiments provide supportive evidence of the affinity of the antibody towards the phosphorylated version of the epitope under evaluation.

2.2.5. Inhibitors of Kinases and Phosphatases

Compounds that inhibit kinases and specific signaling pathways can be powerful tools to further elucidate the specificity of a phosphosite-specific antibody. Development and study of kinase inhibitors has been spurred by the critical roles of kinases in diseases like cancer and diabetes. For instance, a compound such as STI571 (Gleevec) will bind to the BCR-Abl kinase, and related kinases inhibiting its catalytic activity, protein phosphorylation, and leukemia (42). In another example, U0126, an inhibitor of MEK, can be used to block the ERK1/2 signaling pathway (43). Pretreatment of cells with this inhibitor prior to cell activation will both inhibit MEK and diminish the phosphorylation of ERK1/2, RSK, p70 S6 Kinase, and other downstream phosphoproteins. A wide range of different kinase and phosphatase inhibitors are commercially available from different companies such as Sigma and Cell Signaling Technologies, among others. The expected decreases in specific protein phosphorylations can be extremely valuable in helping with the validation of phosphosite-specific antibodies. Lysates or cells from inhibitor-treated cells can be prepared by different methods and probed with the desired phosphosite-specific antibodies.

Protein phosphorylation is also regulated by the activity and rate of de-phosphorylation by protein phosphatases (1, 4). There are many different phosphatase inhibitors; for instance, okadaic acid specifically inhibits protein phosphatase 1, 2A, and 2B, while other inhibitors block the activity of many phosphatases nonspecifically (44). Vanadate ions are potent inhibitors of phosphotyrosine-specific protein phosphatases, and have been used extensively to preserve the integrity of pTyr (45). In its oxidized form, pervanadate is membrane-permeable and used to treat cells, or intact animals, prior to activation, inhibiting intracellular pTyr-phosphatases and resulting in increased levels of pTyr-containing proteins (46).

2.2.6. Cell Transfectants and Site-Specific Mutations

Although not always available, site-directed mutagenesis of the target phosphosite is an elegant and direct way to validate and confirm the specificity of an antibody. This method requires the creation of transfected cells carrying an expression system with the gene of interest with or without the desired amino acid substitution. Investigators can generate either transient or stable transfectants. Ideally, the cell selected to carry the artificial gene should not contain an endogenous copy of it. Once the phosphorylation site has been modified, the protein is expressed and subjected to ligand stimulation inducing the desired phosphorylation. Then, the lysates are probed with the phosphosite-specific antibodies, either by direct Western blotting or immunoprecipitation, flow cytometry, or other techniques, and compared with lysates from control cells expressing the unmodified phospho-protein.

Although cells transfected with unique amino acid substitutions are extremely useful for determining the specificity of a phosphosite-specific antibody, they should not be used as a routine method to select antibodies. The rationale is that in the antibody-selection process, it is relatively easy to detect low-affinity antibodies that will recognize the protein of interest if provided in excess amounts; these same antibodies may not be suitable for recognizing the phospho-protein expressed at normal levels in cells and tissues.

2.2.7. Immunocytochemistry

Methods employing microscopy either with cultured cells or tissue sections are also useful for antibody validation and contribute information regarding the spatial localization of any given antigen. For example, if the cellular localization or cell-type distribution of a phospho-protein is known, then microscopy, in its many facets, can be useful in evaluating the specificity of an antibody. Typically, information regarding the expected distribution of a phospho-epitope can be found in literature describing the modification and should be used to evaluate any given antibody. Many proteins, on phosphorylation, change their subcellular and spatial location (47, 48). One of the main advantages of microscopy over other methods is its ability to monitor events in a structurally complete cell. This is especially true when using electron microscopy with tissue sections stained with gold-labeled antibodies. Immuno-gold staining has great potential to help with the elucidation of specificity with phosphosite-specific antibodies; however, we are not aware of descriptions of this technique using phosphosite-specific antibodies. The ability to correlate immunoreactivity with a specific cell-type, among many others in a tissue, provides another level of confidence in the antibody in question.

There are many protocols for imnunostaining of cells and tissues already available in the literature and described in other chapters in this book. Typically, antibodies labeled with fluorochromes or enzymes for detection are applied to fixed and permeabilized cells to monitor intracellular phospho-proteins and the results are monitored with a microscope. However, visual identification with a phosphosite-specific antibody within an intact cell must be viewed with healthy skepticism until the specificity of the antibody is supported by other methods such as subcellular fractionation and Western blotting (47, 48). Nonspecific reactions and antibody bindings are extremely common in immunological methods with intact cells and tissues.

Interpretation of negative and positive results using phosphosite-specific antibodies in tissue sections must be critical and cautious. When negative results with antibodies are obtained, we should always reference the tissue's collection protocol, as many phosphorylations do not survive the opening of tissues and phosphatase activation for very long. In addition, some kinases and stress pathways are activated on dissection and cell lysis and may remain active for some time, depending on the isolation protocols (49, 50). Furthermore, many phosphorylations trigger protein degradation and, if the fixation is not timed correctly, the immunocytochemical results with a phosphosite-specific antibody will be negative. Depending on the tissue, the rate of penetration of the fixative may affect the survival of the phosphorylation in the inner part of the sections. Therefore, careful interpretation of immuohistochemical data is required due to potential artifactual changes in phosphorylations well after tissue collection. Because every phosphorylation is modulated by different kinase(s) and inputs, and regulated by unique phosphatases, there is no uniformity on the stability of the different phosphorylation sites, even within the same protein.

Another possible factor contributing to the lack of reactivity by a phosphosite-specific antibody in the expected cell or tissue is steric hindrance of the phospho-epitope by a binding protein. Phosphorylation sites are often used as binding domains for adaptors and scaffold proteins (51). In theory, if an adaptor or scaffold protein binds tightly to a phosphorylated amino acid in a protein, this adaptor may prevent the antibody's reaction with its phospho-epitope. If the results are negative, it is advisable to explore other extraction and retrieval protocols using saponin and Triton-X-100 as permeabilizing agents and/or other fixatives such as methanol that may solve some steric hindrance problems.

Positive staining by immunocytochemistry using phosphosite-specific antibodies should also be viewed with caution. First, because antibodies are large proteins, they can interact with many cellular components through their different regions in a nonspecific manner (i.e., Fc region) depending on the conditions and the tissue target. Second, a low-affinity antibody can bind to many different phospho-proteins of similar charge or conformation, giving a false-positive result. This could be exacerbated by the formation of neo-epitopes after cross-linking of intracellular proteins with aldehydes. Fortunately, kinases, phosphatases, and proteases are inactivated by low temperatures so the use of frozen sections and staining at low temperatures to preserve the phosphorylations can be very useful in determining the true status of a phospho-protein. Frozen sections are often superior to formalin-fixed tissues in preserving the integrity of the cellular components (52); many antibodies that normally do not work in formalin-fixed and paraffin-embedded tissues may work well in frozen sections. At the very minimum, adsorption controls with phospho-proteins or phospho-peptides should be included in the immunocytochemical validation of a phosphosite-specific antibody.

As of recently, there have been several reviews addressing the suitability and the overall quality of commercially available antibodies for immunohistochemistry, including phosphosite-specific antibodies (53–56). This is not surprising given the recent explosion of large and small companies supplying biomedical tools like antibodies, often providing the same poorly characterized and validated products under different labels. The overall consensus is the wide range of quality and reproducibility of commercially available, and our recommendations echo these previously stated sentiments. Regardless of whether the results are negative or positive, the reactivity of a phosphosite-specific antibody by immunocytochemical methods should be supported by other applications to ensure specificity. For example, tissues and cells from the same origin should be processed and tested by Western blotting to determine the correlation between immunocytochemical and Western blot results. Other controls, such as phosphorylation kinetics, phosphatase treatment of cells, kinase inhibition, siRNA, and knockout mice will be critical in evaluating imaging data.

2.2.8. Flow Cytometry

The use of phosphosite-specific antibodies in flow cytometry creates similar validation issues to that seen with immunocytochemical methods. Results obtained by flow cytometry should similarly also be supported by other means whenever possible. In recent years, we have witnessed an increased number of publications using flow cytometry and phosphosite-specific antibodies with as many different protocols used by this application (57–59).

Initially, flow cytometers were mainly used to investigate blood cells in the immunology field. Currently, however, flow cytometry is routinely utilized for all types of cells and different applications. Flow cytometry requires an expensive instrument but offers the unique advantage of selecting for the cells of interest based on their phenotype (gating), even in a complex cell population. Cell phenotype can be defined in multiple ways, for instance, phenotype can be defined by a unique cell surface set of proteins (CD markers), DNA content, cell granularity, cell viability, apoptosis, or cell size (60). Most of these parameters can be monitored and quantified by fluorescence as the cells are injected into the cytometer. The high sensitivity of flow cytometry coupled with their ability to preserve cell-specific information makes them a superb tool to determine the specificity of antibody. If a particular protein is expected to be present in a unique cell type, this can be gated with an antibody to the cell type coupled with a fluorochrome of unique emission wavelength, and the antibody in question labeled with a different fluorochrome, must coincide with the cell-specific antibody (57). For example, an antibody to phospho-ZAP70 should stain only T cells with the proper activation, but not B cells or other cellular blood components, because ZAP70 is expressed only in T cells. Figure 3 provides an example of an experiment inducing the phosphorylation of STAT6 (Y641) on rhIL-4 stimulated Daudi cells but not on the unstimulated Daudi control cells. The phosphorylation of STAT6 (Y641) was monitored by the increased fluorescence of the polyclonal anti-phospho STAT6 (Y641) antibody labeled with phycoerythrin fluorochrome.

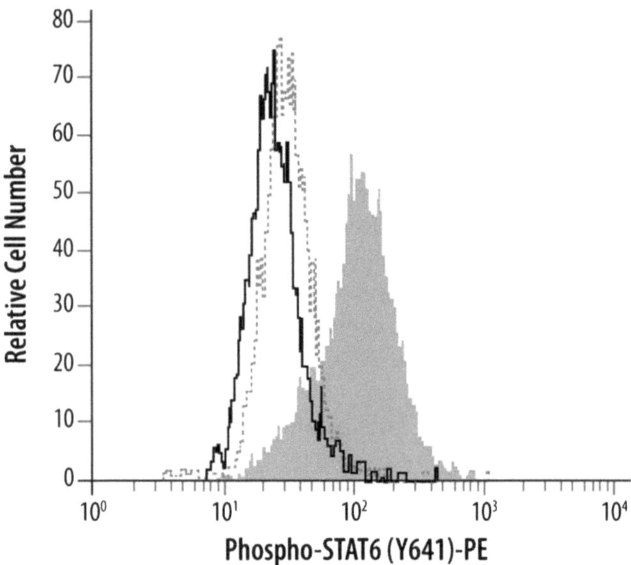

Fig. 3. Anti-phospho-STAT6 (Y641) stains IL-4 treated DAUDI cells. Intracellular staining of unstimulated cells (*open histogram-dotted line*) or rhIL-4-stimulated DAUDI cells (*filled histogram*) with PE-conjugated polyclonal anti-Phospho-STAT6 (Y641) (R&D Systems Catalog #IC3717P) or with isotype control antibody (*open histogram*).

Flow cytometry is not without some obstacles to overcome. Experiments must be carefully designed and optimized for consistent results. First, most phospho-proteins are intracellular and their distribution is seldom unique to a cell type. Second, fixation and permeabilization protocols may introduce nonspecific binding of immunoglobulins to cellular components. Defined and unique protocols describing fixatives and permeabilization steps are available from primary publications or technical books (57–59). As a validation technique, flow cytometry also depends on other methods to support, or refute, the specificity of a phosphosite-specific antibody. Therefore, it is advisable to always have other supporting evidence, such as kinetics and ligand-dependent phosphorylation, inhibitor and siRNA evidence, and Western blotting data correlating with the flow cytometry results. This will ensure a better data interpretation. As with immunocytochemistry, careful sample collection and preparation are important because of the instability of phosphorylations and phosphatase activity. In our experience, the less the cells are manipulated, from collection until fixation, the more consistent are the results.

2.2.9. Immunopurification and Protein Analysis

Once in a while, there are phosphosite-specific antibodies with specificities that cannot be determined by routine methods. Phosphorylations on proteins, like other posttranslational modifications, often induce minor changes of 1–2 kDa in molecular weight (61). Sometimes, however, phosphorylations induce unexpectedly large mobility changes (5–20 kDa) in SDS-PAGE (62).

Any unexpected large mobility shift in a phospho-protein by SDS-PAGE, as detected with a phosphosite-specific antibody, needs to be validated further. Peptide competition or phosphatase treatments of lysates and membranes may not be of sufficient rigor to establish the identity of a phospho-protein. Immunoprecipitation of the phospho-protein with a validated antibody to the total protein, followed by Western blotting with the phosphosite-specific antibody, takes us a step closer to defining antibody specificity; however, we should analyze carefully to uncover co-immunoprecipitation artifacts. Sometimes, antibodies to total proteins fail to recognize the phospho-protein because of low abundance or other structural changes induced by post-translational modifications. A more definitive result can be accomplished by immunopurification of enough phospho-protein to determine the identity of the target protein by N-terminal sequencing or mass spectrometry (63).

3. Methods

3.1. Applications

The main goal for developing high-affinity and discriminating phospho-specific antibodies is to have the right tools to monitor and study key protein phosphorylations in normal and diseased cells. One of the major appeals to monitor protein phosphorylation is its close correlation with protein activity. Therefore, this post-translational modification is very tempting as a potential biomarker for normal cell development and disease. While it is important to monitor the overall changes in expression of a protein, during a cellular process or disease, having the information about changes on the protein's activity provide a better picture of the cellular process. While phospho-specific antibodies are not the only reagents available to investigate phosphorylation events and signaling, these antibodies have proven extremely useful and versatile in advancing the field of signal transduction research. There are many different applications and assays where phospho-specific antibodies have been incorporated, and these applications are virtually the same as for a total protein antibody. Below we will mention some applications that have appeared in the last few years and rapidly advanced the cell-signaling field.

3.1.1. Drug Discovery

There are many different assays described in the literature that utilize phospho-specific antibodies. The majority of these assays are ELISAs designed to rapidly examine the phosphorylation and activation status of key signaling molecules. In the basic format, a cell or tissue lysate is added to one well of a 96-well plate, incubated with a phospho-specific primary antibody, and layered with a secondary antibody for detection and signal amplification

(64, 65). The secondary or detection antibody is typically conjugated to an enzyme, followed by incubation with chemiluminescent or colorimetric detection reagents. In the sandwich format, the well is coated with an antibody to the total protein, incubated with the cell or tissue lysate and layered with a phospho-specific antibody to the same protein followed by the detection system.

Most of the established detection systems for ELISA are fluorescent or colorimetric. Often, the detection systems use an enzyme conjugated second antibody followed by the addition to the enzyme's substrate. Therefore, in addition to the incubation with lysate, ELISA is a multistep protocol, requiring washing steps after the addition of each reagent, and relatively lengthy. More recently, companies like Perkin-Elmer have introduced alternative detection methods for ELISA using a pair of antibodies, each to a different epitope of a molecule (Alpha Screen), for instance, an anti-total and a phosphosite-specific antibody to the same protein. Each antibody is labeled with a different tag, that in close proximity emits at a defined wavelength. The Alpha Screen is homogeneous with no washes required, ideal for large number of samples in large screens initiatives in the pharma industry. For instance, during the search for kinase inhibitors from chemical compounds libraries. Potentially, we could envision some variations to current the detection systems that could enhance sensitivity and expedite the whole assay; for example, the attachment of unique oligonucleotides to the antibodies, resulting in signal amplification.

Advantages of the ELISA format include its relatively low cost, rapid development, and high throughput capability, provided that all needed antibodies are available. Most of the ELISA's that use phospho-specific antibodies examine one phospho-protein at the time and will continue to be an important initial experiment used to monitor the effects of ligands, activators, and inhibitors in cell-signaling pathways. On the other hand, the main limitation of ELISAs is their lack of multiplexing, or the ability to measure multiple analytes simultaneously.

Cell-Based ELISAs or In-Cell Westerns are another modification of the ELISA platform. Both these techniques are commercially available from different vendors such as Cell Signaling Technologies, R&D Systems, and LI-COR, and offer the ability to monitor two parameters at the same time without having to make cell lysates from the individual wells (66, 67). After plating and activation in wells, the cells are fixed and permeabilized before staining with two antibodies. One antibody detects the total protein, or a normalizing control like actin, and the second antibody detects a phospho-epitope. The key requirement for this assay is for the antibodies to be from different species, e.g., one derived from rabbits and the second raised in mice.

Cell-Based ELISAs are, in principle, like immunocytochemistry, but with a quantifiable signal. This assay also allows for a direct comparison of the signal with a control phospho-protein, thereby providing a relative measure of the phosphorylation event. Furthermore, this assay is also amenable for automation and high throughput, making it suitable for drug discovery and validation experiments. Cell-Based ELISAs have the potential to be adapted to fluorescent detection and imaging platforms. An advantage of this platform would be that, in addition to the information on the phosphorylation events, the imaging could also provide the subcellular localization of the phospho-protein in question. Spatial and temporal information would also be extremely valuable in designing and understanding the biology of effective cell activators and kinase inhibitors.

3.1.2. Phospho-Protein Expression, Profiling, and Proteomics

Signaling pathways are intrinsically dynamic and complex, involving hundreds of biochemical reactions and phosphorylation cycles in numerous intracellular proteins. In recent years, we have witnessed the discovery of some of the main players in key cellular pathways, including the MAP kinases and Akt (68, 69). These discoveries have led us to formulate working hypotheses surrounding cellular events like growth and apoptosis, but much more information is still missing. At the same time, there is an urgent need for the development of rapid, sensitive, and quantitative methods to measure the protein phosphorylation of many proteins simultaneously within a pathway, or cellular process, to better profile normal and diseased cells.

Techniques that probe the cells with phospho-specific antibodies, such as Western blot and immunohistochemistry, have helped directly in the characterization of signaling events downstream of a myriad of different conditions and diseases. Most of these studies have been done using a single phospho-specific antibody evaluating different cellular conditions. One of the end results from >10 years of phospho-specific antibody-driven research is an explosion of signaling pathways profiling the stepwise kinase cascades controlling cellular events, including growth factor-activated cells, tyrosine kinase receptors, and the many transcription factors inside the nucleus that become phosphorylated in diseases or during cell development. However, immunocytochemistry, ELISA, and Western blot are somehow limited in their multiplexing capabilities. At most, if phospho-specific antibodies are of good quality, these may be mixed, provided their targets are of different molecular weight, and used to probe a single lane in Western blots, enhancing the data output from the experiment (70–72).

Phospho-protein enrichment with antibodies to pTyr, or metal-affinity chromatography followed by Mass Spectrometry has accelerated the discovery and identification of novel phosphorylations under many different conditions (73, 74). Improvements and variations to the method are frequently described in the literature. For instance, immunoaffinity purification and strong cation-exchange chromatography identified 10,655 unique phosphorylation sites in T cells on ligand activation (74). These phosphorylations corresponded to proteins involved in discrete modules of cellular function like endocytosis, microtubule polarization, and cytokine production.

This unbiased approach has been used to elucidate novel tyrosine phosphorylations in cells activated by oncogenic kinases and other cellular conditions. According to Phosphosite (http://www.phosphosite.org), one of the most comprehensive databases for protein modifications, there are approximately 81,000 different phosphorylations already described. Many of these phosphorylations have been recently derived from immunopurification and mass spectrometry methods. This protocol has accelerated the discovery of previously unknown phosphorylations and paved the way to understand their role in normal and diseased cells.

Discovery-based experiments are extremely useful in the initial identification of a phosphorylation event. These experiments generate massive amounts of data needing thoughtful data analysis. Multiple steps are needed, from the initial experiment to the identification of the phosphorylation sites, and mass spectrometry runs are still expensive and not suitable for analysis of many samples on a routine basis. Thus, once the novel phosphorylations have been identified, phospho-specific antibodies are, perhaps, better suited for use in day-to-day experiments, whether alone or in a multiplex format.

More recently, phospho-protein analysis by microfluidics has been made commercially available (75). In the current format, this assay first separates phospho-proteins by isoelectric focusing and then the proteins are adsorbed, based on their charge, onto small capillary tubes. Next, the tubes are incubated with phospho-specific antibodies followed by secondary antibodies with a chemiluminescent detection system. Microfluidics has the advantage of great sensitivity, and very small amounts of cells and tissues are necessary for the assay. However, with current instrumentation, the sample throughput is limited.

3.1.2.1. Multiplexing and Antibody Arrays

The ever increasing complexity of signaling pathways somehow limits the use of phospho-specific antibodies by Western blotting, immunocytochemistry, and ELISA. First, immunocytochemistry and ELISA can only measure a couple analytes at the same time. Second, although it is possible to multiplex the Western blot, this technique is still time demanding requiring multiple steps. Ideally, we should be able to monitor multiple phosphorylations, and the corresponding total levels of key proteins in a signaling network encompassing the activity of a membrane receptor, intermediary kinases, adaptors, and nuclear transcription factors executing the signal initiated at the cell surface. And this analysis should be accomplished with a minimal time, quantitatively, and with rapid data acquisition and reproducibility. Arrays are the closest platform, or technology, that nearly meets the requirements described above.

Protein and antibody arrays evolved from the DNA array technology used to monitor gene expression. In their basic and initial format, DNA arrays or gene profiling were created by placing up to thousands of genes, in minute amounts, on glass slides (76). The slides were then incubated with cells containing labeled RNA for hybridization and the presence of signal on an individual spot indicated gene expression.

Antibody arrays follow a similar design but on a reduced scale. The arrays are comprised with a few or up to several tens of individual antibodies spotted in a very small area of a solid matrix. Basically, there are two main antibody array configurations. In the first configuration, the microarray, only one antibody per analyte is needed. These first antibody arrays were created by depositing nanoliter amounts of antibodies onto derivatized glass slides as individual spots (77). Each spot represents an individual antibody and each slide contains several hundred different antibodies. Then, the arrays are incubated with a cell extract, with its protein content labeled with a tag or a fluorochrome, like Cy3 or Cy5, and read on an instrument. Provided the antibodies are monospecific, the data obtained with these arrays is extremely valuable because it eliminates the need for multiple Western blots, or other analyses that monitor all the analytes one at the time.

Several companies offer arrays with single antibodies spotted on membrane-coated or glass slides detecting up to about 500 different analytes. There are three major limitations to this approach. First is the suitability of each antibody to be immobilized on a surface and still be capable of recognizing only its target of interest. The expansion of the antibody microarrays has been hampered by the lack of suitable reagents for this application. Since most available antibodies have not been screened for microarrays, while in development, many fail this basic requirement. Second, the affinity of each antibody needs to be high enough to collect all of the protein in a sample from a cell extract or fluid. Due to the limiting and transient nature of phosphorylation events, in our experience, the single antibody approach often does not provide a robust and reproducible fluorescent signal from total cell extracts. It is estimated that the lower limit of sensitivity of a fluorochrome is about 1 ng/mL, which may not nearly sensitive enough to detect events of low frequency such as most phosphorylations. Finally, the cell or tissue labeling protocol must be robust and capable of a uniform label of all proteins in an extract or biological fluid. On the other hand, antibody microarrays have the advantage of needing only one antibody that isolates the antigen of interest and reveals its presence in a cell or tissue extract. In addition, the small printing area for microarrays diminishes the sample required for analysis to approximately 20–50 μg.

In spite of its apparent limitations, antibody microarrays have been successfully used as a discovery tool providing comprehensive data on multiple phospho-proteins in unique experimental systems. For instance, microarrays were used in the elucidation of key phosphorylation events during oocyte maturation (78).

The second approach or configuration requires a pair of antibodies recognizing the same protein at two different epitopes without interference, as a sandwich (79). This approach is also called a macroarray. One antibody is deposited, or arrayed, as a spot onto a solid matrix, such as derivatized glass, nitrocellulose, or wells in microplates. Next, arrayed antibodies are incubated with a lysate or biological sample to allow each antibody to capture its intended target. Subsequently, the array is incubated with a cocktail of antibodies, or detectors, against each of the targeted molecules captured in the array. The second antibodies, or detectors, can be labeled with enzymes, or fluorochromes, for a readout using film or the appropriate instrument.

By analogy, single antibody arrays are like multiple Western blots, whereas sandwich-antibody arrays are like immunoprecipitations/Western assays. The sandwich antibody approach has the advantage of superior sensitivity and specificity because it enriches its intended target and minimizes any potential nonspecific binding by the antibodies (80, 81). This format also allows for more detailed analysis than simple expression of a target protein. In a typical configuration, antibodies to the total protein are arrayed in different spots and used to capture their target of interest. The cocktail of detection antibodies could contain phospho-specific antibodies to the captured proteins. This approach can provide a rapid assessment of unique phosphorylations in a multiplex manner. Thus, with a small amount of lysate, such as 50–100 μg, it is possible to monitor protein phosphorylation changes on multiple targets simultaneously with a similar, or better, sensitivity to a direct Western blot (79, 81).

In a recent example, a sandwich antibody array was developed using three different antibodies to the same protein. One antibody to the total protein is used as a capture, a second antibody to the same total protein, but at a different epitope, labeled with one fluorochrome, and a third phospho-specific antibody to the same protein with a different label (82). Thus, by measuring the ratios of the two tags, it was possible to quantify both the total and the phosphorylation levels of the target in the same spot. It will be interesting if this approach can be scaled up with multiple analytes.

However, the development of the sandwich arrays is more complex since it requires the identification and stringent validation of antibody pairs. Phospho-specific antibodies have also been incorporated in recently developed arrays that are commercially available by R&D Systems, Inc, Cell Signaling Technologies, and others, either as capture or detector antibodies. In addition to the limited availability of matched-antibody pairs for arrays, there are a number of considerations to monitor when developing these tools. The potential for nonspecific interactions by antibodies, as well as background and interference, increases with the number of capture antibodies present on the array. False-positive signals are also an issue with sandwich arrays. Many proteins are assembled into oligomeric complexes and their signal may land in a nonspecific spot, requiring unique lysis buffer conditions to break down these large protein assemblies. As with the previously discussed approaches, once a signal, or lack thereof, is observed with an antibody array, it is advisable to confirm it by other methods before proceeding to a large-scale study.

Nonetheless, the availability of antibody arrays has helped scientists obtain answers in less time than before. Arrays measuring multiple phosphorylation events simultaneously have simplified screens for therapeutics and inhibitors that otherwise would have been tested individually. The substrate for the arrays can be nitrocellulose, glass, and the bottom of a 96 or 384 plate, among others. Each substrate has its benefits and disadvantages. Figure 4 illustrates a typical experiment with an antibody array using the sandwich approach and monitoring the phosphorylation of unique activation sites in up to 46 different intracellular kinases. In this particular example, the capture antibodies have been spotted on a nitrocellulose membrane by an automated contact robot for precision and accuracy. The planar arrays on membranes are economical, have low background, and can be used with different types of commercially available detection systems, like chemiluminescence. In addition, the large real state of the membrane allows for the spotting of dozens of different antibodies side by side. The drawback for the planar arrays is the relatively large surface and volumes needed for the reaction to occur in an optimal manner, requiring approximately 50–200 μg of lysate.

Figure 5 illustrates a sandwich array using the bottom of a typical 96 well plate as the substrate for the capture antibodies. Spotting of the antibodies was done with a noncontact printer robot for consistency and accuracy. The main advantages of this type of support are the reduced volumes need for the reactions to occur, typically in the 50 μL range, reduced cell lysate requirements (1.0–50 μg), and the potential for automation. This 96-well format is ideal when screening for phosphorylations in multiples of small and limited samples. We believe these arrays containing phosphosite-specific antibodies, along with anti-total protein can

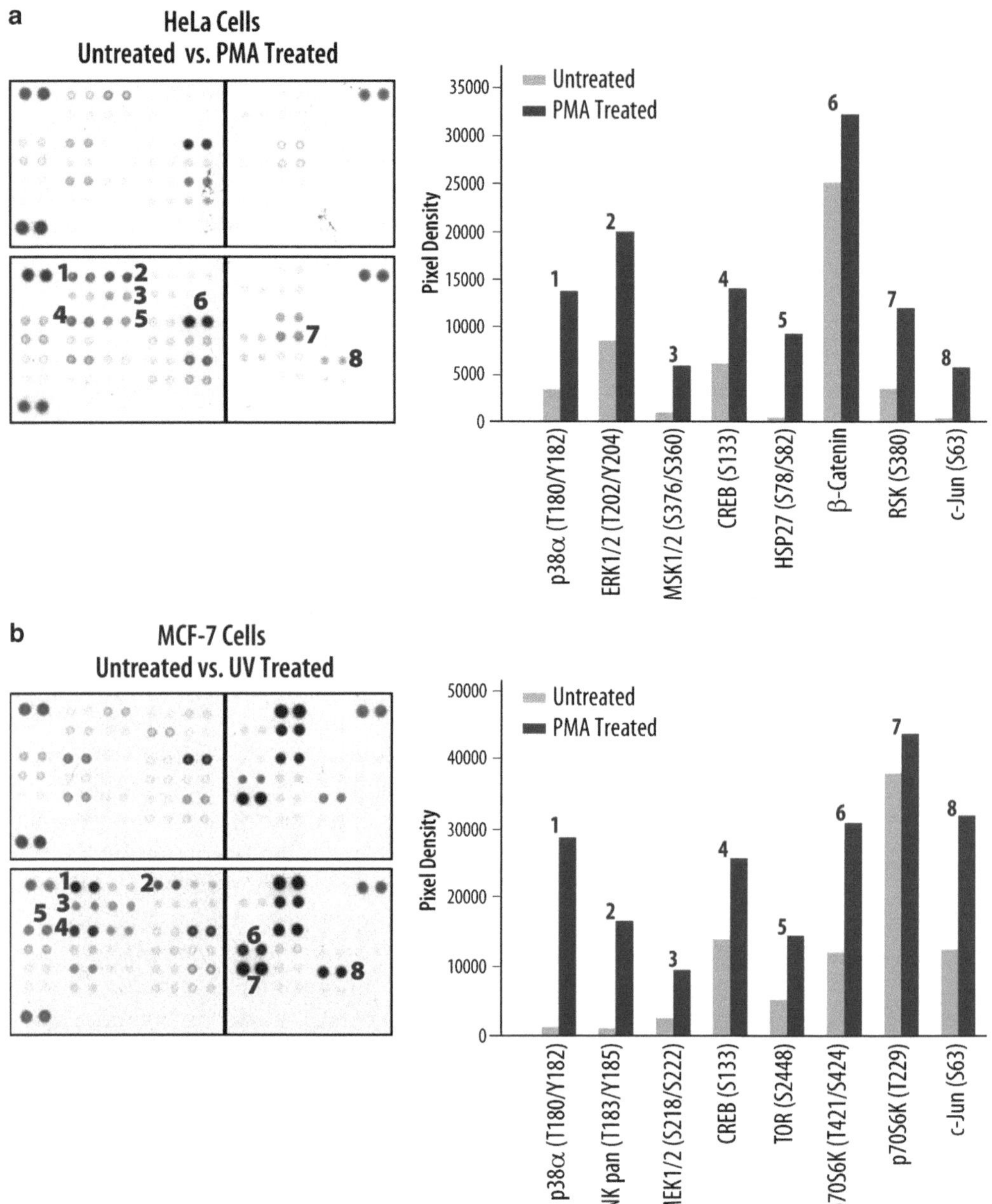

Fig. 4. The Human Phospho-Kinase Array detects phosphorylated proteins in untreated and treated cell lysates. (**a**) HeLa cells were either left untreated or treated with 200 nM PMA for 20 min. The array (R&D Systems Catalog #ARY003) was incubated with 300 mg of cell lysate. (**b**) MCF-7 cells were either left untreated or exposed to 50 J/m^2 of UV light followed by a 4 h recovery period before lysis. The array was incubated with 300 mg of cell lysate.

make a difference when investigating cell signaling pathways in cells and tissues. Arrays provide a rapid screening tool to identify unique changes in cellular protein phosphorylation, during drug and biomarker discovery that can be followed and scrutinized in more detail with more refined tools.

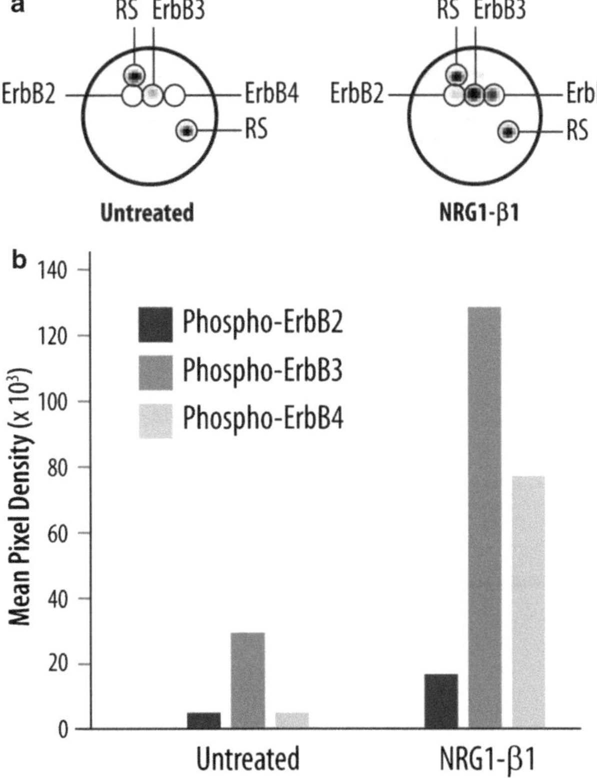

Fig. 5. NRG1-b1 induces phosphorylation of ErbB family receptors in breast cancer cells. (**a**) Cell lysates prepared from MDA-MB-453 human breast cancer cells, untreated or treated with recombinant human NRG1-β1 for 5 min, were assessed for the phosphorylation of 16 different receptor tyrosine kinases using the Proteome Profiler *96* Human Phospho-RTK Antibody Array 1 (R&D Systems Catalog #ARZ001). *RS* reference spot. (**b**) Histogram profiles for RTKs exhibiting significant phosphorylation were generated by quantifying the mean spot pixel densities from individual antibody spots using analytical software.

In a theme variation, instead of antibodies, a multiplicity of lysates have been spotted on membranes and then probed with individual antibodies tagged to an enzyme or fluorochrome. These are also called “reverse-phase arrays” (83–85). The reverse-phase array, coupled with laser tissue microcapture, is a method that places the cellular analysis of diseased cells as close as the in vivo condition. The capture of a selected cell population from a tissue allows for a direct analysis of phospho-proteins and other parameters, like gene expression, in the selectively isolated cells. Nanoliter amounts of a lysate are deposited on a membrane-coated slide and incubated with a tagged antibody detecting either the total protein or a phospho-epitope. Because of the high density of proteins on the array, phospho-specific antibodies are capable of detecting their targets like in a Western blot where all the phospho-protein is concentrated in a band. Although reverse-phase arrays have considerable potential, we have only seen reports of its use when detecting relatively abundant phosphorylations, as with ZAP70, GSK-3β, and ERK (86, 87). Reverse-phase arrays have also been done in epithelial cells from laser capture microdissections from prostate cancer (83). Thus, reverse-phase arrays could have a utility when screening for the effects of kinase inhibitors or activators on abundant signaling proteins, in homogeneous systems such as cell lines. This technique could be expanded to the simultaneous detection approach when investigating one or two phosphorylations by tagging phospho-specific antibodies with nonoverlapping fluorochromes such as Cy3 and Cy5. Reverse-phase arrays have similar limitations of sensitivity to the single antibody array described before; however, this approach may be amenable to potential amplification steps with the detection antibody that may increase the signal-to-noise ratio (88).

Regardless of the configuration used, it is critical to have additional supportive data on the antibody reagents used in the arrays. Researchers should always validate these antibodies by Western blotting and immunoprecipitation under conditions that closely resemble the sample preparation used in the arrays. One needs to be confident that the arrayed antibodies only recognize the phospho-target of interest and there is no interference or nonspecific binding among the different antibodies used.

In addition, supporting evidence illustrating the precision, sensitivity, linearity, and spike recovery from the antibody array vendor is needed to draw the right conclusions from the results.

We can anticipate further improvements and innovation in the area of arrays, from new developments in support to new fluorochromes and detection methods. However, much of the success using this application will be derived from the availability of phosphosite-specific antibodies with the highest affinity that are fully validated for this application. Although there is a continuous demand for applications focusing on large amounts of data points, there is also a stronger need for arrays covering unique pathways from the receptor level and downstream kinases and substrates. More important than the large amounts of data obtained by an array is the quality and reproducibility of the data.

Another variation of antibody multiplexing is the technology developed by Meso Scale Discovery (MSD). Capture antibodies are deposited at discrete locations of a 96 or 384 well plate. Carbon electrodes are incorporated in the bottom of a microplate, and electrochemiluminescent labels, bound to the detection antibody, emit light when stimulated with an electrical field, and the localized signal is captured by the electrodes and decoded by the instrument. The advantages of this approach include low background and interference from very few compounds, which allows the screening of large compound libraries for drug discovery. The signal is amplified with multiple excitation signals and the assay requires no wash steps, as only the labels near the electrode emit light. The main disadvantage of this approach is the need for an instrument to analyze the plates, as other imaging instruments are not equipped to detect the electrochemiluminescent signal (89).

3.1.2.2. Flow Cytometry

The dysregulation of unique signaling events has been implicated in multiple disease states, e.g., tumor progression and oncogenesis, and understanding these signaling abnormalities with the help of unique research tools has yielded remarkable results (90). For instance, the development of Gleevec, a BCR-Abl kinase inhibitor, was aided by an antibody to pTyr (91). Anti-pTyr antibodies have been instrumental in biochemical studies of protein tyrosine phosphorylation of chronic myeloid leukemia cells and many other diseases. Further advances to understand disease are on the horizon. For example, phospho-specific antibodies to several proteins, including CrkL Y207, one of the main targets of the BCR-Abl kinase, have been extremely useful in dissecting the abnormalities in signaling from this oncogenic kinase to further understand chronic myeloid leukemia (92).

As mentioned before, most methodologies, from Western blotting to mass spectrometry, collect data from cellular signaling events that represent the averaged response of a cell tissue like blood, or a homogenized cell culture dish. The signals recorded only represent the pooled responses from different cell types in the tissue, or cells in different states in the cell cycle, apoptotic, anergic, or with different ligand/receptor responses. Protein phosphorylations could either originate entirely from a small number of cells or from the whole cell population, each cell contributing to a small extent in the recorded phosphorylation.

On the other hand, flow cytometry allows the measurement of phosphorylation and signaling events from individual cells distinguished by their phenotype, e.g., cell cycle, cell surface markers, and receptor expression. The marrying and use of both phospho-specific antibodies and flow cytometry has been dubbed as phospho-flow. Phospho-flow can be leveraged in complex cell mixtures like blood and other tissue biopsies to reflect more accurately the in vivo conditions of individual cell types. Furthermore, the ability to tag and identify individual cell phenotypes with specific cell markers allows the monitoring of phosphorylation events in rare cell populations such as stem cells, naive and memory T cells, and regulatory T cells, among many others (93).

The combination of phosphorylation-dependent antibodies and other cellular characteristics such as apoptosis and cell cycle stage by flow cytometry has been documented for a while. Some of the first studies demonstrating the utility and potential of flow cytometry in dissecting signaling pathways were done in the lab of Zbignew Darzynkiewicz (94). These early studies used two antibodies to the Rb protein, one to the total Rb protein and the second to the under-phosphorylated Rb. Each antibody was conjugated to a different fluorochrome to distinguish their individual signals.

Furthermore, cells were stained for DNA with a third fluorochrome, illustrating the potential for the use of multiparameter flow cytometry to monitor changes in protein phosphorylation from normal and diseased cells. Darzynkiewicz's group quickly envisioned the potential of this technique for monitoring the prognostic value of antitumor agents in blood cells from patients. Currently, there are numerous examples of using this multiparameter approach to monitor unique protein phosphorylation changes together with DNA damage, apoptosis, cell cycle, stem cells, and other cellular processes (95–98).

Gary Nolan's lab at Stanford University has dramatically expanded and optimized phospho-flow by simultaneously measuring up to 11 different parameters in blood cells (99). Their early phospho-flow studies included up to eight different surface markers and three different phospho-specific antibodies detecting activated kinases. The number of analytes to be measured simultaneously depends on the selection and availability of the fluorochromes and the instrument capabilities. There have been rapid advances in the performance of flow cytometers and the availability of different sources of fluorochromes and materials such as nanocrystals, with sharp emission and nonoverlapping spectra. More recently, Nolan's lab developed an instrument reading phosphorylation events in the Raman spectra using phosphosite-specific antibodies conjugated to nanoparticles with the potential to expand the multiplexing capabilities (100).

Typically, a phospho-flow assay requires 10,000–100,000 cells/assay. Each assay can be accommodated in a well of a 96-well plate and is thus amenable for automation. Provided that the phospho-specific antibodies have been validated by other methods, the success of the phospho-flow application depends on several factors. These factors include cell fixation and permeabilization, and the careful selection of the fluorochromes. Fixation and permeabilization need to be determined empirically, as not all phospho-epitopes may be readily accessible to the antibodies. Alternatively, some of these phosphorylations may be rapidly dephosphorylated, or may not survive the fixation and permeabilization steps. Ideally, the fluorochromes need to be bright and have nonoverlapping spectra to be used in a multiplex assay. Currently, some of the most successfully used fluorochromes in phospho-flow are Alexa 488, Alexa, 647, PE, and Pacific Blue. Each fluorochrome-conjugated phospho-specific antibody needs to be tested and validated for the assay of interest (57–59).

Recently, multiplexed phospho-flow has elevated the understanding of complex signaling abnormalities in cells from AML, Lupus, and other diseases (101, 102). Phospho-flow has proven useful in the analysis of phosphorylation networks of key proteins like STATs, ERK1/2, and p38 MAPK in cells from AML patients. In unique cell populations, it has been possible to stratify AML patients according to their responsiveness to specific ligands. It has been shown that individual patients have unique differences in signaling pathways at the cellular level that may correlate with resistance or response to some therapeutics. By mapping the intrinsic and active signaling pathways in individual cells, phospho-flow is advancing our understanding on how key protein phosphorylations drive disease and potentially could help in predicting which patient may respond to certain therapeutics and drug regimes. In addition, phospho-flow may be useful when monitoring the off-target effects of kinase inhibitors by examining the different cell types in a tissue. For example, the cell specificity of a kinase inhibitor can be monitored by analyzing its effects on different cell subpopulations like T, B, and NK cells, and determine the side effects, and the safety of the drug or treatment (103).

Overall, the ability to monitor phosphorylations in blood cells makes phosflow one of the least intrusive techniques that may help to advance the understanding of many diseases. Further advances in this area will depend largely on improving the instrumentation and the availability of fully validated phosphosite-specific antibodies.

3.1.2.3. Bead Assays

There is another variation on the use of flow cytometry and phospho-specific antibodies. Two commercially available platforms are on the market, xMAP technology from Luminex and CBA (cytometric bead array) from BD Biosciences (104–106). Both platforms use the same principle of differentially colored micro-beads, with each color-coded bead conjugated to a specific capture antibody. The color-coded and antibody-conjugated beads are used to capture antigens from lysates or biological fluids, followed by incubation with a detection antibody conjugated to a fluorochrome like PE. The detection antibody has to recognize a different epitope of the captured protein, for instance a phospho-specific antibody, and it is possible to use a cocktail of detection antibodies, provided these are specific for their targets and there is no interference when they are mixed together.

This method is an adaptation of a sandwich ELISA performed on beads instead of wells. Potentially, bead assays have several advantages, one being the multiplex capability to analyze close to 100 analytes simultaneously. In addition, the fluorescent signal allows for a semiquantitative estimation of the phosphorylation. This method is ideal for drug or biomarker discovery, when using cell lysates, and its subsequent validation steps. For instance, bead assays can monitor the activation of several key pathways in the same cells, providing useful information on any off-target effects, in un-related pathways, of a particular compound. The sensitivity is very similar to any immunoprecipitation/Western approach because of the enriching step of the capture antibodies.

3.1.2.4. Immunocytochemistry

One of the ultimate goals for the use of phospho-specific antibodies is their potential ability to monitor kinase activity in intact cells and tissues. This goal is particularly challenging with tumors derived from solid tissues. The phosphorylation of a particular residue is either due to the activity of an upstream kinase, autophosphorylation, or phosphatase inactivation. Nonetheless, the analysis of protein phosphorylation in cells and tissues can give us a glimpse of the active pathways in that particular cell type and predict their behavior and the patient's response to therapeutics.

There is a recent explosion in the literature describing the phosphorylation profiles of key kinases and substrates from many different types of human tumors using immunohistochemical protocols. Some of these studies have been very informative and correlations have been found between protein phosphorylation and prognosis. For instance, the phosphorylation of Akt has been found to be a prognostic indicator in a subtype of tongue cancers and breast carcinomas (107, 108). On the other hand, there are numerous reports where phosphorylation of key proteins, like EGFR, do not appear to provide any prognostic benefit for cancer patients (109). This is in sharp contrast to the results obtained using more uniform protocol models such as mice where the availability of knock-outs and point mutations have firmly established the role of protein-phosphorylation in disease (110). There are many conflicting results in studies, sometimes analyzing the same phospho-protein, indicating both the complexity of the tumor environment and a lack of standard protocols for tissue fixation and preservation, in addition to the variety and quality variability of phospho-specific antibodies used in the studies.

Ideally, one should be able to retrieve a small tumor biopsy and rapidly determine the protein phosphorylation status of particular cells before and after drug treatments. If possible, this information must be quantitative; in other words, the protein phosphorylation and kinase activity should be provided in units per tissue section or on a per cell basis. Both immunocytochemistry and reverse-phase arrays have the potential to provide valuable information on signaling pathways from diseased tissues. This only stresses the need for more uniformity on the validation standards of the phosphosite-specific antibodies by immunocytochemistry for the results from these techniques are considered as a reliable prognosticator in biomedicine (111, 112).

Fluorescence resonance energy transfer (FRET) has recently been introduced to monitor and quantify phospho-proteins in cells and tissues. Some of these initial experiments used transfected cell lines where the inducible domain of CREB was flanked by two GFP domains of different wavelengths, that in proximity emitted FRET (113). In this assay, phosphorylation of the CREB domain by PKA reduced the FRET between the two GFP proteins. With a similar approach, the phosphorylation of PKCα and others have been monitored by FRET in fixed or live cultured cells (114). In this format, an antibody to PKCα tagged with Cy3 together with a phosphosite-specific antibody to PKCα T250 labeled with Cy5 emitted FRET when in proximity. Similar approaches have been used successfully to monitor the phosphorylation of many intracellular phosphorylation events in cell lines and tissues resolving the intracellular localization of the phosphorylations (115, 116, 117). Monitoring protein phosphorylation hence activation by FRET could open a simpler avenue to monitor patients' responses to therapeutics and as a prognostic tool. We envision that this methodology will continue to expand in different cell and tissue settings using phosphorylations to monitor drug efficacies and prognosis. The main limiting factor is to have a pair of validated antibodies with nonoverlapping epitopes, and at close enough proximity to transfer energy between two different fluorochromes. Thus, FRET has high potential in tumor immunohistochemistry where relative values of phosphorylation, and thus signal pathway activity, can be used for prognostic and pharmacodynamic purposes.

In conclusion, in recent years we have witnessed an explosion on the availability of phosphosite-specific antibodies as tools for biomedical research. Their exquisite sensitivity and specificity for unique phosphorylation sites, within important intracellular proteins, combined with molecular biology will continue to provide crucial information in basic research. Furthermore, their availability for assays monitoring disease cells from patients could help in the prognosis of disease, drug discovery, and target validation. We are certain that in the years to come, we will continue to witness the expansion on the number of phosphosite-specific antibodies as well as their incorporation into existing and novel applications. However, all of the expectations hinge on the reliability, validation, and reproducibility of these reagents, such that researchers all over the world will reach rapid and valid conclusions from their experiments. We believe that the suggestions listed in this short review may be a minimum standard for the generation and validation of phosphosite-specific antibodies. This is very important because of the myriad of phosphosite-specific antibodies provided by numerous companies, with a wide range in the characterization of their products. Finally, the tools and the strategy outlined in this chapter could be used for the validation of antibodies to other posttranslational modifications like acetylation, methylation, among others.

References

1. Ubersax, J. A., and Ferrell, J. E. Jr. (2007) Mechanisms of specificity in protein phosphorylation. Nat Rev Mol Cell Biol 8, 530–541.
2. Manning, G., Whyte, D. B., Martinez, R., Hunter, T., and Sudarsanam, S. (2002) The protein kinase complement of the human genome. Science 298, 1912–1343.
3. Tarrant, M. K., and Cole, P. A. (2009) The chemical biology of protein phosphorylation. Annu Rev Biochem 78, 797–825.
4. Alonso, A., Sasin, J., Bottini, N., Friedberg, I., Osterman, A., Godzik, A., et al. (2004) Protein phosphatases in the human genome. Cell 117, 699–711.
5. Hunter, T. (2009) Tyrosine phosphorylation: thirty years and counting. Curr Opin Cell Biol 21, 140–146.
6. Boyle, W. J., van der Geer, P., and Hunter T. (1991) Phosphopeptide mapping and phosphoamino acid analysis by two-dimensional separation on thin-layer cellulose plates. Methods Enzymol 201, 201–240.
7. Ross, A. H., Baltimore, D., and Eisen, H. N. (1981) Phosphotyrosine-containing proteins isolated by affinity chromatography with antibodies to a synthetic hapten. Nature 294, 654–656.
8. Nairn, A. C., Detre, J. A., Casnellie, J. E., and Greengard, P. (1982) Serum antibodies that distinguish between the phospho- and dephospho-forms of a phosphoprotein. Nature 299, 734–736.
9. Glenney, J. R. Jr., Zokas, L., and Kamps, M. P. (1988) Monoclonal antibodies to phosphotyrosine. J Immunol Methods 109, 277–285.
10. Kanakura, Y., Druker, B., Cannistra, S. A., Furukawa, Y., Torimoto, Y., and Griffin, J. D. (1990) Signal transduction of the human granulocyte-macrophage colony-stimulating factor and interleukin-3 receptors involves tyrosine phosphorylation of a common set of cytoplasmic proteins. Blood 76, 706–715.
11. Okamoto, M., Karasik, A., White, M. F., and Kahn, C. F. (1990) Epidermal growth factor stimulated phosphorylation of a 120-kilodalton endogenous substrate protein in rat hepatocytes. Biochemistry 29, 9489–9494.
12. Glenney, J. R. Jr. (1989) Tyrosine phosphorylation of a 22-kDa protein is correlated with transformation by Rous sarcoma virus. J Biol Chem 264, 20163–20166.
13. Kanner, S. B., Reynolds, A. B., Vines, R. R., and Parsons, J. T. (1990) Monoclonal antibodies to individual tyrosine-phosphorylated protein substrates of oncogene-encoded tyrosine kinases. Proc Natl Acad Sci USA 87, 3328–3332.
14. Towbin, H., Staehelin, T., and Gordon, J. (1979) Electrophoretic transfer of proteins from polyacrylamide gels to nitrocellulose sheets: procedure and some applications. Proc Natl Acad Sci USA 76, 4350–4354.
15. Glenney, J. R. Jr. (1992) Tyrosine phosphorylated proteins: mediators of signal transduction from the tyrosine kinases. Biochim Biophys Acta 1134, 113–127.
16. Sefton, B. M. (1982) Phosphorylation and metabolism of the transforming protein of Rous sarcoma virus. J Virol 41, 813–820.
17. Shankaran, H., Ippolito, D. L., Chrisler, W. B., Resat, H., Bollinger, N., Opresko, L. K., et al. (2009) Rapid and sustained nuclear-cytoplasmic ERK oscillations induced by epidermal growth factor. Mol Syst Biol 5, 1–13.
18. Lemeer, S., and Heck, A. J. (2009) The phosphoproteomics data explosion. Curr Opin Chem Biol 13, 414–420.
19. Kehoe, J. W., Velappan, N., Walbolt, M., Rasmussen, J., King, D., Lou, J., et al. (2006) Using phage display to select antibodies recognizing post-translational modifications independently of sequence context. Mol Cell Proteomics 5, 2350–2363.
20. Harlow, E., and Lane, D. (1988) Antibodies: a laboratory manual. New York: Cold Spring Harbor Laboratory Press.
21. Weng, Q-P., Kozlowski, M., Belham, C., Zhang, A., Comb, M. J., et al. (1995) Regulation of the p70 S6 kinase by phosphorylation *in vivo*. J Biol Chem 273, 16621–16629.
22. Yung, Y., Dolginov, Y., Zao, Z., Rubinfeld, H., Michael, D., Hanoch, T., et al. (1997) Detection of ERK activation by a novel monoclonal antibody. FEBS Lett 408, 292–296.
23. Campos-Gonzalez, R., and Glenney, J. R. Jr. (1991) Immunodetection of the ligand-activated receptor for epidermal growth factor. Growth Factors 4, 305–316.
24. Sternberger, L. A., and Sternberger, N. H. (1983) Monoclonal antibodies distinguish phosphorylated and nonphosphorylated forms of neurofilaments in situ. Proc Natl Acad Sci USA 80, 6126–6130.
25. Heffetz, D., Fridkin, M., and Zick, Y. (1991) Generation and use of antibodies to phosphothreonine. Methods Enzymol 201, 44–52.

26. Wang, J. Y. (1991) Generation and use of anti-phosphotyrosine antibodies raised against bacterially expressed abl protein. Methods Enzymol 201, 53–65.
27. Briand, J. P., Muller, S., and Van Regenmortel, M. H. V. (1985) Synthetic peptides as antigens: pitfalls of conjugation methods. J Immunol Methods 78, 59–69.
28. Epstein, R. J., Druker, B. J., Roberts, T. M., and Stiles, C. D. (1992) Synthetic phosphopeptide immunogens yield activation-specific antibodies to the c-erbB-2 receptor. Proc Natl Acad Sci USA 89, 10435–10439.
29. Edbauer, D., Cheng, D., Batterton, M. N., Wang, C.-F., Duong, D. M., et al. (2009) Identification and characterization of neuronal mitogen-activated protein kinase substrates using a specific phosphomotif antibody. Mol Cell Proteomics 8, 681–695.
30. Shi, Y., Dodson, G. E., Mukhopadhyay, P. S., Shanware, N. P., Trinh, A. T., and Tibbetts, R. S. (2007) Identification of carboxyl-terminal MCM3 phosphorylation sites using polyreactive phosphospecific antibodies. J Biol Chem 282, 9236–9243.
31. Tam, J. P., and Zavala, F. (1989) Multiple antigen peptide: a novel approach to increase detection sensitivity of synthetic peptides in solid-phase immunoassays. J Immunol Methods 124, 53–61.
32. Spieker-Polet, H., Sethupathi, P., Yam, P. C., and Knight, K. L. (1995) Rabbit monoclonal antibodies: generating a fusion partner to produce rabbit-rabbit hybridomas. Proc Natl Acad Sci USA 92, 9348–9352.
32a. Babcook, J. S., Leslie, K. B., Olsen, O. A., Salmon, R. A., and Schrader, J. H. (1996) Proc Natl Acad Sci USA 93, 7843–7848.
33. Ruff-Jamison, S., Campos-Gonzalez, R., and Glenney, J. R. Jr. (1991) Heavy and light variable region sequences and antibody properties of anti-phosphotyrosine antibodies reveal both common and distinct features. J Biol Chem 266, 6607–6613.
34. Ruff-Jamison, S., and Glenney, J. R. Jr. (1993) Requirements for both H and L chain V regions, VH and VK joining amino acids, and the unique H chain D region for the high affinity binding of an anti-phosphotyrosine antibody. J Immunol 150, 3389–3396.
35. Ruff-Jamison, S., and Glenner, J. R. Jr. (1993) Molecular modeling and site-directed mutagenesis of an anti-phosphotyrosine antibody predicts the combining site and allows the detection of higher affinity interactions. Protein Eng 6, 661–668.
36. Tuckey, C. D., and Noren, C. J. (2002) Selection for mutants improving expression of an anti-MAP kinase monolconal antibody by filamentous phage display. J Immunol Methods 270, 247–257.
37. Campos-Gonzalez, R., and Glenney, J. R. Jr. (1991) Temperature-dependent tyrosine phosphorylation of microtubule-associated protein kinase in epidermal growth factor-stimulated human fibroblasts. Cell Regul 2, 663–673.
38. Vaughan, M. H., Xia, X., Wang, X., Chronopoulou, E., Gao, G. J., Campos-Gonzalez, R., et al. (2007) Generation and characterization of a novel phospho-specific monoclonal antibody to p120-catenin serine 879 Hybridoma 26, 407–415.
39. Borrebaeck, C. A. K., Malmborg, A. C., Furebring, C., Michaelsson, A., Ward, S., Danielsoon, L., et al. (1992) Kinetic analysis of recombinant antibody-antigen interactions: relation between structural domains and antigen binding. Nat Biotechnol 10, 697–698.
40. Michalewski, M. P., Kaczmarski, W., Golabek, A., Kida, E., Kaczmarski, A., and Wisniewski, K. E. (2002) Immunoblotting with anti-phosphoamino acid antibodies: importance of the blocking solution. Anal Biochem 276, 254–257.
41. Song, K. S., Tang, Z., and Lisanti, M. P. (1997) Mutational analysis of the proteperties of caveolin-1. A novel role for the C-terminal domain in mediating homo-typic caveolin-caveolin interactions. J Biol Chem 271, 4398–4403.
42. Heinrich, M. C., Griffith, D. J., Druker, B. J., Wait, C. L., Ott, K. A., and Zigler, A. J. (2000) Inhibition of c-kit receptor tyrosine kinase kinase activity by STI571, a selective tyrosine kinase inhibitor. Blood 96, 925–932.
43. Nelson, E. A., Walker, S. R., Kepich, A., Gashin, L. B., Hideshima, T., Ikeda, H., et al. (2008) Nifuroxazide inhibits survival of multiple myeloma cells by directly inhibiting STAT3. Blood 112, 5095–5102.
44. Hardie, D. G., Haystead, T. A. J., and Sim, A. T. R. (2001) Use of okadaic acid to inhibit protein phosphatases in intact cells. Methods Enzymol 201, 531–538.
45. Gordon, J. A. (2001) Use of vanadate as protein-phosphotyrosine phosphatase inhibitor. Methods Enzymol 201, 581–586.
46. Evans, G. A., Garcia, G. G., Erwin, R., Howard, O. M., and Farrar, W. L. (1994) Pervanadate stimulates the effects of interleukin-2 (IL-2) in human T cells and provides evidence for the activation of two distinct tyrosine kinase pathways by IL-2. J Biol Chem 269, 23407–23412.

47. Ruff, S. J., Chen, K., and Cohen, S. (1997) Peroxovanadate induces tyrosine phosphorylation of multiple signaling proteins in mouse liver and kidney. J Biol Chem 272, 1263–1267.
48. Yang, T. T., Yu, R. Y., Agadir, A., Gao, G. J., Campos-Gonzalez, R., Tournier, C., and Chow, C. W. (2008) Integration of protein kinases mTOR and extracellular signal-regulated kinase 5 in regulating nucleocytoplasmic localization of NFATc4. Mol Cell Biol 28, 3489–3501.
49. Espina, V., Edmiston, K. H., Heiby, M., Pierobon, M., Sciro, M., Merritt, B., Banks, S., Deng, J., VanMeter, A. J., Geho, D. H., Pastore, L., Sennesh, J., Petricoin, E. F., and Liotta, L. A. (2008) A portrait of tissue phosphoprotein stability in the clinical tissue procurement process. Mol Cell Proteomics 7, 1998–2018.
50. Gilbert, C., Rollet-Labelle, E., Con, A. C., and Naccache, P. H. (2002) Immunoblotting and sequential lysis protocols for the analysis of tyrosine phosphorylation-dependent signaling. J Immunol Methods 271, 185–201.
51. Skolnik, E. Y., Lee, C. H., Batzer, A., Vicentini, L. M., Zhou, M., Daly, R., et al. (1993) The SH2/SH3 domain-containing protein GRB2 interacts with tyrosine-phosphorylated IRS1 and Sch: implications for insulin control of ras signaling. EMBO J 12, 1929–1936.
52. Barbareschi, M., Girlando, S., Mauri, F. M., Eccher, C., Mauri, F. A., Togni, R., et al. (1994) Quantitative growth fraction evaluation with MIB1 and Ki67 antibodies in breast carcinomas. Am J Clin Pathol 102, 171–175.
53. Mandell, J. W. (2003) Phosphorylation state-specific antibodies. Applications in investigative and diagnostic pathology. Am J Pathol 163, 1687–1698.
54. Bordeaux, J., Welsh, A. W., Agarwal, S., Killiam, E., Baquero, M. T., Hanna, J. A., Anagnostou, V. K., and Rimm, D. L. (2010) Antibody validation. Biotechniques 48, 197–209.
55. Mandell, J. W. (2008) Immunohistochemical assesment of protein phosphorylation state: the dream and the reality. Histochem Cell Biol 130, 465–471.
56. Kalyuzhny, A. E. (2009) The dark side of the immunohistochemical moon: industry. J Histochem Cytochem 57, 1099–1101.
57. Krutzik, P. O., Irish, J. M., Nolan, G. P., and Perez, O. D. (2004) Analysis of protein phosphorylation and cellular signaling events by flow cytometry: techniques and clinical applications. Clin Immunol 110, 206–221.
58. Perez, O. D., Mitchell, D., Campos, R., Gao, G-J., Li, L., and Nolan, G. P. (2005) Multiparameter analysis of intracellular phosphoepitopes in immunophenotyped cell populations by flow cytometry. Curr Protoc Cytom 6, 1–22.
59. Chow, S., Patel, H., Hedley, D. W. (2001) Measurement of MAP kinase activation by flow cytometry using phospho-specific antibodies to MEK and ERK: potential for pharmacodynamic monitoring of signal transduction inhibitors. Cytometry 46, 72–78.
60. Lombardi Givan, A. (2001) Flow cytometry. First principles. 2nd edition. New York: Wiley-Liss.
61. Smith, C. L., Debouk, C., Rosenberg, M., and Culp, J. S. (1988) Phosphorylation of ferine residue 89 of human adenovirus E1A proteins is responsible for their characteristic electrophoretic mobility shits, and its mutation affects biological fuction. J Virol 63, 1569–1577.
62. Wegener, A. D., and Jones, L. R. (1984) Phosphorylation-induced mobility shift in phospholamban in sodium dodecyl sulfate-polyacrylamide gels. Evidence for a protein structure consisting of multiple identical phosphorylatable subunits. J Biol Chem 259, 1834–1841.
63. Jorgensen, C. S., Jagd, M., Sorensen, B. K., McGuire, J., Barkholt, V., Hojrup, P., et al. (2004) Efficacy and compatibility with mass spectrometry of methods for elution of proteins from sodium dodecyl sulfate-polyacrylamide gels and polyvinyldifluoride membranes. Anal Biochem 330, 87–97.
64. Forrer, P., Tamaskovic, R., and Jaussi, R. (1998) Enzyme-linked immunosorbent assay for measurement of JNK, ERK and p38 kinase activities. Biol Chem 379, 1101–1111.
65. Suzuki, S., Tamai, K., and Yoshida, S. (2002) Enzyme-linked immunosorbent assay for distinct cyclin-dependent kinase activities using phosphorylation-site-specific anti pRB monoclonal antibodies. Anal Biochem 301, 65–74.
66. Offterdinger, M., and Bastiaens, P. I. (2008) Prolonged EGFR signaling by ERBB2-mediated sequestration at the plasma membrane. Traffic 9, 147–155.
67. Loos, T., Mortier, A., Gouwy, M., Ronsee, I., Put, W., Lenaerts, J-P., et al. (2008) Citrullination of CXCL10 and CXCL11 by peptidylarginine diminase: a naturally occurring posttranslational modification of chemockines and new dimension of immunoregulation. Blood 112, 2648–2656.

68. Ramos, J. W. (2008) The regulation of extracellular signal-regulated kinase (ERK) in mammalian cells. Int J Biochem Cell Biol 40, 2707–2719.

69. Gonzalez, E., and McGraw, T. E. (2009) The Akt kinases: isoform specificity in metabolism and cancer. Cell Cycle 8, 2502–2508.

70. Ribeiro-Oliveira, A. Jr., Franchi, G., Kola, B., Dalino, P., Pinheiro, S. V., Salahuddin, N., et al. (2008) Protein western array analysis in human pituitary tumors: insights and limitations. Endocr Relat Cancer 15, 1099–1114.

71. Pelech, S., Sutter, C., and Zhang, H. (2003) Kinetworks protein kinase multiblot analysis. Methods Mol Biol 218, 99–111.

72. Ciaccio, M. F., Wagner, J. P., Chuu, C.-P., Lauffenburger, D. A., and Jones, R. B. (2010) Systems analysis of EGF receptor signaling dynamics with microwestern arrays. Nat Methods 7, 148–155.

73. Rikova, K., Guo, A., Zeng, Q., Possemato, A., Yu, J., Haack, H., et al. (2007) Global survey of phosphotyrosine signaling identifies oncogenic kinases in lung cancer. Cell 131, 1190–1203.

74. Mayya, V., Lundgren, D. H., Hwang, S.-I., Rezaul, K., Wu, L., Eng, J. K., Rodionov, V., and Han, D. K. (2009) Quantitative phosphoproteomic analysis of T cell receptor signaling reveals system-wide modulation of protein-protein interactions. Sci Signal 2, ra46 1–ra46 16.

75. Fan, A., Deb-Basu, D., Orban, M. W., Gotlib, J. R., Natkunam, Y., O'Neill, R., et al. (2009) Nanofluidic proteomic assay for serial analysis of oncoprotein activation in clinical samples. Nat Med 15, 566–571.

76. Hughes, T. R., and Shoemaker, D. D. (2001) DNA microarrays for expression profiling. Curr Opin Chem Biol 5, 21–25.

77. Andersson, O., Kozlowski, M., Garachtchenko, T., Nikoloff, C., Lew, N., Litman, D. J., et al. (2005) Determination of relative protein abundance by internally normalized ratio algorithm with antibody arrays. J Proteome Res 4, 758–767.

78. Pelech, S., Jelinkova, L., Susor, A., Zhang, H., Shi, X., Pavlok, A., et al. (2008) Antibody microarray analyses of signal transduction protein expression and phosphorylation during porcine oocyte maturation. J Proteome Res 7, 2860–2871.

79. MacBeath, G. (2002) Protein microarrays and proteomics. Nat Genet 32, 526–532.

80. Russo, G., Zegar, C., and Giordano, A. (2003) Advantages and limitations of microarray technology in human cancer. Oncogene 22, 6497–6507.

81. Nielsen, U. B., Cardone, M. H., Sinskey, A. J., MacBeath, G., and Sorger, P. K. (2003) Profiling receptor kinase activation by using Ab microarrays. Proc Natl Acad Sci USA 100, 9330–9335.

82. Liu, X., Kim, P., Kirkland, R., Magonova, K., Liu, L., Zhang, I., et al. (2009) Prevalence of activated & total p95HER2 and other receptor tyrosine kinases in breast cancer. AACR San Antonio Breast Cancer Symposium Abstract #3053.

83. Paweletz, C. P., Charboneau, L., Bichsel, V. E., Simone, N. L., Chen, T., Gillespie, J. W., et al. (2001) Reverse phase protein microarrays which capture disease progression show activation of pro-survival pathways at the cancer invasion front. Oncogene 20, 1981–1989.

84. Espina, V., Woodhouse, E. C., Wulfkuhle, J., Asmussen, H. D., Petricoin, E. F., and Liotta, L. A., (2004) Protein microarray detection strategies: focus on direct detection technologies. J Immunol Methods 290, 121–133.

85. Spurrier, B., Ramalingam, S., and Nishizuka, S. (2008) Reverse-phase protein lysate microarrays for cell signaling analysis. Nat Protoc 3, 1796–1808.

86. Chan, S. M., Ermann, J., Su, L., Fathman, C. G., and Utz, P. J. (2004) Protein microarrays for multiplex analysis of signal transduction pathways. Nat Med 10, 1390–1396.

87. Natarajan Mendes, K., Nicorici, D., Cogdell, D., Tabus, I., Ylf-Harga, O., Guerra, R., et al. (2007) Analysis of signaling pathways in 90 cancer cell lines by protein lysate array. J Proteome Res 6, 2753–2767.

88. Schweitzer, B., Roberts, S., Grimwade, B., Shao, W., Wang, M., Fu, Q., et al. (2002) Multiplex protein profiling on microarrays by rolling-circle amplification. Nat Biotechnol 20, 359–365.

89. Dahut, W. L., Scripture, C., Posadas, E., Jain, L., Gulley, J. L., Arlen, P. M., et al. (2008) A phase II clinical trial of sorafenib in androgen-independent prostate cancer. Clin Cancer Res 14, 209–214.

90. Tan, C. S. H., Bodenmiller, B., Pascualescu, A., Jovanovic, M., Hengartner, M. O., Jorgensen, C., et al. (2009) Comparative analysis reveals conserved protein phosphorylation networks implicated in multiple diseases. Sci Signal 2, ra39 1–ra39 13.

91. Drucker, B. J. (2009) Perspectives on the development of imatinib and the future of cancer research. Nat Med 10, 1149–1152.

92. Jilani, I., Kanttarjian, H., Gorre, M., Cortes, J., Ottmann, O., Bhalla, K., et al. (2008) Phosphorylation levels of BCR-ABL, CrkL, AKT, and STAT5 in imatinib-resistant

chronic myeloid leukemia cells implicate alternative pathway usage as a survival strategy. Leuk Res 32, 643–649.

93. Irish, J. M., Kotecha, N., and Nolan, G. P. (2006) Mapping normal and cancer cell signaling networks: towards single-cell proteomics. Nat Rev Cancer 6, 146–155.

94. Juan, G., Gruenwald, S., and Darzynkiewicz, Z. (1998) Phosphorylation of retinoblastoma susceptibility gene protein assayed in individual lymphocytes during their mitogenic stimulation. Exp Cell Res 239, 104–110.

95. Juan, G., Traganos, F., and Darzynkiewicz, Z. (1999) Histone H3 phosphorylation in human monocytes and during HL-60 cell differentiation. Exp Cell Res 246, 212–220.

96. Zell, T., Khoruts, A., Ingulli, E., Bonnevier, J. L., Mueller, D. L., and Jenkins, M. K. (2001) Single-cell analysis of signal transduction in CD4 T cells stimulated by antigen in vivo. Proc Natl Acad Sci USA 98, 10805–10810.

97. Krutzik, P. O., Hale, M. B., and Nolan, G. P. (2005) Characterization of the murine immunological signaling network with phosphospecific flow cytometry. J Immunol 175, 2366–2373.

98. Lu, X. P., Alpdogan, O., Lin, J., Balderas, R., Campos-Gonzalez, R., Wang, X., et al. (2008) STAT-3 and ERK1/2 phosphorylation are critical for T-cell activation and graft-versus-host disease. Blood 112, 5254–5258.

99. Perez, O. D., and Nolan, G. P. (2002) Simultaneous measurement of multiple active kinase states using polychromatic flow cytometry. Nat Biotechnol 20, 155–162.

100. Shachaf, C. M., Elchuri, S. V., Koh, A. L., Zhu, J., Nguyen, L. N., Mitchell, D. J., et al. (2009) A novel method for detection of phosphorylation in single cells by surface enhanced raman scattering (SERS) using composite organic-inorganic nanoparticles. PLoS One 4, e5206 1–e5206 12

101. Irish, J. M., Hovland, R., Krutzik, P. O., Perez, O. D., Bruserud, O., Gjertsen, B. T., and Nolan, G.P. (2004) Single cell profiling of potentiated phospho-protein networks in cancer cells. Cell 118, 217–228.

102. Hale, M. B., Krutzik, P. O., Samra, S. S., Crane, J. M., and Nolan, G. P. (2009) Stage dependent aberrant regulation of cytokine-STAT signaling in murine systemic lupus erythematosus. PLoS One 4, e6756 1–e6756 10.

103. Krutzik, P. O., and Nolan, G. P. (2006) Fluorescent cell barcoding in flow cytometry allows high-throughput drug screening and signaling profiling. Nat Methods 3, 361–368.

104. Pritchard, J. R., Cosgrove, B. D., Hemann, M. T., Griffith, L. G., Wands, J. R., and Lauffenburger, D. A. (2009) Three-kinase inhibitor combination recreates multipathway effects of a geldanamycin analogue on hepatocellular carcinoma cell death. Mol Cancer Ther 8, 2183–2192.

105. Morgan, E., Varro, R., Sepulveda, H., Ember, J. A., Apgar, J., Wilson, J., et al. (2004) Cytometric bead array: a multiplexed assay platform with applications in various areas of biology. Clin Immunol 110, 252–266.

106. Chen, L., Huynh, L., Apgar, J., Tang, L., Rassenti, L., Weiss, A., and Kipps, T. J. (2008) ZA-70 enhances IgM signaling independent of its kinase activity in chronic lymphocytic leukemia. Blood 111, 2685–2692.

107. Massarelli, E., Liu, D. D., Lee, J. J., El-Naggar, A. K., Lo Muzio, L., Staibano, S., et al. (2005) Akt activation correlates with adverse outcome in tongue cancer. Cancer 104, 2430–2436.

108. Smitz, K. J., Otterbach, F., Callies, R., Levkau, B., Holscher, M., Hoffmann, O., et al. (2004) Prognostic relevance of activated Akt kinase in node-negative breast cancer: a clinicopathological study of 99 cases. Mol Pathol 17, 15–21.

109. Okamoto, I., Kenyon, L. C., Emlet, D. R., Mori, T., Sasaki, J., Hirosako, S., et al. (2003) Expression of activated EGFRvIII in small cell lung cancer. Cancer Sci 94, 50–56.

110. D'Andrea, M. R., Mel, J. M., Tuman, R. W., Galemmo, R. A., and Johnson, D. L. (2005) Validation of in vivo pharmacodynamic activity of a novel PDGF receptor tyrosine kinase inhibitor using immunohistochemistry and quantitative image analysis. Mol Cancer Ther 4, 1198–1204.

111. Kong, A., Leboucher, P., Leek, R., Calleja, V., Winter, S., Harris, A., et al. (2006) Prognostic value of an activation state marker for epidermal growth factor receptor in tissue microarrays of head and neck cancer. Cancer Res 66, 2834–2843.

112. VanMeter, A. J., Rodriguez, A. S., Bowman, E. D., Jen, J., Harris, C. C., Deng, J., et al. (2008) Laser capture microdissection and protein microarray analysis of human non-small cell lung cancer. Mol Cell Proteomics 7, 1902–1924.

113. Nagai, Y., Miyasaki, M., Akoi, R., Zama, T., Inouye, S., Hirose, K., et al. (2000) A fluorescent indicator for visualizing cAMP-induced phosphorylation in vivo. Nat Biotechnol 18, 313–316.

114. Ng, T., Squire, A., Hansra, G., Bornancin, F., Prevostel, C., Hanby, A., et al. (1999)

Imaging protein kinase Calpha activation in cells. Science 283, 2085–2089.

115. Tomida, T., Takekawa, M., O'Grady, P., and Saito, H. (2009) Stimulus-specific distinctions in spatial and temporal dynamics of stress-activated protein kinase kinase kinases revealed by a fluorescence resonance energy transfer biosensor. Mol Cell Biol 29, 6117–6127.

116. Ting, A. Y., Kain, K. H., Klemke, R. L., and Tsien, R. Y. (2001) Genetically encoded fluorescent reporters of protein tyrosine kinase activities in living cells. Proc Natl Acad Sci USA 98, 15003–15008.

117. Kelleher, M. T., Fruhwirth, G., Patel, G., Ofo, E., Festy, F., Barber, P. R., et al. (2009) The potential of optical proteomic technologies to individualize prognosis and guide rational treatment for cancer patients. Target Oncol 4, 235–252.

Chapter 2

Selection and Validation of Antibodies for Signal Transduction Immunohistochemistry

Juraj Bodo and Eric D. Hsi

Abstract

The in situ expression levels and subcellular localization of molecules involved in signal transduction using specific antibodies can be useful for prognosis and diagnosis of human diseases such as cancer. In addition, it has the potential to be helpful in monitoring biologic response to targeted therapies. The increasing availability of such antibodies makes these studies feasible. However, compared to typical immunohistochemical stains in which stabile molecules such as cytokeratins are targeted, additional validation may be required for signal transduction immunohistochemistry.

Key words: Antibody validation, Phosphoprotein, Immunohistochemistry

1. Introduction

The study of signal transduction pathways is nearly synonymous with the study of phosphoproteins (1). Phosphoproteins are involved in regulating nearly all cellular functions. Phosphorylation states can determine key properties of proteins including enzyme activity, protein–protein physical interactions, protein–nucleic acid physical interactions, and subcellular localization. The first antibody against phosphoproteins was discovered almost 30 years ago (2). Ten years later, researchers successfully developed antibodies specific for phosphorylated tyrosine and threonine (3, 4), but these antibodies were mostly useful in Western blotting (WB) analyses where different phosphoproteins could be determined by their molecular weight. Nevertheless, researchers applied these antibodies in immunohistochemical studies of human cancer tissues and discovered a significant increase in phosphorylated proteins in these tissues (5).

Alexander E. Kalyuzhny (ed.), *Signal Transduction Immunohistochemistry: Methods and Protocols*, Methods in Molecular Biology, vol. 717, DOI 10.1007/978-1-61779-024-9_2,

The current commercial availability of phosphorylation–state-specific antibodies makes the in situ study of signal transduction molecules possible and is opening many opportunities in diagnostic pathology and targeted therapeutic monitoring. However, an extra degree of validation may be required for the selection of the antibody suitable for signal transduction immunohistochemistry (IHC), because the detection of the phosphoproteins is highly dependent on the antibody sensitivity and specificity, as well as tissue integrity.

2. Materials

2.1. Cell Controls for Validation of Anti-pAKT Antibody

1. HT-29 cell line (ATCC, Manassas, VA).
2. McCoy's medium (Lonza, Basel, Switzerland) supplemented with 10% fetal bovine serum (Invitrogen, Carlsbad, CA).
3. Hydrogen peroxide.
4. Wortmannin (Cell Signaling Technology, Danvers, MA).
5. 10% Buffered formalin.

2.2. Automated IHC

1. HistoGel system (Richard-Allan Scientific, Kalamazoo, MI).
2. Immunostainer Discovery (Ventana Medical Systems, Tucson, AZ).
3. Cell Conditioning 1 (CC1, pH = 8, Ventana Medical Systems).
4. Reaction buffer (Ventana Medical Systems).
5. Background Sniper (Biocare Madical, Concord, CA).
6. Endogenous Biotin Blocking Kit (Ventana Medical Systems).
7. Anti-pAKT (S473) (736E11) antibody (Cell Signaling Technology).
8. OmniMap anti-rabbit HRP (Ventana Medical Systems).
9. ChromoMap kit (Ventana Medical Systems).
10. Hematoxylin.
11. 95% Dehydrant, 100% dehydrant, xylenes (Richard-Allan Scientific).
12. Cytoseal XYL (Richard-Allan Scientific).

3. Methods

3.1. Search Primary Antibodies: Vendors and Primary Literature

There are some free online search tools available that allow the user to search multiple companies at once. For example, www.antibodybeyond.com, www.linscottsdirectory.com, www.biocompare.com, or www.antibodydirectory.com are good places to start. With these tools, the antibody search can often be narrowed by antigen, species, type (monoclonal, polyclonal), or application. Importantly, one of the crucial requirements is that the chosen specific antibody must be completely described by the commercial vendor. The species, type (monoclonal vs. polyclonal), subclass, the structure of the immunizing antigen, and specificity must be known (6). This can often eliminate potential problems of using less-specific antibodies or antibodies not suitable for IHC. Furthermore, if peer-reviewed literature exists, this can give a realistic review of the performance of the antibody in a particular setting. Such literature can also provide the basis for optimizing immunoreactivy of the antibody (7).

3.1.1. Monoclonal and Polyclonal Antibodies

Based on the type of production, antibodies are divided into monoclonal and polyclonal. In general, monoclonal antibodies are produced by a single B-cell clone using hybridoma techniques. This provides excellent specificity; the antibody binds to a single epitope and it is less likely to cross-react with other proteins. The first monoclonal sequence-specific phosphoprotein antibody was successfully produced in the early 1990s (8). The phosphopeptide immunization approach was later applied but with the production of polyclonal instead of monoclonal antibodies (9).

Monoclonal antibodies have very high homogeneity in comparison to polyclonal antibodies that are prone to "batch-to-batch" variability. Usually, if experimental conditions are kept constant, results will be highly reproducible between experiments. Thus, one of the disadvantages of using polyclonal antibodies, especially for quantitative analysis of signal transduction proteins, is that every new lot must be reevaluated. Polyclonal antibodies contain a mix of antibodies recognizing multiple epitopes on any one antigen, and thus may cause higher nonspecific background staining and be less specific than monoclonal antibodies. On the other hand, polyclonal antibodies may demonstrate high affinity allowing higher dilutions when compared to monoclonal antibodies. Finally, polyclonal antibodies are also more tolerant to changes in the antigen induced by sample processing. Although it seems that monoclonal antibodies are more suitable for signal transduction IHC, a few examples of particularly well-suited polyclonal antibodies are exceptions to this generalization.

3.2. Validation of the Antibody

In the clinical laboratory, we are "at the mercy" of manufacturers. Regulations such as those governing analyte-specific reagents are meant to ensure that reagents are correctly manufactured and labeled as to content. For most antibodies in diagnostic use, we have some idea, based on the results of positive and negative control tissues, that antibodies are specific. Since most applications in routine IHC are lineage or cell-of-origin assignments, normal tissues with stable targets are often suitable. However, signal transduction proteins are very labile by nature; therefore, we believe that an extra degree of diligence may be required (see Note 1). Although the exact method for this type of antibody validation is not established, several independent methods can be used to validate the antibody specificity, including immunostaining of different stimulated and unstimulated cultured cells (see Note 2), immunostaining of tissue with peptide preincubation controls, and genetic (knockout) controls (see Notes 3 and 4). Archival primary human tissues are difficult to use as controls since there may be substantial variability in tissue processing, most importantly delays in fixation, that will affect phosphoprotein levels.

3.2.1. Automated Immunohistochemistry of pAKT

1. 2×10^7 cells are treated with 5 mM hydrogen peroxide for 15 min and 1 μM wortmannin for 1 h, harvested, and fixed in 50 ml of 10% buffered formalin overnight at 4°C with gentle shaking.
2. Paraffin-embedded cell blocks are prepared using the "HistoGel" system for IHC. All samples are processed overnight using the conventional histological techniques and embedded in paraffin, using an automatic apparatus (Tissue-TEK VIP, Miles Scientific). The melted wax temperature does not exceed 60°C.
3. IHC is performed using an automated immunostainer. After deparaffinization and heat-induced epitope retrieval (HIER) using standard Cell Conditioning 1, slides are incubated in Background Sniper for 30 min. Subsequently, Avidin and Biotin block is applied for 20 min.
4. Samples are incubated with 1:50 dilution of the anti-pAKT (S473) (736E11) for 2 h at room temperature (see Notes 5 and 6).
5. OmniMap anti-rabbit HRP is added and incubated for 30 min. Staining is then visualized by using ChromoMap DAB kit (see Notes 7 and 8).
6. Finally, cells are counterstained with hematoxylin for 1 min. After rinsing, the slides are submerged for 2 min twice in 95% dehydrant, 100% dehydrant, and in xylenes and slides are mounted in Cytoseal XYL.

If the antibody is suitable for WB, paired WB and IHC can be performed (Fig. 1a). In this case, WB should result in detection of

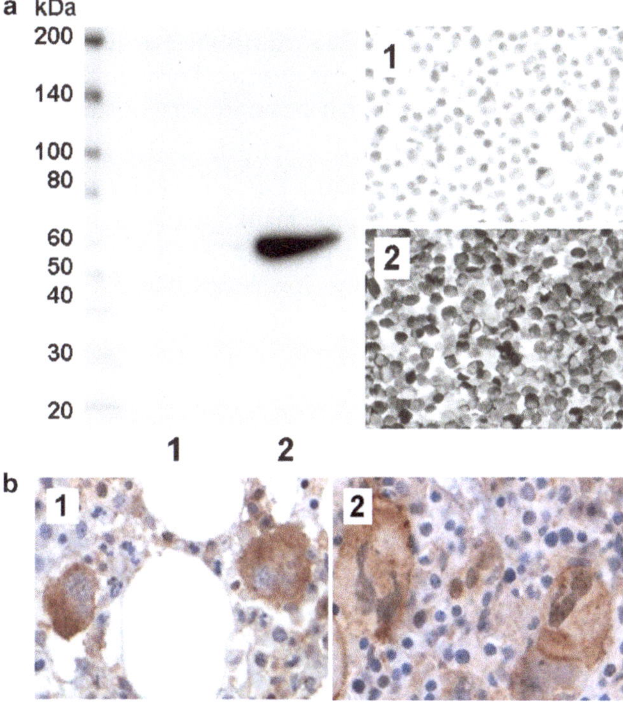

Fig. 1. (**a**) Western blot analysis of pAKT (S473) of HT-29 culture cells treated with (*1*) AKT inhibitor, wortmannin and (*2*) hydrogen peroxide. Negative (*1*) and positive (*2*) IHC stainings correspond to lanes of WB. (**b**) pSTAT5 (Y694/9) IHC staining of (*1*) normal bone marrow showing cytoplasmic localization of pSTAT5 in megakaryocytes and of (*2*) refractory anemia with ringed sideroblasts associated with thrombocytosis sample with nuclear positive pSTAT5 megakaryocytes.

a single band (or multiple bands if family members share the same motifs) of appropriate molecular weight. Alternatively, methods such as ELISA, intracellular flow cytometry (10), reverse phase protein microarray (11), or mass spectrometry (12, 13) can prove the status of signal transduction protein levels in the cells that can be prepared as controls for IHC. Using these known positive and negative controls, the specificity of the antibody can be validated followed with further confirmation in particular tissues.

4. Notes

1. Since little data of signal transduction protein patterns in human tissues exist, one would need to use judgment, informed by knowledge of active biologic processes in various cell types (Fig. 1b), as to whether the antibody remains specific and is sensitive enough to be used in studied tissues. This can be even more complicated, because some signal transduction proteins can oscillate between different cell

compartments in a very short time after their activation (14). Antibodies against different sites of the same protein or against total protein that produce the same staining pattern may be an important strategy for establishing specificity.

2. For semiquantitative analysis of signal transduction proteins using IHC, a further method of validation is needed. Cell lines manipulated to produce different levels of intended targets (Fig. 2a) or cell lines known to express high, medium, and low levels of (Fig. 2b) a particular signal transduction

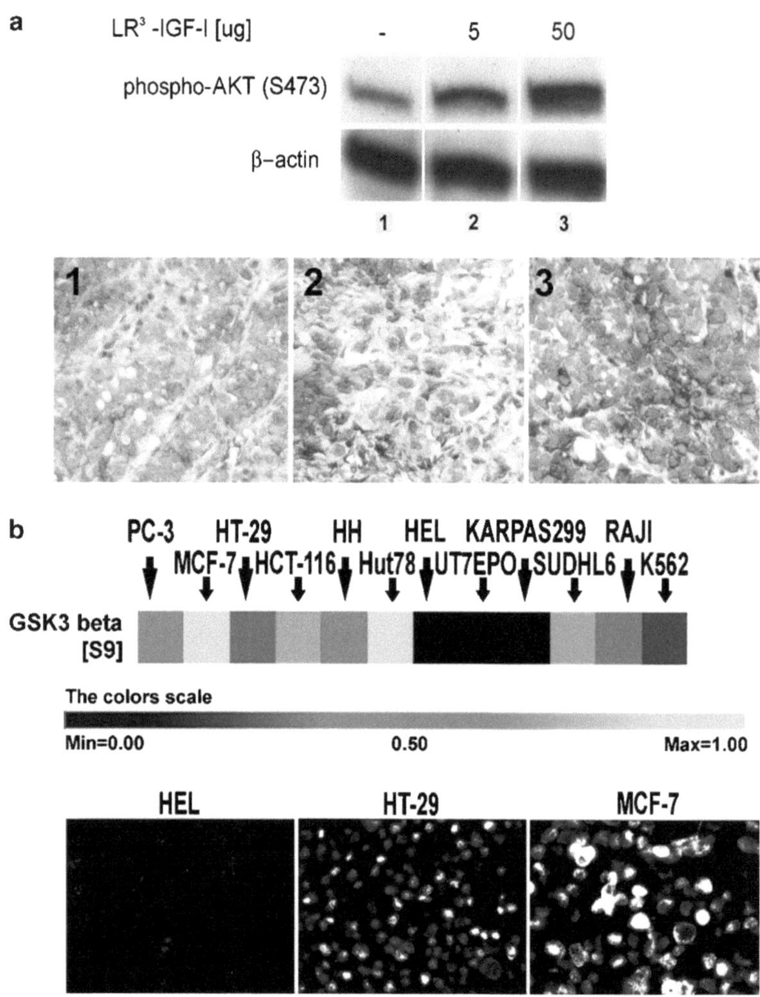

Fig. 2. (**a**) Western blot and IHC analysis of pAKT (S473) in HT-29 mouse xenografts treated with LR3-IGF1. Numbers correspond to lanes of WB. (**b**) The expression of pGSK3 beta (S9) was determined by using Kinetworks phosphoprotein screen and Quantum dot based immunofluorescence staining in the panel of various human cell lines. Phosphoproteins are normalized to the cell line with the highest expression of corresponding phosphoprotein, and relative intensities are shown on the heat map on a 0–1 scale. Three human cell lines with low, medium, and high expression of pGSK3beta (S9) represent different intensities of Quantum dot based staining.

protein may be useful to confirm the specificity and sensitivity. Such information is available for some phosphoproteins (12, 15, 16), but more published data are needed. Based on the same principles, validated antibody can be used for the truly quantitative analysis, using immunofluorescence assay (Fig. 2b).

3. Besides validation methods mentioned above, antigen adsorption, where the antibody is mixed with the appropriate purified antigen before application to the tissue section or cells, is a powerful way to look for nonspecific reactivity of the antibody. Using this approach, we can confirm, especially with monoclonal antibodies, that the cloning process was performed efficiently and only one antigen epitope is recognized. However, this does not provide information on whether other tissue proteins may cross-react with the tested antibody.
4. The last method, especially for validation of antiphosphoprotein antibody, is using (as negative control) tissue sections or cells pretreated by alkaline phosphatase. Although this method excludes only cross-reaction with nonphosphorylated proteins, it is still more useful than using omission controls (staining without primary antibody) that can be only used as a control for evaluation of the specificity of the secondary antibody.
5. The rate of binding between antigen and antibody is dependent on the affinity constant. This constant can be affected by many factors such as temperature, pH, and buffer type. Further, changing the antibody concentrations can also control the amount of antibody–antigen complex formation. Often, the manufacturing company has guidelines for starting dilutions. Typically, two to tenfold dilutions above and below the manufacturer's recommended dilution provide a good starting point. Depending on the type of antibody and type of tissue, a range of 1–5 μg/ml should be used for an initial titration. Moreover, it is important to realize that antibody dilutions may vary between different tissues.
6. If an antibody has not been tested yet, a systematic approach using a wide range of dilutions, as well as different antigen-retrieval methods, is needed. A checkerboard design combining all combinations of chosen retrieval conditions and primary antibody concentrations will allow one to quickly identify promising conditions that one can focus on during a more detailed study. For example, a broad checkerboard experiment may identify that low pH heat-induced epitope retrieval stains positive and negative control tissue appropriately. Focusing secondary experiments on more detailed antibody concentrations, epitope retrieval, incubation times, and temperatures can then be systematically tested. Incubation

for most routine IHC protocols is 30–90 min at room temperature. However, for detection of signal transduction proteins, it is better to use longer incubation times with higher dilutions in order to eliminate nonspecific staining.

7. Primary antibodies can be directly labeled, but this staining is rarely used, usually due to low sensitivity. Secondary antibodies, labeled with the first step in a detection (such as biotin) system, are widely available from commercial sources and are generally of good quality. Because high quality secondary antibodies are widely available, selection is only based on the type of the primary antibody. The most important criterion is that secondary antibody must be directed against the species in which the primary antibody was raised. Furthermore, if the primary antibody is monoclonal, the secondary antibody should match the class (isotype) of the primary antibody. For example, if the primary antibody is rabbit IgG, an anti-rabbit IgG should be used.
8. Selecting an optimal secondary antibody and optimizing the immunoreactivity can improve immunostaining and reduce false-positive or -negative staining. However, especially for poorly expressed signal transduction proteins, special polymer detection systems (such as EnVision (DakoCytomation, Carpinteria, CA), ImmPRESS (Vector Laboratories, Burlingame, CA), or MACH4 (Biocare Medicals, Concord CA)) that amplify signal may be more useful.

Acknowledgments

The authors acknowledge Lisa Durkin for her technical expertise in immunostaining.

References

1. Mandell JW. (2003) Phosphorylation state-specific antibodies: applications in investigative and diagnostic pathology. *Am J Pathol* **163**, 1687–1698.
2. Ross AH, Baltimore D, Eisen HN. (1981) Phosphotyrosine-containing proteins isolated by affinity chromatography with antibodies to a synthetic hapten. *Nature* **294**, 654–656.
3. Glenney JR, Jr., Zokas L, Kamps MP. (1988) Monoclonal antibodies to phosphotyrosine. *J Immunol Methods* **109**, 277–285.
4. Heffetz D, Fridkin M, Zick Y. (1991) Generation and use of antibodies to phosphothreonine. *Methods Enzymol* **201**, 44–53.
5. Ogawa R, Ohtsuka M, Sasadaira H et al. (1985) Increase of phosphotyrosine-containing proteins in human carcinomas. *Jpn J Cancer Res* **76**, 1049–1055.
6. Saper CB. (2005) An open letter to our readers on the use of antibodies. *J Comp Neurol* **493**, 477–478.

7. Hsi ED. (2001) A practical approach for evaluating new antibodies in the clinical immunohistochemistry laboratory. *Arch Pathol Lab Med* **125**, 289–294.
8. Yano T, Taura C, Shibata M et al. (1991) A monoclonal antibody to the phosphorylated form of glial fibrillary acidic protein: application to a non-radioactive method for measuring protein kinase activities. *Biochem Biophys Res Commun* **175**, 1144–1151.
9. Czernik AJ, Girault JA, Nairn AC et al. (1991) Production of phosphorylation state-specific antibodies. *Methods Enzymol* **201**, 264–283.
10. Krutzik PO, Irish JM, Nolan GP, Perez OD. (2004) Analysis of protein phosphorylation and cellular signaling events by flow cytometry: techniques and clinical applications. *Clin Immunol* **110**, 206–221.
11. Espina V, Edmiston KH, Heiby M et al. (2008) A portrait of tissue phosphoprotein stability in the clinical tissue procurement process. *Mol Cell Proteomics* 7, 1998–2018.
12. Rikova K, Guo A, Zeng Q et al. (2007) Global survey of phosphotyrosine signaling identifies oncogenic kinases in lung cancer. *Cell* **131**, 1190–1203.
13. Zheng H, Hu P, Quinn DF, Wang YK. (2005) Phosphotyrosine proteomic study of interferon alpha signaling pathway using a combination of immunoprecipitation and immobilized metal affinity chromatography. *Mol Cell Proteomics* **4**, 721–730.
14. Shankaran H, Ippolito DL, Chrisler WB et al. (2009) Rapid and sustained nuclear-cytoplasmic ERK oscillations induced by epidermal growth factor. *Mol Syst Biol* **5**, 332.
15. Bodo J, Durkin L, Hsi ED. (2009) Quantitative in situ detection of phosphoproteins in fixed tissues using quantum dot technology. *J Histochem Cytochem* **57**, 701–708.
16. Fantin VR, Loboda A, Paweletz CP et al. (2008) Constitutive activation of signal transducers and activators of transcription predicts vorinostat resistance in cutaneous T-cell lymphoma. *Cancer Res* **68**, 3785–3794.

Chapter 3

An Overview of Western Blotting for Determining Antibody Specificities for Immunohistochemistry

Biji T. Kurien, Yaser Dorri, Skyler Dillon, Anil Dsouza, and R. Hal Scofield

Abstract

Despite its overall simplicity, protein blotting or Western blotting has been proven to be a powerful procedure for the immunodetection of proteins, especially those that are of low abundance, following electrophoresis. The usefulness of this procedure stems from its ability to provide simultaneous resolution of multiple immunogenic antigens within a sample for detection by specific antibodies. Protein blotting has evolved greatly since its inception and researchers have a variety of ways and means to carry out this transfer. This procedure is used in combination with other important antibody-based detection methods such as enzyme-linked immunosorbant assay and immunohistochemistry to provide confirmation of results both in research and diagnostic testing. Specificity of antibodies used for immunohistochemistry is of critical importance and therefore Western blot is a "must" to address antibodies' specificity.

Key words: Western blotting, SDS-PAGE, Nitrocellulose

1. Introduction

The process of protein or nucleic acid transfer to microporous membranes is referred to as "blotting." This term includes both manual sample deposition (spotting) and transfer from planar gels. In this procedure, termed as protein blotting or Western blotting (WB) (1, 2), proteins resolved on sodium dodecyl sulfate polyacrylamide gel electrophoresis (SDS-PAGE) gels are typically transferred to adsorbent membrane supports under the influence of an electric current. Nucleic acids are transferred from agarose gels to a membrane support, in a procedure known as Southern blotting, through capillary action. The protein blotting procedure evolved from DNA (Southern) blotting (3) and RNA

Alexander E. Kalyuzhny (ed.), *Signal Transduction Immunohistochemistry: Methods and Protocols*, Methods in Molecular Biology, vol. 717, DOI 10.1007/978-1-61779-024-9_3,

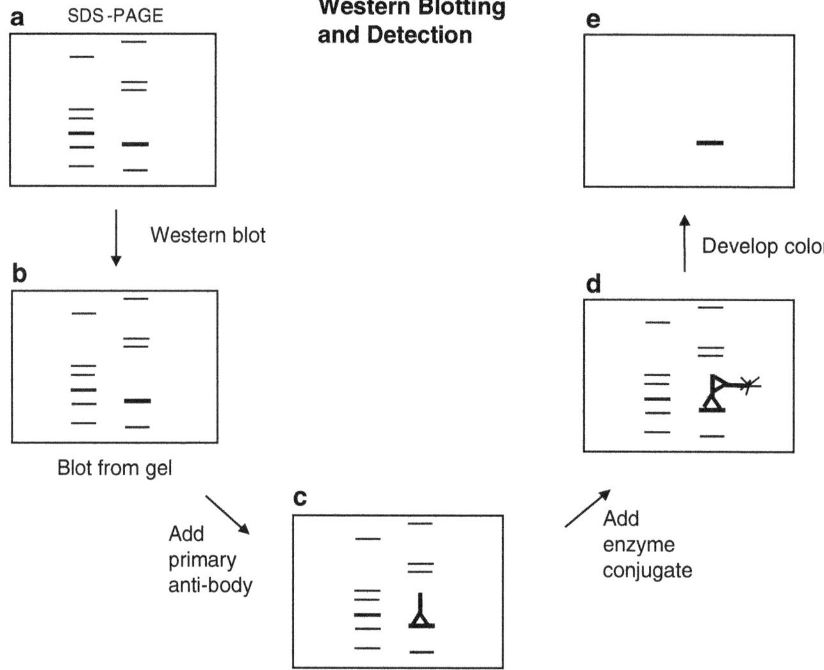

Fig. 1. Schematic representation of Western blotting and detection procedure. (**a**) Unstained SDS-PAGE gel prior to Western blot. The bands shown are hypothetical. (**b**) Exact replica of SDS-PAGE gel obtained as a blot following Western transfer. (**c**) Primary antibody binding to a specific band on the blot. (**d**) Secondary antibody conjugated to an enzyme (alkaline phosphatase or horse radish peroxidase) binding to primary antibody. (**e**) Color development of specific band (reproduced from ref. (10) with permission from Elsevier).

(Northern) blotting (4). "Western blotting" was coined to describe (5) this procedure to retain the "geographic" naming tradition initiated by Southern (3). The blotted proteins form an exact replica of the gel and have proved to be the starting step for a variety of experiments. The subsequent employment of antibody probes directed against the membrane-bound proteins (immunoblotting) has revolutionized the field of immunology (Fig. 1). Dot blotting refers to the analysis of proteins applied directly to the membrane rather than after transfer from a gel.

Until the introduction of protein blotting, the usefulness of the immense resolving power of SDS-PAGE (6) was limited. The prime reason for this was that the separated proteins in the gel matrix were difficult to access with molecular probes. Protein transfer followed by immunodetection has found wide application in biomedical research. This method (1, 2) is a powerful tool to detect and characterize a multitude of proteins, especially proteins that are of low abundance. WB offers the following specific advantages: (i) moist membranes are pliable and are easy to handle compared to gels, (ii) there is easy accessibility of the proteins immobilized on the membrane to different ligands, (iii) requires only small amount of reagents for transfer analysis (d) provides multiple replicas of a gel, (iv) allows prolonged storage of transferred patterns prior to use, and (v) the same protein transfer can be utilized for multiple successive analyses (7–9).

Since its introduction, protein blotting has been evolving constantly. Researchers now have a number of options for transferring proteins (10). Western blot sensitivity, however, is dependent on efficiency of blotting or transfer, retention of antigen during processing, and the final detection/amplification system used. Results are compromised if there are deficiencies at any of these steps (11).

1.1. Blotting Efficiency

Transfer of proteins efficiently from a gel to a solid membrane support depends on the nature of the gel, the molecular mass of the proteins being transferred, and the type of membrane used. The best option would be to run the softest gel, in terms of acrylamide and cross-linker, which yields the required resolution. Transfer becomes more complete and faster when thinner gels are used. The use of ultrathin gels, however, may cause handling problems and a 0.4-mm thickness represents the lower practical limit (12). High molecular proteins blot poorly following SDS-PAGE, resulting in low levels of detection on immunoblots. The use of heat, special buffers, and partial proteolytic digestion of the proteins prior to transfer (11, 13–17), however, has facilitated the efficient transfer of such proteins.

2. Immobilizing Supports for Protein Transfer

The most common solid microporous phases used for protein blotting comprise microporous surfaces and membranes such as cellulose, nitrocellulose (NC), polyvinylidine diflouride, cellulose acetate, polyethane sulfone, and nylon. The unique properties of microporous surfaces that make them suitable for "protein blotting" are (i) large volume to surface area ratio, (ii) high binding capacity, (iii) short- and long-term storage of immobilized molecules, (iv) ease of processing by allowing a solution phase to interact with the immobilized molecule, (v) lack of interference with the detection strategy, and (vi) reproducibility. These properties are useful for the high-thoroughput assays used in the postgenomic era as well (2, 4, 14, 18, 19).

These microporous surfaces, typically, are used in the form of membranes or sheets with a thickness of 100 μm and possessing an average pore size that ranges from 0.05 to 10 μm in diameter. The interaction of biomolecules with each of these membranes is not completely understood, except for the fact that it is generally known to interact in a noncovalent fashion (20, 21) (http://www.ncbi.nlm.nih.gov/sites/entrez).

2.1. Nitrocellulose Membranes

NC is used in high-throughput array, immunodiagnostic as well as mass-spectrometry coupled proteomic applications, filtration/concentration, ion exchange, and amino acid sequencing in addition to traditional protein blotting procedures. Southern first demonstrated (in 1975) the usefulness of NC to capture nucleic acids. Towbin et al. (1) and Burnette (5) showed that they could also be used for proteins. NC continues to be useful in the post-genomic era technology (19), since high-thoroughput methodologies for proteomics and genomics rely heavily on traditional concepts of molecular immobilization followed by hybridization binding or analysis.

2.1.1. Mechanism of Immobilization

The exact mechanism by which biomolecules interact with NC is unknown. However, several lines of evidence suggest that the interaction is noncovalent and hydrophobic. One evidence favoring hydrophobic interaction is the fact that since most proteins at pH values above 7 are negatively charged, it is surprising that NC, which is also negatively charged, can bind proteins efficiently. An additional fact is that nonionic detergents, such as Triton X-100, are effective in removing bound antigens from NC (8). High salt concentrations and low methanol concentrations increase immobilization efficiency (22).

NC can be stained with amido black (4), Coomassie brilliant blue (CBB) (1), aniline blue black, Ponceau S, fast green, or toluidine blue. Amido black staining can detect a 25 ng spot of bovine serum albumin readily with acceptable background staining. The background staining tends to be higher with CBB while Ponceau S gives a very clean pattern but with slightly less sensitivity than Amido black.

2.1.2. Disadvantages of Nitrocellulose Membrane

NC cannot be stripped and reprobed multiple times owing to its fragile nature. It also has a tendency to become brittle when dry. Also, small proteins tend to move through NC membranes and only a small fraction of the total amount actually binds. Using membranes with smaller pores can obviate this (12). Gelatin-coated NCs have been used for quantitative retention (10, 23). In supported-NC (e.g., Hybond-C Extra), the mechanical strength of the membrane has been improved by incorporating a polyester support web, thereby making handling easier.

2.2. Polyvinylidene Difluoride

Polyvinylidene difluoride (PVDF) is a linear polymer with repeating $-(CF_2-CH_2)-$ units. The use of "di" in PVDF is redundant (including its use here) and its use needs to be discouraged (2). The membrane was renamed as Immobilion-P™ Transfer Membrane after being initially referred to as Immobilon™ PVDF transfer membrane to differentiate it from other PVDF and non-PVDF-based blotting membranes referred to collectively as Immobilon family and marketed by Millipore. Immobilon-P^{SQ}

membrane with a 0.2 μm pore size suitable for proteins with a molecular weight less than 20 kDa (to prevent blow through) and immobilon-FL membrane optimized for all fluorescence applications also form part of the Immobilon family of PVDF membranes, added recently. Sequelon (24), a PVDF-based sequencing membrane, sold by Milligen/BioSearch, a Millipore subsidiary, is advantageous because of high protein-binding capacity, physical strength, and chemical stability.

2.2.1. Mechanism of Immobilization

Proteins transferred to the Immobilon-P membrane during WB are retained efficiently on its surface throughout the immunodetection process via a combination of dipole and hydrophobic interactions. The antigen-binding capacity of the membrane is 170 μg/cm^2 (for bovine serum albumin) and this is proportionate with that of NC. Also, the immobilon-P membrane has very good mechanical strength and like Teflon™ (a related fluorocarbon polymer) is compatible with a range of chemicals and organic solvents (acetonitrile, trifluoroacetic acid, hexane, ethylacetate, and trimethylamine) (2, 25).

It is important to prewet the PVDF membrane in either methanol or ethanol before using with aqueous buffers. Except for this, the blotting mechanics are not different from that seen with NC. This is because PVDF is highly hydrophobic and there is no added surfactant in PVDF.

2.2.2. Advantages of PVDF

A major advantage of electroblotting proteins onto PVDF membranes is that replicate lanes from a single gel can be used for various purposes such as N-terminal sequencing, proteolysis/peptide separation/internal sequencing along with Western analysis. Proteins blotted to PVDF membranes can be stained with amido black, India ink, and silver nitrate (26). These membranes are also amenable to staining with CBB, thus allowing excision of proteins for N-terminal protein sequencing, a procedure first demonstrated by Matsudaira (25) and Xu and Shivley (27).

2.3. Activated Paper

Activated paper (diazo groups) binds proteins covalently, but is disadvantageous in that the coupling method is incompatible with many gel electrophoresis systems. Linkage is through primary amines and therefore systems that use gel buffers without free amino groups must be used with this paper. In addition, the paper is expensive and the reactive groups have a limited half-life once the paper is activated.

2.4. Nylon Membranes

Nylon-based membranes are thin and smooth surfaced as NC, but with much better durability. Two kinds of membranes are available commercially, Gene Screen and Zetabind (ZB). ZB is a nylon matrix (polyhexamethylene adipamine or Nylon 66) modified by the addition of numerous tertiary amino groups during the manufacturing process (extensive cationization). It has excellent mechanical strength and also offers the potential of very significant (yet reversible) electrostatic interactions between the membrane and polyanions. Nylon shows a greater protein-binding capacity compared to NC (480 vs. 80 μg BSA bound/cm^2). In addition, nylon offers the advantages of more consistent transfer results and a significantly increased sensitivity compared to other membranes (7, 18). This effect is possible owing to the extra potential difference created by the positive charge of ZB.

2.4.1. Disadvantages of Nylon

The high binding capacity of these membranes, however, produces higher nonspecific binding. Another problem with using nylon membranes is the fact that they bind strongly to the commonly used anionic dyes such as Coomassie blue, amido black 10B (18), aniline blue black, Ponceau S, fast green, or toluidine blue. SDS, dodecyl trimethylammonium bromide, or Triton X-100 at low concentrations (0.1% in water) remove the dyes from the membrane while simultaneously destaining the transferred proteins, with SDS being the best. Destaining of this membrane is thus not possible, unlike NC, and therefore the background remains as high as the signal (8). On account of these problems, NC membranes have remained the best compromise for most situations. However, an immunological stain and India ink have been used to detect proteins on ZB (28–30) and NC membranes.

Nylon membranes have been found very useful in binding the negatively charged DNA, especially the positively charged ZB membranes. As a consequence, it has been used more for DNA blotting than for protein blotting.

3. Antibody Considerations

In many instances, protein blots are used in combination with enzyme-linked immunosorbant assays or immunohistochemistry, which are important alternative antibody-based detection methods. Immunoblotting is a "must" to determine specificity of antibodies used for immunohistochemistry. A major feature with any successful Western blot is the highly specific interaction between an antigen and an antibody. The actual point of interaction occurs between a small portion of the antigen (an epitope) and the sites of recognition found on the Fab region of the antibody molecule (a paratope). Antibodies that are selected for immunodetection protocols should be tested by Western blot analysis, when possible, and experimental conditions recommended by the antibody supplier must be adhered to (31). Since protein electrophoresis is carried out under denaturing conditions, the Western blots derived from a protein SDS-PAGE gel would contain its replica of denatured proteins. Western blot positive antibodies normally recognize a short linear segment of amino acids found in the nonlinearized target protein, which become available for binding under denaturing and reducing conditions, such as found in WB.

However, antibodies recognizing conformational epitopes, regions forming a three-dimensional structural configuration of amino acids, would lose its binding ability on denaturation of the protein. However, WB protocols are flexible. Since an investigator can choose gel electrophoresis and protein blotting conditions, it is possible to modify buffers to retain sufficiently higher-order protein structure for detection by some antibodies. That datasheet obtained with the antibody normally provides information about buffers best-suited for specific antigen–antibody interaction (31).

3.1. Polyclonal Antibodies vs. Monoclonal Antibodies

Polyclonal antibodies are normally made in experimental animals such as mice, rabbits, sheep, goats, and donkeys by immunization with a specific protein or peptide. These antibodies contain a pool of immunoglobulin molecule that bind to different epitopes found on a single protein.

Monoclonal antibodies, on the other hand, bind only to a single epitope within an antigen. These contain homogeneous cloned immunoglobulins and are made by fusing antibody-producing B cells from the spleen of the immunized animal (rat or mouse) with an immortalized cell line.

Both kinds of antibodies are used in protein blotting, and the choice should be made depending on the downstream application. Polyclonal antibodies can give higher background and cross-reactivity, compared to monoclonal antibodies, owing to detection of multiple epitopes. However, polyclonal antibodies are more sensitive than monoclonals since the signal is amplified as a result of binding of several antibodies per antigenic target (31).

4. Methods to Transfer Proteins from Gel to Membrane

Protein transfer from SDS-PAGE or native gels to NC or PVDF membranes has been achieved by (a) simple diffusion; (b) vacuum-assisted solvent flow; and (c) "Western" blotting or electrophoretic elution (4, 12, 32–39).

4.1. Simple Diffusion

Diffusion blotting was originally developed for transferring proteins separated by iso-electric focusing on thin gels to membranes and this was later expanded to other gel systems (32, 40–46). In this method, a membrane is placed on the gel surface with a stack of dry filter papers on top of the membrane. A glass plate and an object

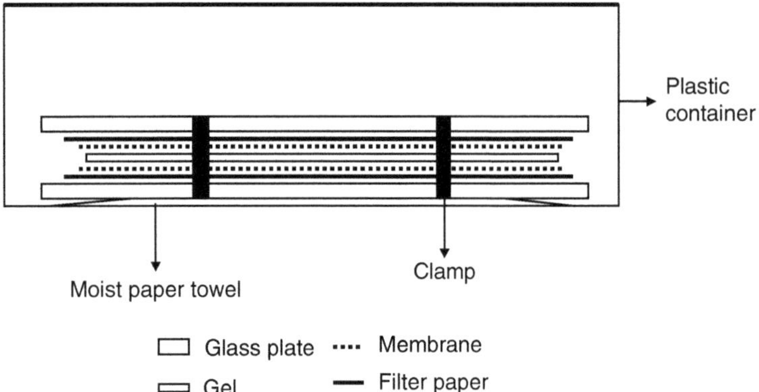

Fig. 2. Bi-directional, nonelectrophoretic transfer of proteins from SDS-PAGE gels to NC membranes to obtain up to 12 blots. The PAGE gel is sandwiched between two membranes, filter paper and glass plates, and incubated at 37°C for varying periods of time to obtain up to 12 blots (reproduced from ref. (10) with permission from Elsevier).

with a certain weight are usually placed on this assembly to enable the diffusion process. However, since quantitative transfer of protein was lacking, this protocol had not gained widespread acceptance. Interest began to pick up, when it was demonstrated that it was possible to obtain up to 12 blots from a single gel by sandwiching it between two membranes sequentially (Fig. 2) (32).

Nonelectrophoretic membrane lifts from SDS-PAGE gels for immunoblotting, obtained by this procedure, provides a useful way for identifying proteins by mass spectrometry (47, 48). The gel can be stained with Coomassie following diffusion blotting. The antigens on the blot are detected by immunostaining and the immunoblotted target band can be compared with the Coomassie-stained gel by superimposing the blot and the stained gel, allowing the identification of the band to be excised for tryptic digestion and subsequent matrix-assisted laser desorption time-of-flight mass spectrometric analysis. The main advantage of diffusion blotting compared to electroblotting is that several transfers or imprints can be obtained from the same gel and different antisera can be tested on identical imprints.

Subsequently, quantitative information regarding protein transfer during diffusion blotting was obtained using ^{14}C labeled proteins. A 3-min diffusion blotting procedure was shown to allow a transfer of 10% compared to electroblotting. Diffusion blotting of the same gels carried out multiple times for prolonged periods at 37°C causes the gel to shrink. This was overcome by using gels cast on plastic supports (44, 45).

Activity gel electrophoresis or Zymography has also been studied with regard to the utility of diffusion. This procedure involves the electrophoresis of enzymes (either nucleases or proteases) through discontinuous polyacrylamide gels containing enzyme substrate (either type III gelatin or β-casein). After electrophoresis, SDS is removed from the gel by washing in 2.5% Triton X-100. This allows the enzyme to renature, and the substrate to be degraded. Staining of the proteins with CBB allows the bands of enzyme activity to be detected as clear bands of lysis against a blue background (49). An additional immunoblotting analysis using another gel is often required in this procedure to examine a particular band that is involved. Diffusion blotting has been used to circumvent the use of a second gel for this purpose (45). The activity gel was blotted onto PVDF for immunostaining and the remaining gel after blotting was used for routine "activity staining." Since the blot and the activity staining are derived from the same gel, the signal localization in the gel and the replica can be easily aligned for comparison.

Diffusion blotting transfers 25–50% of the (45) proteins to the membrane compared to electroblotting. However, the advantage of obtaining multiple blots from the same gel could outweigh the loss in transfer and actually could be compensated for by using sensitive detection techniques. The gel remains on its plastic support, which prevents stretching and compression; this ensures identical imprints and facilitates more reliable molecular mass determination. If only a few imprints are made, sufficient protein remains within the gel for general protein staining. These advantages make diffusion blotting the method of choice when quantitative protein transfer is not required.

4.2. Vacuum Blotting

Vacuum blotting was developed (50) as an alternative to diffusion blotting and electroblotting. The suction power of a pump connected to a slab gel dryer system drives the separated polypeptides from the gel to the NC membrane. Both low and high molecular weight proteins could be transferred using this method. Since small molecular weight proteins (±14,000 Da) are not well adsorbed by the 0.45 μm membrane NC, membranes with a small pore size (0.2 or 0.1 μm) should be used when using low molecular weight proteins.

The gel can dry out if the procedure is carried out over 45 min and in such a scenario enough buffer should be used. In some instances, low concentration polyacrylamide gels stuck to the membrane following transfer. Rehydrating the gel helps detaching the NC membrane from the gel remnants in such a scenario.

4.3. Electroblotting

This is the most commonly used procedure to transfer proteins from a gel to a membrane. The major advantages are speed and the completeness of transfer compared to diffusion or vacuum blotting. Electroblotting can be achieved either by (a) complete immersion of a gel–membrane sandwich (Fig. 3) in a buffer (wet transfer) or by (b) placing the gel–membrane sandwich between absorbent paper soaked in transfer buffer (semidry transfer).

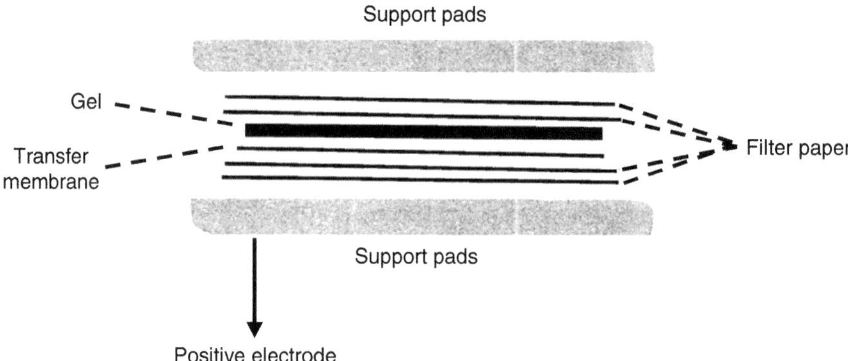

Fig. 3. The Western blot transfer assembly (reproduced from ref. (10) with permission from Elsevier).

The conditions for transfer are dependent on gel type, the immobilization membrane, the transfer apparatus used as well as the protein themselves. SDS gels, urea gels (4), lithium dodecyl sulfate-containing gels, nondenaturing gels, two-dimensional gels (51), and agarose gels have been used for protein electrophoretic blotting (18). The electric charge of the protein should be determined and the membrane should be placed on the appropriate side of the gel. When using urea gels, the membrane should be placed on the cathode side of the gel (4). Proteins from SDS-PAGE gels are eluted as anions and therefore the membrane should be placed on the anode side of the gel.

4.3.1. Wet Transfer

In the wet transfer procedure, the sandwich is placed in a buffer tank with platinum wire electrodes. A large number of different apparati are available to efficiently transfer proteins (or other macromolecules) transversely from gel to membrane. Most of these, however, are based on the design of Towbin et al. (1): i.e., they have vertical stainless steel/platinum electrodes in a large tank.

4.3.2. "Semidry" Transfer

In this procedure, the gel–membrane sandwich is placed between carbon plate electrodes. "Semidry" or "horizontal" blotting uses two plate electrodes (stainless steel or graphite/carbon) for uniform electrical field over a short distance, and sandwiches between these up to six gel/membrane/filter paper assemblies, all well soaked in transfer buffer. The assembly is clamped or otherwise secured on its side, and electrophoretic transfer effected in this position, using as transfer buffer only the liquid contained in the gel and filter papers or other pads in the assembly.

The advantages to this procedure over the conventional upright protocol are (a) gels can be blotted simultaneously; (b) electrodes can be cheap carbon blocks; and (c) less power is required for transfer (and therefore a simpler power pack).

5. Conclusion

Protein blotting has been evolving constantly and now the scientific community is faced with a number of ways and means of transferring and detecting proteins. The usefulness of protein blotting stems from its ability to provide simultaneous resolution of multiple immunogenic antigens within a sample for detection by specific antibodies. This has made it a very valuable method, especially for testing the specificity of antibodies to be used in immunohistochemistry experiments.

References

1. Towbin, H., Staehelin, T., and Gordon, J. (1979) Electrophoretic transfer of proteins from polyacrylamide gels to NC sheets: procedure and applications. *Proc Natl Acad Sci USA.* **76**, 4350–4354.
2. LeGendre, N. (1990). Immobilon-P transfer membrane: applications and utility in protein biochemical analysis. *Biotechniques* **9**(6 Suppl), 788–805. Review.
3. Southern, E.M. (1975). Detection of specific sequences among DNA fragments separated by gel electrophoresis. *J Mol Biol.* **98**, 503–517.
4. Alwine, J.C., and Kemp, D.J., and Stark, G.R. (1977) Method for detection of specific RNAs in agar gels by transfer to diazobenzyloxymethyl-paper and hybridization with DNA probes. *Proc Natl Acad Sci* USA. **74**, 5350–5354.
5. Burnette, W.N. (1981) "Western Blotting": electrophoretic transfer of proteins from sodium dodecyl sulfate–polyacrylamide gels to unmodified NC and radiographic detection with antibody and radioiodinated protein A. *Anal Biochem.* **112**, 195–203.
6. Laemmli, U.K. (1970) Cleavage of structural proteins during assembly of the head of bacteriophage T4. *Nature* **227**, 680–685.
7. Kost, J., Liu, L-S., Ferreira, J., and Langer, R. (1994) Enhanced protein blotting from PhastGel media to membranes by irradiation of low-intensity. *Anal Biochem.* **216**, 27–32.
8. Gershoni, J.M., and Palade, G.E. (1982) Electrophoretic transfer of proteins from sodium dodecyl sulfate-polyacrylamide gels to a positively charged membrane filter. *Anal Biochem.* **124**, 396–405.
9. Gershoni, J.M. (1988) Protein blotting: a manual. *Methods Biochem Anal.* **33**, 1–58. Review.
10. Kurien, B.T., and Scofield, R.H. (2006) Western blotting. *Methods* **38**, 283–293.
11. Karey, K.P., and Sirbasku, D.A. (1989) Glutaraldehyde fixation increases retention of low molecular weight proteins (growth factors) transferred to nylon membranes for Western blot analysis. *Anal Biochem.* **178**, 255–259.
12. Harlow, E., and Lane, D. (1988) Immunoblotting. In: Antibodies. A laboratory manual. Cold Spring Harbor Laboratory, New York, p. 485.
13. Renart, J., Reiser, J. and Stark, G.R. (1979) Transfer of proteins from gels to diazobenzyloxymethyl paper and detection with anti-sera: a method for studying antibody specificity and antigen structure. *Proc Natl Acad Sci USA.* **76**, 3116–3120.
14. Elkon, K.B., Jankowski, P.W., and Chu, J.L. (1984) Blotting intact immunoglobulins and other high-molecular-weight proteins after composite agarose-polyacrylamide gel electrophoresis. *Anal Biochem.* **140**, 208–213.
15. Gibson, W. (1981). Protease-facilitated transfer of high-molecular-weight proteins during electrotransfer to NC. *Anal Biochem.* **118**, 1–3.
16. Bolt, M.W., and Mahoney, P.A. (1997) High efficiency blotting of proteins of diverse sizes following sodium dodecyl sulfate-polyacrylamide gel electrophoresis. *Anal Biochem.* **247**, 185–192.
17. Kurien, B.T., and Scofield, R.H. (2002) Heat mediated, ultra-rapid electrophoretic transfer of high and low molecular weight proteins to NC membranes. *J Immunol Methods* **266**, 127–133.
18. Gershoni, J.M., and Palade, G.E. (1983) Protein blotting: principles and applications. *Anal Biochem.* **131**, 1–15.

19. Thornton, D.J., Carlstedt, I., and Sheehan, J.K. (1996) Identification of glycoproteins on nitrocellulose membranes and gels. *Mol Biotechnol.* **5**, 171–176.
20. Tonkinson, J.L., and Stillman, B. (2002) NC: a tried and true polymer finds utility as a post-genomic substrate. *Front Biosci.* 7, c1–c12. Review.
21. Lauritzen, E., Masson, M., Rubin, I., Bjerrum, O.J., and Holm, A. (1993) Peptide dot immunoassay and immunoblotting: electroblotting from aluminum thin-layer chromatography plates and isoelectric focusing gels to activated NC. *Electrophoresis* **14**, 852–859.
22. Masson, M., Lauritzen, E., and Holm, A. (1993) Chemical activation of NC membranes for peptide antigen-antibody binding studies: direct substitution of the nitrate group with diaminoalkane. *Electrophoresis* **14**, 860–865.
23. Too, C.K., Murphy, P.R., and Croll, R.P. (1994) Western blotting of formaldehyde-fixed neuropeptides as small as 400 daltons on gelatin-coated NC paper. *Anal Biochem.* **219**, 341–348.
24. Coull, J.M., Dixon, J.D., Laursen, R.A., Koester, H., and Pappin, D.J.C. (1989) Development of membrane supports for the solid-phase sequence analysis of proteins and peptides. In: B. Witmann-Liebold (Ed.) Methods in protein sequence analysis. Springer, Berlin, pp. 69–78.
25. Matsudaira, P. (1987) Sequence from picomole quantities of proteins electroblotted onto polyvinylidene difluoride membranes. *J Biol Chem.* **262**,10035–10038.
26. Pluskal, M.F., Przekop, M.B., Kavonian, M.R., Vecoli, C., and Hick, D.A. (1986) Immobilon™ PVDF transfer membrane. A new membrane substrate for western blotting of proteins. *Biotechniques* **4**, 272–282.
27. Xu, Q.Y., and Shively, J.E. (1988) Microsequence analysis of peptides and proteins. VIII. Improved electroblotting of proteins onto membranes and derivatized glass-fiber sheets. *Anal Biochem.* **170**, 19–30.
28. Kittler, J.M., Meisler, N.T., Viceps-Madore, D., Cidlowski, J.A., and Thanassi, J.W. (1984) A general immunochemical method for detecting proteins on blots. *Anal Biochem.* **137**, 210–216.
29. Hughes, J.H., and Mack, K., and Hamparian, V.V. (1988) India ink staining of proteins on nylon and hydrophobic membranes. *Anal Biochem.* **173**, 18–25.
30. Tovey, E.R., and Baldo, B.A. (1989) Protein binding to NC, nylon and PVDF membranes in immunoassays and electroblotting. *J Biochem Biophys Methods* **19**, 169–183.
31. Moore, C. (2009) Introduction to western blotting. AbD serotec. www.abdserotec.com/uploads/WesternBlottingBrochure.pdf.
32. Kurien, B.T., and Scofield, R.H. (1997) Multiple immunoblots after non- electrophoretic bidirectional transfer of a single SDS-PAGE gel with multiple antigens. *J Immunol Methods* **205**, 91–94.
33. Kyhse-Andersen, J. (1984) Electroblotting of multiple gels: a simple apparatus without buffer tank for rapid transfer of proteins from polyacrylamide to nitrocellulose. *J Biochem Biophys Methods* **10**, 203–209.
34. Otter, T., King, S.M., and Witman, G.B. (1987) A two-step procedure for efficient electro transfer of both high-molecular weight (greater than 400,000) and low-molecular weight (less than 20,000) proteins. *Anal Biochem.* **162**, 370–377.
35. Harper, D.R., Kit, M.L., and Kangro, H.O. (1990) Protein blotting: ten years on. *J Virol Methods* **30**, 25–39. Review.
36. Egger, D., and Bienz, K. (1994) Protein (western) blotting. *Mol Biotechnol.* **1**, 289–305.
37. Wisdom, G.B. (1994) Protein blotting. *Methods Mol Biol.* **32**, 207–213.
38. Kurien, B.T., and Scofield, R.H. (2003) Protein blotting: a review. *J Immunol Methods* **274**, 1–15. Review.
39. Kurien, B.T., and Scofield, R.H. (2005) Blotting techniques. In: P.J. Worsfold, A. Townshend, and C.F. Poole (Eds.) Encyclopedia of analytical science, Second edition. Elsevier, Oxford, p 425.
40. Reinhart, M.P., and Malamud, D. (1982). Protein transfer from isoelectric focusing gels: the native blot. *Anal Biochem.* **123**, 229–235.
41. Jagersten, C., Edstrom, A., Olsson, B., and Jacobson, G. (1988) Blotting from PhastGel media after horizontal sodium dodecyl sulfate-polyacrylamide gel electrophoresis. *Electrophoresis* **9**, 662–665.
42. Kazemi, M., and Finkelstein R.A. (1990) Checkerboard immunoblotting (CBIB): an efficient, rapid, and sensitive method of assaying multiple antigen/antibody cross-reactivities. *J Immunol Methods* **128**, 143–146.
43. Heukeshoven, J., and Dernick, R. (1995). Effective blotting of ultrathin polyacrylamide gels anchored to a solid matrix. *Electrophoresis* **16**, 748–756.
44. Olsen, I., and Wiker, H.G. (1998) Diffusion blotting for rapid production of multiple identical imprints from sodium dodecyl sulfate polyacrylamide gel electrophoresis on a solid support. *J Immunol Methods* **220**, 77–84.

45. Chen, H., and Chang, G.D. (2001) Simultaneous immunoblotting analysis with activity gel electrophoresis in a single polyacrylamide gel. *Electrophoresis* **22**, 1894–1899.
46. Bowen B., Steinberg J., Laemmli U.K., and Weintraub H. (1980) The detection of DNA-binding proteins by protein blotting. *Nucleic Acids Res.* **8**, 1–20.
47. Kurien, B.T., and Scofield, R.H. (2000) Association of neutropenia in systemic lupus erythematosus with anti-Ro and binding of an immunologically cross-reactive neutrophil membrane antigen. *Clin Exp Immunol.* **120**, 209–217.
48. Kurien, B.T., Matsumoto, H., and Scofield, R.H. (2001) Purification of tryptic peptides for mass spectrometry using polyvinylidene fluoride membrane. *Indian J Biochem Biophys.* **38**, 274–276.
49. Bischoff, K.M., Shi, L., and Kennelly, P.J. (1998) The detection of enzyme activity following sodium dodecyl sulfate-polyacrylamide gel electrophoresis. *Anal Biochem.* **260**, 1–17. Review.
50. Peferoen, M., Huybrechts, R., and De Loof, A. (1982) Vacuum-blotting: a new simple and efficient transfer of proteins from sodium dodecyl sulfate-polyacrylamide gels to NC. *FEBS Lett.* **145**, 369–372.
51. Dorri, Y., Kurien, B.T., and Scofield, R.H. (2009) A simpler and faster version of two-dimensional gel electrophoresis using vertical, mini SDS-PAGE apparatus. *Iran J Chem Chem Eng.* **28**, 51–56.

Chapter 4

Optimized Protocol to Make Phospho-Specific Antibodies that Work

Amy J. Archuleta, Crystal A. Stutzke, Kristin M. Nixon, and Michael D. Browning

Abstract

Phosphoproteins are considered to be among the most important proteins in the body. They are the proteins that regulate almost all cell processes from cell division in cancer to neuronal signal transduction in learning and memory. This review will describe the development of a revolutionary immunochemical technique that produces antibodies that bind to target proteins only when the protein is in the phosphorylated state. These phospho-specific antibodies can thus be used to track the activity of a protein, not simply its level of expression. In this review, we will discuss both the design of the phosphopeptide immunogen and immunization. The affinity purification of the phospho-specific antibody as well as the methods most suitable for characterizing the phosphospecificity of the antibody will be described here. Taken together, these methods will cover the key procedures and protocols required to produce a phospho-specific antibody that works.

Key words: Phosphoprotein, Antibodies, Western blots, Affinity purification, Antibody characterization

1. Introduction

Phosphoproteins are considered to be among the most important proteins in the body. They are the proteins that regulate almost all cell processes from cell division in cancer to neuronal signal transduction in learning and memory. A protein becomes phosphorylated by an enzymatic, reversible post-translational modification in which a phosphoryl group is covalently attached to or removed from specific serine, threonine, or tyrosine residues in the protein. This changes the conformation and hence the function of the protein. Since phosphoproteins regulate virtually every important

Alexander E. Kalyuzhny (ed.), *Signal Transduction Immunohistochemistry: Methods and Protocols*,
Methods in Molecular Biology, vol. 717, DOI 10.1007/978-1-61779-024-9_4,

cellular function, we like to say that "phosphoproteins are the verbs of the proteomic language."

This review will describe the development of a revolutionary immunochemical technique that produces antibodies that bind to target proteins only when the protein is in the phosphorylated state. These phospho-specific antibodies can thus be used to track the activity of a protein and not simply its level of expression.

1.1. History of Phosphoprotein Detection

Protein phosphorylation is the principal cellular mechanism used to regulate protein function. The stoichiometry of phosphorylation (percentage of the protein that is phosphorylated) of a given site is controlled by the relative activities of a cell's protein kinases and phosphatases, and can often generate extremely rapid and reversible changes in the activity of target proteins. The ability to assay the state of phosphorylation of specific proteins is of great utility in the quest to establish the function of a given protein and how that activity is influenced by cellular signals. Such assays are also critical for the identification of drugs that can influence the phosphorylation and hence the function of specific proteins.

In early studies, most methods commonly used to measure protein phosphorylation and dephosphorylation in cell preparations employed prelabeling with 32Pi, in vitro phosphorylation with [γ-32P] ATP, or "back" phosphorylation. These methods have several practical and theoretical limitations including the facts that they are very time-consuming and labor-intensive and they provide little in the way of quantitative data on specific phosphorylation events. An immunochemical approach became an attractive alternative for detecting changes in the state of phosphorylation of specific proteins at a specific site based in large part on the successful use of short synthetic peptides to produce epitope-targeted antibodies. The use of phosphorylation state-specific antibodies takes advantage of the sensitivity and selectivity afforded by immunochemical methodology to greatly increase not only the throughput but also quantitative accuracy of phosphoprotein assays.

1.2. Development of Phospho-Specific Antibodies

The first report of the production of phosphorylation-dependent antibodies appeared in 1981, when polyclonal antibodies that could detect phosphotyrosine-containing proteins were produced by immunization of rabbits with benzyl phosphonate conjugated to keyhole limpet hemocyanin (KLH) (1). These antibodies became key reagents in oncogenic virus and cancer research, but detected phosphotyrosine on many proteins. Shortly thereafter, Nairn and colleagues reported the production of serum antibodies that distinguished between the phospho- and dephospho-forms of G-substrate, a protein localized to cerebellar Purkinje cells and phosphorylated by cGMP-dependent protein kinase (2). A synthetic heptapeptide, Arg-Lys-Asp-Thr-Pro-Ala-Leu, corresponding to a repeated sequence surrounding two phosphorylated threonyl residues in the intact protein, served as antigen. Rabbit antisera against a peptide–KLH conjugate were specific for the dephospho-form of G-substrate. Phospho-specific antibodies were prepared by immunization of rabbits with the purified phosphoprotein, phosphorylated in vitro to a stoichiometry of 2 mol/mol with cGMP-dependent protein kinase. Despite this initial success, other attempts to produce phospho-specific polyclonal antisera by immunization with the phospho-form of intact proteins were not very successful, probably because of two significant factors. First, many phosphorylated proteins are believed to undergo rapid dephosphorylation during immunization, regardless of the route of injection, leading to the loss of the desired phospho-epitope. Second, holoproteins generally contain multiple immunogenic epitopes. This decreases the probability that clonal dominance for a phospho-specific epitope will be obtained.

Taking a more direct approach utilizing phosphorylated and unphosphorylated forms of synthetic peptides, a general protocol for the production of phosphorylation state-specific antibodies for substrates with established site(s) of phosphorylation was developed (3). In early stages of the development of this methodology, phosphopeptides were routinely prepared by enzymatic phosphorylation. At the same time, advances were being made in the chemical synthesis of phosphopeptides (4), and such phosphopeptides were being used to produce phospho-specific antibodies (5). Chemically phosphorylated peptides were also produced and contributed to the refinement of postsynthesis global phosphorylation to produce phospho-specific antibodies (6). These enzymatic and chemical approaches remain perfectly valid today. However, the use of commercially available, high-quality, affordable *O*-benzyl-protected Fmoc derivatives of phosphoamino acids has become state of the art in the preparation of synthetic phosphopeptides (7).

The production and use of these phospho-specific antibodies has become an area of intense interest. Therefore, we describe below some of the salient features that underlie production of phospho-specific antibodies.

1.3. Overall Strategy for Antibody Production

Antibodies are glycoproteins that are produced by an organism in response to the presence of a foreign substance. Foreign substances capable of eliciting immune responses are known as antigens. Each antibody has a unique and specific affinity for the antigen that stimulated its synthesis. This affinity is for a specific site located within the antigen known as the epitope.

Antibodies are typically produced by immunizing a mammal with the desired antigen. The first and quite possibly most crucial step in producing an antibody that works is choosing the antigen. Without the proper immunogenic antigen, your efforts in making the antibody you want are not likely to be successful. There are generally two ways to achieve this: one using endogenous or recombinant protein, and the other producing synthetic peptides. We will focus here on the production and applications of phospho-specific antibodies. As mentioned above, phospho-specific antibodies were first successfully prepared by immunization of rabbits with native purified protein. However, virtually all phospho-specific antibodies are now produced using phosphopeptides as antigens. In using synthetic peptides to produce antibodies, the key step is the design of the antigen. Ensuring that you have an immunogenic sequence is of utmost importance because the specificity and utility of the antibody depends critically on the design of the immunizing peptide. Important considerations in choosing an antigen include selecting a sequence that is specific to your targeted protein to ensure that there is no cross-reactivity with other similar proteins and that it also has amino acid homology with the species of interest in which you will utilize the antibody.

When producing an antibody that is specific to certain post-translational modifications, such as phosphorylation sites, the antigen design choices are constrained by the sequence directly surrounding the phosphorylated residue of interest. Antigen sequences used for generating these phospho-specific antibodies are generally short. This forces the phosphorylated residue into the epitope recognized by the antibody, as the recognition site of an antigen is typically only about six amino acids wide. There are numerous algorithms available that seek to improve selection of antigenic peptides. We rely on a database of successful epitopes we have chosen during our nearly three decades of phospho-specific antibody production to guide our antigen selection.

When using synthetic peptides as antigens, they need to be conjugated to larger carrier molecules in order to enhance the immunogenicity of the antigen. Common carrier proteins include KLH, bovine thyroglobulin, and BSA. KLH is often the preferred choice as it is a very large molecule produced in mollusks, which are genetically distant from mammals used to produce antibodies, thereby decreasing the risk of cross-reactivity.

Once the antigen has been conjugated, you are ready to begin immunizations in host animals. It is best to immunize at least two animals in order to improve the odds of obtaining the desired antibody, as a single antigen is capable of eliciting antibodies to many different epitopes. Common antibody hosts include rabbits, mice, guinea pigs, sheep, goats, and chickens. Rabbits are frequently the species of choice for many researchers as they have been found to be highly immunogenic, as well as easy to handle and cost-effective, yielding larger quantities of serum than smaller rodents, but without the higher costs involved in care of larger mammals such as goats and sheep.

Following immunization and after you have collected the first few bleeds from your animals, you can begin to screen the antiserum in ELISAs or dot-blots using the phospho-peptide antigen to determine if you are indeed getting the desired immune response. We usually delay initial screening for the antibody until we reach roughly 8 weeks out in the immunization protocol. This is done to ensure that the immune response has fully matured because the quality and character of antibodies produced in response to repeated antigen exposure (immune boosts) changes over time. A mature immune response ideally displays peak titers of the desired higher affinity IgG, whereas lower-affinity IgM is more abundant in earlier immune responses. Once you have obtained positive initial screening results, you will progress to the purification and further characterization of your phospho-specific antibody.

Once the antigen has been selected and used as an immunogen, the production of the antibody is dependent on the animal's immune system. Despite decades of trying to create a comparable protein detection molecule, no other system – be it phage display, aptamers, etc., has ever come close to mimicking the power and specificity of an animal's native immune system. This is perhaps not surprising given the eons devoted to selection of this mechanism for foreign substance detection. However, once the antibody has been produced, the art and science of purifying the antibody under conditions that preserve its activity begin. In the remainder of this chapter, we will focus on how best to isolate a phospho-specific antibody that works.

1.4. Affinity Purification of Phospho-Specific Antibodies

When a phosphorylated antigen is injected into a host, three types of antibodies may be produced (Fig. 1). One possibility is a phospho-specific antibody that recognizes the protein only if it is phosphorylated at the specific amino acid of interest (A). This is usually the desired antibody. A second possibility is a dephospho-specific antibody that recognizes the protein only when the same specific amino acid does not have a phosphate group attached (B). This dephospho-specific antibody may be created if the phosphatases in the animal dephosphorylate the

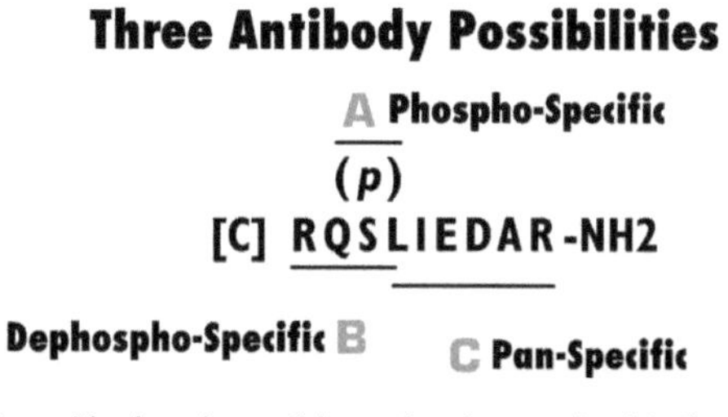

Three Antibody Possibilities

A Phospho-Specific

(p)

[C] RQSLIEDAR-NH2

Dephospho-Specific B

C Pan-Specific

Fig. 1. Immunization with phosphopeptide conjugates can lead to the presence of several antibody types in the antiserum. One can obtain the phospho-specific antibody that is desired (**a**). However, it is also possible to obtain antibodies that are specific for the dephosphopeptide (**b**) as well as antibodies that are pan-specific (**c**) and do not react with the protein in a phospho-specific manner.

phosphopeptide antigen that is injected. A third possibility is a pan-specific antibody that recognizes a region of the sequence that does not contain the specific amino acid residue that has the phosphate group attached (C). Pan antibodies will detect the total amount of the protein that is present regardless of phosphorylation state.

Both pan- and phospho-antibodies are purified using affinity columns. A pan-antibody requires only one column, while a phospho-specific antibody requires two sequential columns. For a phospho-specific purification, a column bound with only the phosphopeptide and a separate column bound with only the dephosphopeptide are used to assure that only the phospho-specific antibody is recovered. If the host's immune response generated a dephospho- or pan-antibody, they may also be recovered if they are desired. In the example below, sera containing all three types of antibody is first sent over an affinity column whose matrix is bound to the phosphopeptide antigen (Fig. 2). The phospho-specific (A) and pan (C) antibodies will recognize the phosphopeptide, so both will stick to the column. The dephospho-antibody (B), which does not stick to the phospho column, will fall through. The phospho and pan antibodies are then eluted from the phospho column, and sent over a column made with the dephosphopeptide (Fig. 3). Once again, the pan antibody (C) will stick to the dephospho column, but this time the phospho antibody (A) will fall through. This flow through is the purified phospho-specific antibody. The pan antibody can then be eluted off the dephospho column if desired. The dephospho antibody may be separated and purified as well by sending the flow through from the phospho column over the dephospho column. The dephospho antibody will stick to the column, and can then be eluted.

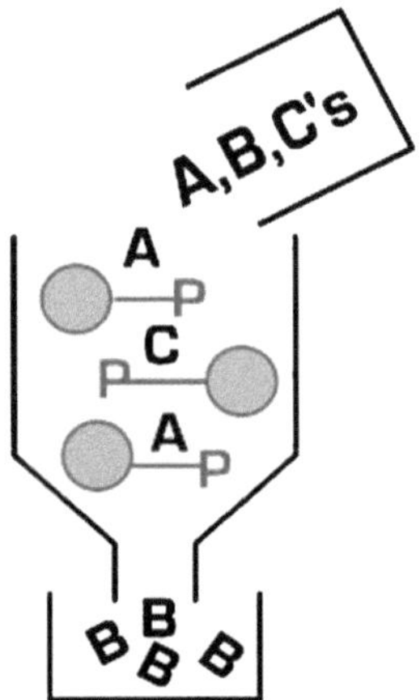

Fig. 2. Sequential affinity chromatography is used to isolate phospho-specific antibodies. The first column used is the column made with the phosphopeptide used as antigen.

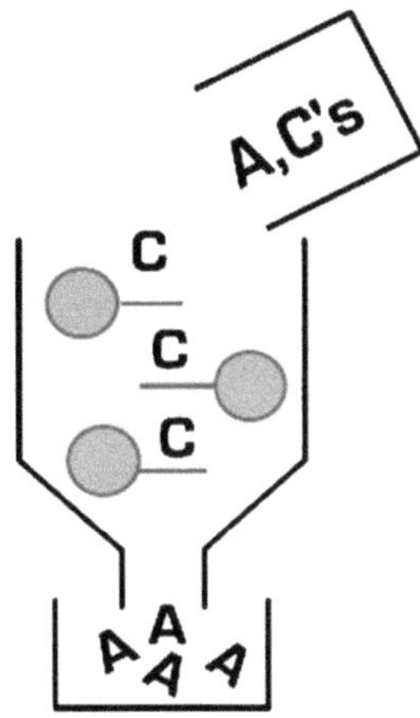

Fig. 3. Sequential affinity chromatography is used to isolate phospho-specific antibodies. The second column used is the column made with the dephosphopeptide form of the antigen.

1.5. Antibody-Detection Assays

1.5.1. Peptide ELISA

A peptide enzyme-linked immunoabsorbent assay (ELISA) is commonly used for screening sera and/or assaying affinity column fractions. It is important to emphasize that such ELISAs have little utility in determining whether the antibody will work for detecting the protein or phospho-protein of interest. When using a peptide immunogen to screen for antibody in sera via peptide ELISA, one almost always obtains positive signal, but such antibodies detect the protein of interest in other assays only 50–75% of the time. Nevertheless, ELISAs can be quite useful for rapid screens for the presence of an immune response or for affinity purification monitoring.

1.5.2. Western Blot

A Western blot (WB) is the most useful application for antibody characterization. It is a powerful tool for a number of reasons. First, the technique demonstrates that the antibody can detect the actual protein of interest and not simply the peptide used as antigen as in ELISAs. Second, because the molecular weight of the protein detected by the antibody can be determined in the WB, one can obtain strong confirmation that the antibody is in fact detecting the protein of interest. Lastly, the technique can indicate whether the antibody only recognizes the protein of interest, or if it cross-reacts with other proteins.

1.6. Antibodies that Work

We have developed a proprietary wash-and-elution protocol that has two goals: to bind only the highest-affinity antibodies due to the stringent washes, and to retain the high activity of the antibodies during the elution. It is necessary to use a rigorous washing technique to remove nonspecific and lower-affinity antibodies as well as any antibodies that were raised to the carrier protein. At the same time, it is important that these washes as well as the elution buffers do not damage the desired antibody. The combination of stringent washes and our highly effective elution cocktail allows for a new definition of specific activity unique to polyclonal antibodies.

There are a number of terms that have been used over the years to describe the quality of an antibody. These include: titer, avidity, affinity, and often simply the amount in micrograms. We have not found any of these terms particularly useful. We prefer to analyze our antibodies in terms of specific activity in much the same way as enzymes are analyzed. Thus, just as with enzymes, we characterize our antibodies in terms of units of activity per milligram. The basic unit we use is a mini WB. A typical antibody that we consider to "work" has a specific activity of 100s of mini WBs per mg. Thus, 1–2 μg of a highly purified antibody should be sufficient to perform a mini WB in most cell lysates. This is only a rule of thumb as expression level of the protein and its phosphorylation stoichiometry are also important in this measure.

However one characterizes an antibody, a key question always remains: Does the antibody work in your application? There is no absolute answer to this question. However, we have tried to describe the processes that we have developed over the years to help us develop antibodies that work for us. Hopefully, these methods will help you to develop antibodies that work in your applications.

2. Materials

2.1. Affinity Column Preparation

The reagents needed for column preparation vary depending on the chemical structure of the antigen. Items 1–4 below are used for all types of purifications. Select the set of items in 5, 6, or 7 based on your antigen.

1. Disposable 5 mL polypropylene columns (Pierce, Rockford, IL).
2. Wash solution: 1.0 M NaCl.
3. Dimethyl sulfoxide (DMSO). Store under a hood at room temperature.
4. Storage solution: 1× PBS containing 0.05% NaN_3.
5. Peptide antigens with terminal cysteines:
   (a) SulfoLink® Coupling Gel: (Pierce, Rockford, IL). Store at 4°C.
   (b) Coupling buffer: 50 mM Tris, 5 mM EDTA-Na, pH 8.5.
   (c) Ellman's Reagent Solution, 4 mg/mL in coupling buffer (5,5′-Dithio-bis(2-Nitrobenzoic Acid) (Sigma, St. Louis, MO)
   (d) Quench solution: 50 mM l-Cysteine (Use pH 10.0 environment if the ligand to be coupled is stable at 10.0. If it is not stable, or the stability is unknown, use pH 7.2 environment.) HCl in coupling buffer.
   (e) Reducing agent: Tris(2-Carboxyethyl)-Phosphine Hydrochloride (TCEP) (Pierce, Rockford, IL).

6. Peptide antigens with no terminal cysteines:
   (a) UltraLink® Biosupport beads (Pierce, Rockford, IL).
   (b) Coupling buffer: 0.6 M sodium citrate ($C_6H_5Na_3O_7$) pH 6.0.
   (c) Quench solution: 3.0 M ethanolamine pH 9.0 (Note 1).
7. Fusion protein antigens:
   (a) AminoLink® Coupling Gel – (Pierce, Rockford, IL). Store at 4°C.
   (b) Coupling buffer pH 7.2: 0.1 M sodium phosphate (Na_2HPO_4), 0.15 M NaCl, **or** pH 10.0: 0.1 M sodium citrate ($C_6H_5Na_3O_7$), 0.05 M sodium carbonate (Na_2CO_3).
   (c) 5 M Cyanoborohydride ($NaCNBH_3$) in 0.01 M NaOH.
   (d) 1 M Tris, pH 7.4.

2.2. Affinity Purification

1. Precipitating agent: Ammonium sulfate $(NH_4)_2SO_4$.
2. 200 mM Benzamidine (100× stock solution) (Kodak, Rochester, NY). Store at 4°C.
3. 100 mM Pefablock (500× stock solution) 4-(2-Aminomethyl) benzenesulfonyl fluoride hydrochloride, AEBSF (Sigma, St. Louis, MO). Store at −20°C.
4. Wash solution: Borate buffered saline (BBS): 1 M NaCl, 100 mM boric acid (H_3BO_3), 20 mM sodium tetraborate, decahydrate ($B_4Na_2O_7{\cdot}10H_2O$), 0.1% Tween-20.
5. Wash solution: Phosphate buffered saline (PBS), pH 7.6: 137 mM NaCl, 280 mM Na_2HPO_4, 5.4 mM KCl, 2.9 mM KH_2PO_4.
6. Elution buffer: This is dependent on the affinity of the antibodies.
7. Dialysis buffer: 10 mM Hepes, 150 mM NaCl, pH 7.5.
8. Column storage buffer: PBS + 0.05% sodium azide (NaN_3).
9. Purging buffer: 6.0 M NaSCN, 0.05 M Hepes.
10. Dialysis tubing: 3500 MW Snakeskin Pleated Dialysis Tubing (Pierce, Rockford, IL).
11. Polyethylene glycol: 8,000 MW.

2.3. Antibody Characterization

2.3.1. Peptide ELISA

1. Immunolink-Amino 96 well plate (NUNC, Denmark). Store at 4°C.
2. Immunolink Peptide Coupling Buffer – 100 mM Na_2CO_3, pH 9.6.
3. PBS: 136 mM NaCl, 28 mM Na_2PO_4, 5 mM KCl, 3 mM KH_2PO_4, pH 7.6.
4. Wash solution: PBS + 0.05% Tween-20.

5. Antibody dilution solution: Wash solution with 0.5% BSA. Store at 4°C.
6. Secondary antibody: Goat Anti-Rabbit-HRP Conjugate (Bio-RAD, Hercules, CA). Store at 4°C.
7. Substrate solution: 1:1 mixture of TMB Peroxidase Substrate and Peroxidase Substrate Solution B (KPL, Gaithersburg, MD). Store at 4°C.
8. Quenching reagent: 2 N H_2SO_4.

2.3.2. Western Blot

1. 4× Sample buffer: 0.25 M Tris, 8% sodium dodecyl sulfate, 40% glycerol, 10% beta-mercapto-ethanol. Adjust pH to 6.8. Trace of bromophenol blue.
2. Lower polyacrylamide gel: Varying percentages of 7.5–12% of 30% Acrylamide/Bis solution, 37.5:1 (Bio-RAD, Hercules, CA), 0.375 M TRIS pH 8.8, 0.001% SDS, 0.001% ammonium persulfate [$(NH_4)_2S_2O_8$], 0.1% N,N,N′,N′ Tetra-methyl-ethylenediamine (TEMED) (Bio-RAD, Hercules, CA).
3. Upper polyacrylamide gel: 4% of 30% Acrylamide/Bis solution, 37.5:1, 0.125 M Tris pH 6.8, 0.001% SDS, 0.001% $(NH_4)_2S_2O_8$, 0.1% TEMED.
4. Gel Electrophoresis Apparatus: Mini-Protean Tetra Cell (Bio-RAD, Hercules, CA).
5. Gel Transfer Apparatus: Transfer Genie (Idea Scientific Company, Minneapolis, MN).
6. Membrane: Polyscreen PVDF Hybridization Transfer Membrane (Perkin Elmer, Waltham, MA).
7. Running buffer: 25 mM Tris, 80 mM glycine, 3.5 mM SDS.
8. Transfer buffer: 25 mM Tris, 190 mM glycine, 10% methanol.
9. Membrane wash solution: 1× TTBS: 14 mM NaCl, 2 mM Tris. Adjust pH to 7.6, 0.1% Tween-20.
10. Membrane blocking solution: 5% Nonfat dry milk (Carnation (Nestle), Solon, OH) in 1× TTBS.
11. Primary and secondary antibody solution: 1% Nonfat dry milk in 1× TTBS.
12. Secondary antibody: goat-anti-rabbit IgG (H+L)-HRP Conjugate (Bio-RAD, Hercules, CA).
13. ECL detecting solution: 1:1 mixture of SuperSignal West Dura Stable Peroxide Buffer (Thermo-Fischer, Rockford, IL) and SuperSignal West Dura Luminol/Enhancer Solution (Thermo-Fischer, Rockford, IL).
14. Imaging Station: FluorCHEM (Alpha Innotech, San Leanandro, CA).
15. Lambda Protein Phosphatase (Sigma, St. Louis, MO).

3. Methods

3.1. Affinity Column Preparation

The optimal column preparation method is dependent on the chemical structure of your antigen. Refer to Notes 2–4 before proceeding.

3.1.1. Ellman's Assay

This is necessary for SulfoLink® column preparations only.

1. Make approximately 1 mL of peptide at 1 mg/mL in SulfoLink® coupling buffer. If peptide does not fully dissolve, weigh out fresh peptide and add 50 µL DMSO, or no more than 10% DMSO. Once peptide has dissolved in DMSO, bring peptide to a final concentration of 1 mg/mL with coupling buffer.
2. Prepare fresh Ellman's Reagent, in SulfoLink® coupling buffer. This should be made fresh each time columns are prepared.
3. Combine one part Ellman's Reagent (60 µL), one part peptide solution (60 µL), and eight parts ddH_2O (480 µL). Make a control using 60 µL of H_2O instead of peptide solution. Aliquot 250 µL of control and each unknown and control in duplicates into an untreated 96-well plate. Read the absorbance at Abs_{412}.
4. If the peptide is 100% reduced, then the actual micromoles of dry peptide and micromoles of free sulfhydryls are the same. The ratio of the actual absorbance to the extinction coefficient for -SH at Abs_{412} will be 1. Determine the volume of stock needed to have 0.2 µmol of free sulfhydryl (-SH) groups for the column using the equation below:

$$\left(\frac{Abs_{412}}{2.72}\right)10 = x$$

$$\frac{1}{x} = \mu\text{L stock}$$

µL stock = volume of 1 mg/mL peptide stock needed for 0.2 µmol peptide. Add coupling buffer to calculated volume of peptide stock to reach final volume of 1 mL. This is your peptide solution for incubation on the Sulfo-Link beads.

5. If Ellman's absorbance reading is low, <0.300, or if the volume needed for 0.2 µmol is more than the volume of the prepared peptide solution, then the peptide should be reduced using TCEP-HCl as described below:
   (a) Use the molecular weight of the peptide to determine what volume of the 1 mg/mL stock is equal to 0.2 µmol. Assume the peptide is 75% pure (Only 750 µg peptide in 1 mL of 1 mg/mL stock).
   (b) Aliquot the calculated volume of 1 mg/mL peptide stock solution into a tube and add TCEP to a final concentration

of 20 mM. Add coupling buffer to a final volume of 1 mL. Incubate at room temperature for 30 min. Assume peptide is 100% reduced. This is your peptide solution for incubation on the SulfoLink® beads.

6. Perform peptide coupling with SulfoLink® as per manufacturer's instructions.

3.2. Affinity Purification

Steps 1–4 detail a preliminary $(NH_4)_2\ SO_4$ precipitation, which is not necessary in all cases. Refer to Note 5 before proceeding with affinity purification.

1. Pour sera to be precipitated into a beaker with a stir bar and place on a stir plate at 4°C. Stir slowly to avoid bubbles. Very slowly add $(NH_4)_2SO_4$ to sera to 40% saturation (2.26 g/10 mL).
2. Wait until $(NH_4)_2SO_4$ is completely dissolved and allow to stir for an additional 30 min.
3. Centrifuge at 10,000 × *g* for 20 min.
4. Discard supernatant and resuspend pellet in 10 mL PBS.
5. If frozen, thaw sera or ammonium sulfate precipitated sera. If sera that have not been precipitated look cloudy, spin down at 2,0000 × *g* for 10 min to clarify.
6. If using unprecipitated sera, filter with 1.0 μm Whatman filter disc into a clean tube and spike with 500× Pefabloc to a final concentration of 0.2 mM–2 μL per 1 mL. Spike with 100× Benzamidine to a final concentration of 2 mM–10 μL per 1 mL.
7. Remove sodium azide storage buffer from previously made phospho or pan affinity column and wash column with at least 25 mL of PBS.
8. Cycle sera through column three times and save collected flow through (Note 6). The flow through can be used to purify the dephospho antibody (Note 7).
9. Wash with 100 mL of BBS, then 50 mL PBS.
10. Elute the antibody from column. For a column bed of 1 mL, use 1 mL of elution buffer for each 1 mL of sera put over the column (Notes 8–11).
11. Wash column with 50 mL PBS and 5 mL of column storage buffer. Cap the bottom of column, add 2 mL of column storage buffer, and cap the top of the column (Note 12). Store at 4°C.
12. Place eluate in dialysis tubing and dialyze against 2 L of buffer overnight at 4°C.
13. Concentrate the dialyzed antibody in the dialysis tubing down to around 1–3 mL with Polyethylene Glycol MW 8000. Rinse tubing in H_2O and dialyze a second time in 2 L of dialysis buffer for at least 3 h, up to overnight at 4°C.

14. Remove antibody from dialysis tubing and place in a conical tube. If purifying a pan antibody, the purification process is complete. If purifying a phospho antibody, continue to the next step.
15. Remove the sodium azide storage buffer from the nonphospho column. Wash with 25 mL PBS.
16. Send the dialyzed antibody over the nonphospho column (Note 13). Collect the first milliliter of the flow through and discard (for columns with 1 mL bed). This is just the displaced PBS that was in the column bed. Collect remaining flow through. Push any antibody remaining on the column out into collection tube by running 1 mL of the dialysis buffer over the column. The flow through is the affinity purified phospho antibody.
17. Elute the pan antibody off the nonphospho column into a separate tube. Dialyze this antibody if desired.
18. Wash column with 25 mL PBS and 5 mL of column storage buffer. Cap the bottom of column, add 2 mL of column storage buffer, and cap the top of the column. Store at 4°C (Note 12).

3.3. Antibody Characterization

3.3.1. Peptide ELISA Method – Serum Screen

1. Measure out approximately 1 mg of phosphopeptide. Make a 1 mg/mL solution using Peptide Coupling Buffer. Make a 1 μg/mL stock solution from the 1 mg/mL solution.
2. Using 1 μg/mL stock solution, make serial dilutions of phosphopeptide of 500, 100, 10, 2.5 ng/mL. Refrigerate additional 1 mg/mL peptide solutions for further ELISA testing.
3. Aliquot 100 μL of each dilution of phosphopeptide into the wells of an Immunolink Amino plate (Notes 14–15). There should be one well for each sera sample at each dilution. Incubate while rocking at room temperature for 2 h or overnight at 4°C. Remove peptide solution and rinse out wells with wash solution three times (Note 16).
4. Dilute sera in Antibody Dilution Solution at 1:1,000 (Note 17). Aliquot 100 μL of the diluted sera into the wells. Incubate while rocking at room temperature for 2 h or overnight at 4°C. Remove antibody solution and rinse three times with wash solution.
5. Dilute secondary antibody in Antibody Dilution Solution at manufacturer's recommended dilution (Note 18). Aliquot 100 μL into each well and incubate while rocking at room temperature for 1 h. Remove solution and rinse three times with wash solution.
6. Remove excess solution by inverting plate and patting against paper towels.

7. Aliquot 100 μL of TMB liquid substrate into each well. Use a multichannel pipettor to apply solution quickly so that reaction happens simultaneously. Lightly agitate the plate to ensure even development and allow substrate to incubate until a blue color appears and variation can be seen. This should take approximately 15 s to 2 min.
8. Aliquot 100 μL of a quenching reagent such as sulfuric acid into each well. The reaction is complete when the solution is yellow. Again, lightly agitate the plate to ensure homogeneity.
9. Read plate at Abs_{450}.

3.3.2. Peptide ELISA: Affinity Purification Screen

In testing the various fractions collected during the affinity purification process, use both the phospho and dephosphopeptides. Measure out the dephosphopeptide if that was not done when measuring the phosphopeptide for the sera screen. Measure approximately 1 mg and dissolve in peptide-coupling buffer to a 1 mg/mL solution.

1. Using peptide-coupling buffer, dilute 1 mg/mL peptide stock (both phospho and dephospho) to the optimal concentration determined in the sera screen. The optimal concentration (Note 19) will give a mid-range signal (0.4–0.6). Two wells of the Immunolink Amino plate will be reserved for each fraction taken from the purification process. One will be for phosphopeptide, and the other for dephosphopeptide. Make enough of each diluted peptide to aliquot 100 μL per well.
2. Aliquot 100 μL of each peptide for each sample to be tested into the wells. Incubate while rocking at room temperature for 2 h or overnight at 4°C. Remove peptide solution and rinse wells with wash solution three times.
3. Dilute purification samples in Antibody Dilution Solution at 1:1,000. Aliquot into wells, each fraction into one well with phosphopeptide and one with dephosphopeptide. Incubate while rocking at room temperature for 2 h or overnight at 4°C. Remove antibody solution and rinse three times with wash solution.
4. Dilute secondary antibody in Antibody Dilution Solution at manufacturer's recommended dilution. Aliquot 100 μL into each well. Incubate while rocking at room temperature for 1 h. Remove secondary solution and rinse three times with wash solution.
5. Remove excess solution by inverting plate and patting against paper towels.
6. Aliquot 100 μL of TMB liquid substrate into each well. Use a multichannel pipettor to apply solution quickly so that the reaction happens simultaneously. Lightly agitate the plate to ensure even development and allow the substrate to incubate

until a blue color appears and variation can be seen. This should take approximately 15 s to 2 min.

7. Aliquot 100 μL of a quenching reagent such as sulfuric acid into each well. The reaction is complete when the solution is yellow. Again, lightly agitate the plate to ensure homogeneity.
8. Read plate at Abs_{450}.

3.3.3. WB Protocol

1. Clean all glass and utensils with ethanol.
2. Determine the acrylamide percentage needed to effectively separate your proteins of interest based on their molecular weights and pour lower (resolving) SDS-PAGE gels. Layer the top of the gels with water and polymerize for 45 min.
3. Remove water, pour upper (stacking) gel, and insert desired comb. Polymerize for 20–30 min.
4. Remove combs and excess polymerized gel, attach gels to running gel apparatus per manufacturer's recommendation. Fill the chamber with 1× running buffer.
5. Prepare lysate sample(s) by adding 4× sample buffer and boil sample(s) for 5 min at 100°C. Remove and cool for 5 min.
6. Load gels with samples and molecular weight marker. Refer to manufacturer's recommendation for necessary volume of marker.
7. Attach electrodes to power source, turn on source, and run gels per manufacturer's recommendation for voltage and time, or until the dye front is at the bottom of the gel.
8. Turn off power source, remove gel holder, remove plates from gel holder, then pull apart the glass so that the gel stays on one side. Cut away excess gel, carefully remove the gel of interest, and place in 1× transfer buffer until ready to place in transfer apparatus.
9. Cut membrane to gel size and activate per manufacturer's recommendation. Rinse with water and then place in 1× transfer buffer until time to transfer.
10. Prepare transfer apparatus as per manufacturer's recommendation, making sure to add membrane and gel in the right orientation. Add 1× transfer buffer.
11. Attach electrodes to transfer apparatus. Set the voltage and run time recommended by manufacturer. Turn on power source.
12. Turn off power source and remove blot. Rinse blot with water (Note 20). Allow to dry overnight to set proteins into membrane.
13. Reactivate membrane, rinse with water and block membrane in 5% milk in TTBS for 30 min, rocking, at room temperature.

14. Prepare primary antibody solution: dilute antibody in 1% milk in TTBS.
15. Remove blocking solution from membrane, and add primary antibody solution (Note 21). Incubate membrane overnight, rocking at 4°C.
16. Remove primary antibody solution and wash membrane three times for 5 min each by adding enough wash solution to cover membrane while rocking.
17. Incubate membrane in secondary antibody solution while rocking at room temperature for 1 h.
18. Wash membrane three times for 5 min by adding enough wash solution to cover the membrane while rocking.
19. Add ECL detecting solution to membrane and incubate the blot as per the manufacturer's recommended time. Remove solution and develop the membrane using film or an imaging station (Note 22).

4. Notes

1. When performing the Ultra Link® protocol, bring the 3 M ethanolamine to the correct pH slowly under a vented hood. Noxious fumes can be produced when adding the concentrated HCl.
2. Antigen conjugation is a step that often receives little attention. However, it is a critical step in the production of phospho-specific antibodies. We go to great lengths to ensure that both the conjugation of the peptide to carrier protein and coupling of the peptides to affinity-purification resin are optimized. This enables production of columns with the optimal ratio of peptide to resin in the column. When the peptide ratio is too high, it may be very difficult to elute high-affinity antibodies and when too low the capacity of the column will be compromised.
3. The phosphopeptide used as the antigen is bound to the affinity column matrix using various methods. A peptide antigen with a terminal cysteine can be bound to SulfoLink® Coupling Gel produced by Pierce Biotechnology. The SulfoLink® system relies on covalent immobilization through reduced cysteine residues. This binding depends on available free sulfhydryl groups, so we typically synthesize antigens that have a terminal cysteine. The peptide must be reduced as much as possible for optimal binding to the gel matrix, so it is important to ensure that the peptides are not dimerized or oxidized. To determine how well the peptide is reduced, the peptide can

be tested using dithionitrobenzoic acid (DNTB), commonly known as Ellman's Reagent. The free sulfhydryls on the peptide cleave the disulfide bond of DNTB. The resulting 2-nitro-5-thiobenzoate (TNB^{2-}) ion has a yellow color that can be read on a spectrophotometer at Abs_{412}. A very weak yellow color indicates too few free sulfhydryls, which will result in suboptimal binding to the column matrix. If this is the case, the peptide should be incubated with a reducing agent such as TCEP before being bound to the column.

4. If the amino acid sequence of the protein of interest has cysteine residues around the phosphorylation site, a cysteine-dependent binding matrix like SulfoLink® is not the best purification option. These internal cysteines may bind to the column matrix as well, resulting in a suboptimal binding environment for the phospho-specific antibody. When internal cysteines are present, terminal cysteines should not be added to the synthesized peptide antigen, and a different purification method must be used. An example is UltraLink® Biosupport by Pierce Biotechnology, which uses amide bonds to attach the antigen to the column matrix.
5. It is sometimes useful to purify the IgG fraction from the serum before applying it to the phospho and dephospho affinity columns if a rapid degradation of the phospho columns is observed. This may indicate that agents in the serum are degrading the antigen bound to the column. Using an ammonium sulfate precipitation as a preliminary purification method can reduce these agents and result in a cleaner, more concentrated serum. Steps 1–4 in the Affinity Purification Method detail an ammonium sulfate precipitation to achieve this end. If no preliminary purification is required, begin the purification protocol at Step 5.
6. It is recommended to collect and save flow throughs as well as small aliquots of all washes. These samples can then be tested for loss of antibody using SDS-PAGE gels or Western blotting.
7. The flow through from the phospho column contains any dephospho antibody generated. This flow through can be used to purify the dephospho antibody by sending it over the nonphospho column, or can be used to check the binding efficiency of the phospho column.
8. It may be necessary to test several different elution techniques in order to maximize antibody specificity and activity.
9. Use a higher volume of elution buffer if you are familiar with your system and know that the column has a high binding capacity and there is a high titer of antibody in the sera. A larger volume of elution buffer may also be used for larger bed volumes.

10. If an acidic elution buffer is used, add one bed volume of 1 M Tris pH 8.0 to the collection tube to neutralize the acid in the eluate.
11. A slower flow rate may be observed, especially when using older columns. The column may be placed on a long spinal needle (~4 inches) during wash and elution steps to allow capillary action to speed the flow rate. Older columns may have collected debris in their frits (filter discs), so flow rate may be improved by replacing these frits.
12. Columns may be used for repeated purifications. It is recommended that the column be washed well with purging buffer after 5–10 purifications. After purging, wash column extremely well with PBS before beginning purification.
13. When purifying a novel antibody, only send ½ of the phospho column eluate over the nonphospho column and then test both the reserved phospho eluate as well as the nonphospho flow through for antibody activity. This will check for antibody loss over the nonphospho column.
14. Most ELISAs are performed in 96- or 384-well microtiter plates. The bottoms of the wells are treated or coated to create a specific binding environment. There are many types of plates available, so it is important to make sure that your peptide can bind to the plate's surface. An example of a plate used for peptide ELISAs is the Immobilizer Amino plate created by NUNC. The binding chemistry of this plate requires the peptide to have a free amine. The binding surface uses a patented photochemical method that covalently couples ligands to polymer materials. This photocoupling uses an ethylene glycol spacer and a stable electrophilic group to react with nucleophiles such as free amines located at the end of peptide sequences. The spacer design and the density of the electrophilic group are optimized for peptide-based immunodiagnostic assays. The bound peptide is oriented to facilitate antibody recognition and binding.
15. When performing a peptide ELISA, different solutions will be used for binding, blocking, washing, incubating, and detecting. The manufacturer of the plate should specify the best coupling buffer to ensure the optimal binding environment. Once the peptide is bound, the manufacturer may recommend a blocking solution that will bind to the remaining active sites on the plate. This blocks your antibody from creating a false signal by binding to the plate's active sites rather than the peptide.
16. Thorough washing of the plate is very important because it removes both unbound antibody as well as other nonspecific binding.

17. The dilution of the primary and detection (secondary) antibodies will have to be optimized for your system, but these incubations are generally done at high dilutions. Common antibody dilutions for peptide ELISAs are 1:1,000 for the primary antibody and 1:10,000 for the secondary antibody.
18. Secondary antibodies are host-specific, and will bind to any antibody present from that host. For example, goat-anti-rabbit secondary antibody was produced in a goat, and will recognize all rabbit polyclonal antibodies. There are many commercially available secondary antibodies that have been tagged with common enzyme labels to allow for colorimetric or ECL detection methods.
19. It is important to find the optimal peptide concentration to use in the ELISA. Initially, a wide range of concentrations should be tested. Perform an initial dilution curve until a linear decrease in signal can be observed. Using this information, create a dilution curve with a smaller range that can be used on a variety of antibodies to save space and time.
20. WBs are the best way to demonstrate the phosphospecificity of an antibody. The most rigorous method for demonstrating an antibody's phosphospecificity involves treating a blot with phosphatase such as lambda phosphatase. Incubating the blot in the presence of the phosphatase often removes the phosphate groups from proteins in the blot. When WB analysis is performed on the untreated and treated portions of the blot, a phospho-specific antibody will show signal on the untreated section, but not on the phosphatase-treated section. When performing this treatment, a portion of the blot is cut away immediately after the transfer is complete, and is incubated in a buffer containing diluted protein lambda phosphatase enzyme. For the protein lambda phosphatase method to work, the phosphate group has to be accessible on the protein that has been transferred in the membrane. In our experience, this type of treatment is successful only about 50% of the time.
21. An alternate method to determine phosphospecificity is by performing a peptide block. In this method, the phospho and dephosphopeptides are each separately incubated with the diluted antibody prior to the antibody being introduced to the membrane. Ideally, the phosphoantibody will be absorbed only to the phospho-peptide antigen and thus fail to bind to the protein in the blot. Figure 4 shows an example of the lambda phosphatase methods being used to demonstrate the phosphospecificity of a phosphoantibody for ERK.
22. It is important to emphasize that a positive WB result does not guarantee that the antibody will work in IHC/IF/ICC applications. However, a positive and specific WB result adds

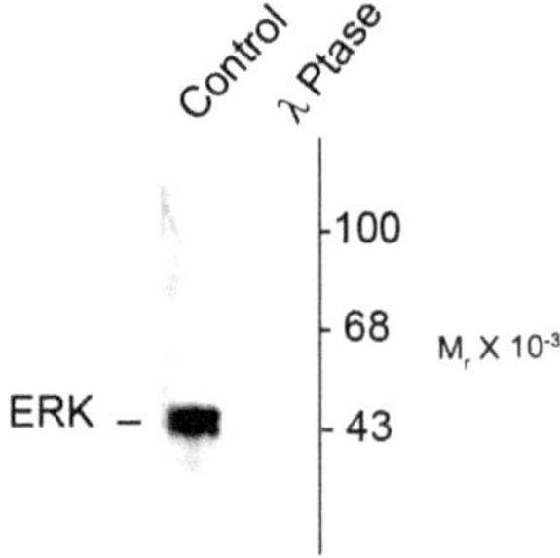

Fig. 4. Western blot of human T47D cell lysates showing specific immunolabeling of ~42 to 44k ERK/MAPK protein phosphorylated at Thr[202] and Tyr[204] (control). The phospho-specificity of this labeling is shown in the second lane (*lambda*-phosphatase: λ-Ptase). The blot is identical to the control except that it was incubated in λ-Ptase (1,200 units for 30 min) before being exposed to the Anti-Thr[202]/Tyr[204] ERK/MAPK. The immunolabeling is completely eliminated by treatment with λ-Ptase.

significant credibility to such imaging applications. It is certainly possible that an antibody may work in IHC and not in WB. But if the antibody cannot be shown to be specific for the protein of interest in a Western blot, it becomes much more difficult to establish the specificity of the IHC signal.

References

1. Ross, A.H., Baltimore, D., and Eisen, H.N. (1981) Phosphotyrosine-containing proteins isolated by affinity chromatography with antibodies to synaptic hapten. *Nature* **294**, 654–656.
2. Nairn, A.C., Detre, J.A., Casnellie, J.E., and Greengard, P. (1982) Serum antibodies that distinguish between the phospho- and dephospho-forms of a phosphoprotein. *Nature* **299**, 734–736.
3. Czernik A.J., Girault J-A., Nairn A.C., Chen J., Snyder G., Kebabian J., et al. (1991) Production of phosphorylation state-specific antibodies. *Methods Enzymol.* **201**, 264–283.
4. Perich, J.W. (1991) Synthesis of O-phosphoserine- and O-phosphothreonine-containing peptides. *Methods Enzymol.* **201**, 225–233.
5. Lee, V.M., Balin, B.J., Otvos, L. Jr, and Trojanowski, J.Q. (1991) A68: a major subunit of paired helical filaments and derivatized forms of normal Tau. *Science* **251**, 675–678.
6. Czernik, A.J., Mathers, J., Tsou, K., Greengard, P., and Mische, S.M. (1995) Phosphorylation state-specific antibodies: preparation and applications. *Neuroprotocols* **6**, 56–61.
7. Czernik, A.J., Mathers, J., and Mische S.M. (1997) Phosphorylation state-specific antibodies. *Neuromethods: Regulatory Protein Modification: Techniques & Protocols* **30**, 219–250.

Part II

Preservation and Unmasking of Tissue Antigens

Chapter 5

Methodology and Technology for Stabilization of Specific States of Signal Transduction Proteins

Mats Borén

Abstract

The ability to adequately measure the phosphorylation state of a protein has major biological as well as clinical relevance. Due to its variable nature, reversible protein phosphorylations are sensitive to changes in the tissue environment. Stabilizor T1 is a system for rapid inactivation of enzymatic activity in biological samples. Enzyme inactivation is accomplished using thermal denaturation in a rapid, homogeneous, and reproducible fashion without the need for added inhibitors. Using pCREB(Ser133) as a model system, the applicability of the Stabilizor system to preserve a rapidly lost phosphorylation is shown.

Key words: Denator, Stabilizor, CREB, Phosphorylation, Postmortem change, Immunohistochemistry, Phosphospecific antibody

1. Introduction

Communication within and between cells is a fundamental requisite for multicellular organisms. Signals controlling cellular processes are constantly transmitted though tightly regulated networks, both intra- and intercellular. Reversible protein phosphorylations are important in the reception of intercellular signals and the major agents of intracellular signaling (1, 2). The phosphorylation state of a protein often affects fundamental protein properties, e.g., activity, and/or subcellular distribution (3). The correct regulation of cellular signaling cascades via phosphorylations is vital in order to maintain cellular homeostasis and health. Aberrant phosphorylation of signaling proteins has been implicated in several neurological and cancerous disease states (4). The ability to adequately measure the phosphorylation state of a protein has major biological as well as

Alexander E. Kalyuzhny (ed.), *Signal Transduction Immunohistochemistry: Methods and Protocols*, Methods in Molecular Biology, vol. 717, DOI 10.1007/978-1-61779-024-9_5,

clinical relevance. With the emergence of therapies specifically targeted to particular proteins and protein states, the need for precise molecular phenotyping of disease states have increased (5). Histochemical (HC) and immunohistochemical (IHC) assessment of biopsies are considered standard methods to provide information and classify, as well as manage, disease states. The use of IHC for molecular phenotyping of tissue specimens faces several challenges, especially when determining not just the presence or absence of a protein but also its phosphorylation state (5). Due to its variable nature, reversible protein phosphorylations are sensitive to changes in the tissue environment, e.g., sampling and subsequent anoxia, and kinases and phosphatases react rapidly by altering the state of phosphorylation in response to the external stimuli (6, 7). The use of phosphospecific antibodies in IHC for detection of phosphorylation state is a promising application, but questions have been raised related to rapid alterations of the phosphorylation state during fixation (8–10). In standard IHC practice, samples are submerged in formalin to fixate the tissue and prevent degradation before they are dehydrated and embedded in paraffin. Formalin fixation is a rather slow process that usually requires several hours of formalin incubation at room temperature to be completed, a time period during which enzymes remain active and change the state and composition of proteins and especially phosphorylations (8). Current IHC protocols do not include the addition of enzyme inhibitors or other measures to prevent postsampling changes besides formalin fixation. Thus, in order to preserve transient epitopes, measures must be taken to prevent protein alterations prior to formalin fixation. The Stabilizor system, consisting of the Stabilizor T1 instrument and Maintainor Tissue consumable, rapidly inactive enzymatic activity in biological samples. Enzyme inactivation is accomplished using thermal denaturation in a rapid, homogeneous, and reproducible fashion without the need of additives (11).

In the current study, we have investigated the applicability of the Stabilizor system in combination with IHC procedures in order to assess prevention of protein alterations, especially of reversible protein phosphorylations, as well as preservation of morphology. The phosphorylation on serine 133 of calcium/cAMP responsive element binding (CREB) protein is rapidly lost following sampling and has been used as a model system during this study (Figs. 1–3).

2. Materials

2.1. Sample Acquisition and Preparation

1. Tools to extract mouse brain: Small scissor and spatula.
2. Stabilizor T1 with Maintainor Tissue consumables (Denator AB, Gothenburg, Sweden).

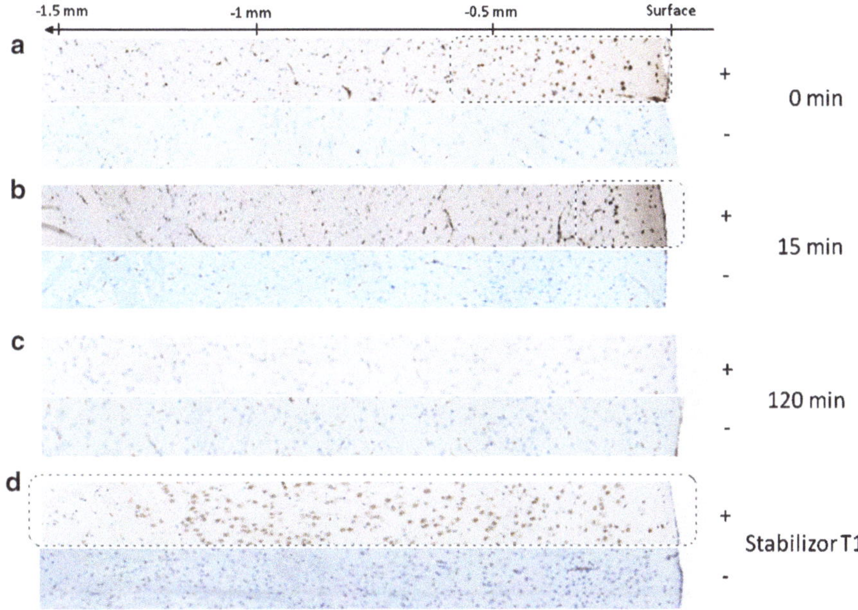

Fig. 1. IHC visualization of the phosphorylated form of CREB, pCREB(Ser133) in formalin-fixed tissue (mouse cortex). (**a**–**c**) 0–120 min postmortem incubation at room temperature before submersion in formalin. (**d**) Stabilizor T1 sample treated directly after extraction and then directly placed in formalin. *Large black dots* are pCREB-positive nuclei and smaller *blue–gray dots* are pCREB-negative nuclei counterstained with Hematoxylin--Eosin stained. A clear gradient from outside (*right*) to interior (*left*) can be seen for 0–120 min samples. Longer postmortem time gives shallower staining. In the Stabilizor-treated sample, black large pCREb-positive nuclei can be seen throughout the whole sample. All brains were cut and embedded as showed in Fig. 4. Panels are divided showing staining with primary antibody (+) and corresponding negative control without primary antibody (–).

3. 10% (v/v) neutral buffered formalin solution (HistoLab Products AB, Västra Frölunda, Sweden).
4. Sakura Tissue Tek VIP E150 Tissue Processor (Sakura Finetek Japan Co., Ltd, Tokyo, Japan).
5. HistoWax 56–58°C (HistoLab Products AB, Västra Frölunda, Sweden).
6. Tissue marking dye (Triangle Biomedical Sciences, Durham, NC, USA).

2.2. Histochemistry and Immunohistochemistry

1. Thermo Scientific Microscope slides (Thermo Fisher Scientific Inc., Waltham, MA)
2. Biobond Tissue Adhesive (BB-20. BioCell, Cardiff, UK)
3. Hematoxylin-Eosin Staining kit (DAKO AS, Copenhagen, Denmark)
4. pCREB (Ser133) antibody (#ab32096, Abcam plc, Cambridge, UK)
5. DAKO Autostainer Plus (DAKO AS, Copenhagen, Denmark)
6. Envision Real Detection kit (DAKO AS, Copenhagen, Denmark)
7. Pertex Mounting Medium (CellPath Ltd, Wales, UK)

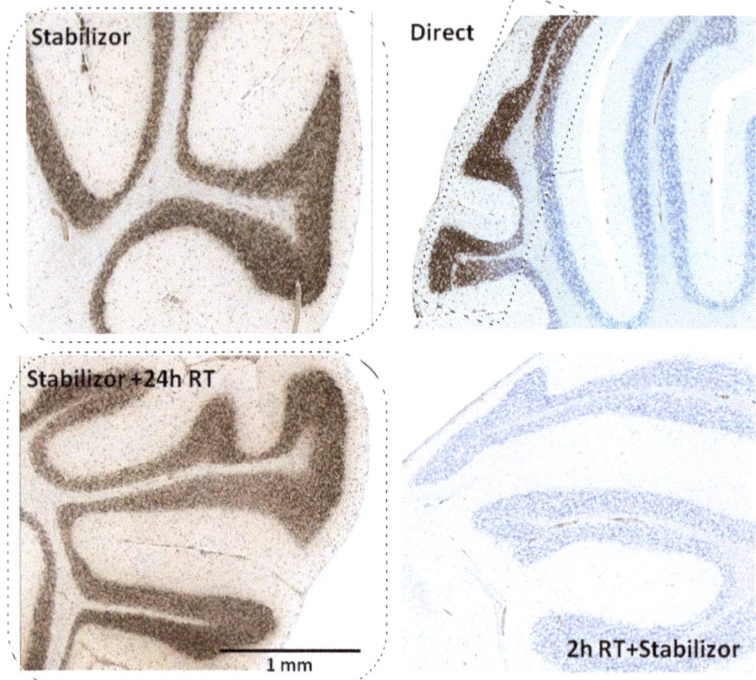

Fig. 2. IHC visualization of the phosphorylated form of CREB, pCREB(Ser133) in formalin-fixed tissue (mouse cerebellum). *Left* panels show stabilized tissue directly placed in formalin (*upper*) and with 24 h of room temperature incubation after treatment and formalin fixation (*lower*), whereas the right panels show tissue that was directly placed in formalin without Stabilizor treatment (*upper*) and tissue incubated at room temperature for 2 h before Stabilizor treatment and then directly placed in formalin (*lower*). Treated tissues show strong staining throughout the tissue (*upper left*) even after having been incubated at room temperature for 24 h prior to stabilization (*lower left*). Tissue placed directly in formalin but without stabilization show a clear outside-in staining (*upper right*). Tissue with 2 h room temperature incubation prior to stabilization shows no pCREB staining, consistent with dephosphorylation prior to stabilization and serves as a negative control for unspecific Stabilizor-induced staining (*lower right*).

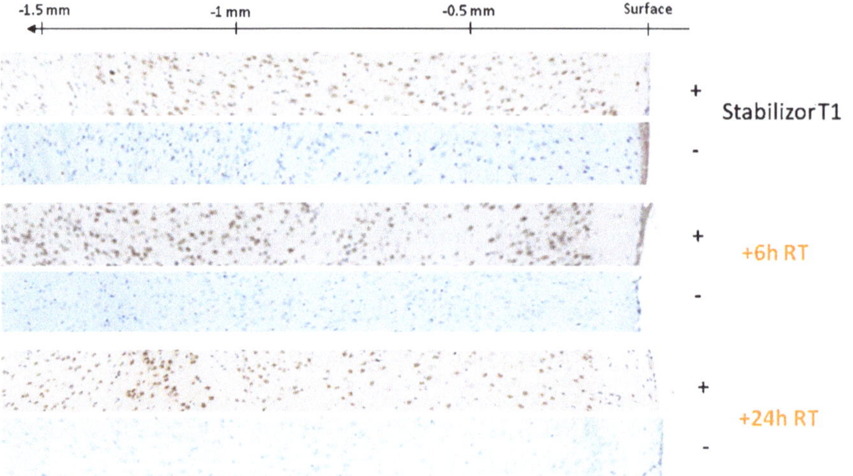

Fig. 3. IHC visualization of the phosphorylated form of CREB, pCREB(Ser133) in formalin-fixed tissue (mouse cortex). Tissue with increasing post-Stabilizor treatment incubations in room temperature. 0 min, 6 h, and 24 h. pCREB-positive nuclei can be seen throughout the tissue even with 24 h incubation at room temperature after treatment. Panels are divided showing staining with primary antibody (+) and corresponding negative control without primary antibody (–).

8. Cover slip (DAKO AS, Copenhagen, Denmark)
9. ScanScope® XT with software (Aperio Technologies Inc., Vista, CA, USA)

3. Methods

On a molecular level, tissue remains alive after sampling and can continue to change as long as conditions are favorable. This is normally not a problem since protein abundance and proteins commonly do not change in the time it takes to fix a sample using formalin. However, this may become a major problem for specific protein phosphorylations as phosphorylation levels are prone to change rapidly during fixation either leading to unnaturally high/low signals or complete loss, due to enzymatic activity during fixation. The Stabilizor instrument has been developed to enable detection of accurate levels of phosphorylations with preserved tissue morphology. This is accomplished by inactivating enzymes prior to fixation, thus precluding change during the fixation step. Stabilization causes protein denaturation, which may change availability of epitopes. This may necessitate optimization of IHC standard protocols developed for nonstabilized tissue specimen. In favorable cases, this can be solved by changing antigen-retrieval protocol or acquiring a new antibody while in extreme cases; it may require raising new antibodies specific for stabilized tissue.

In the following sections, the use of Stabilizor T1 for stabilization of mouse brains for detection of pCREB (Ser133) using IHC with a phosphospecific antibody will be outlined in detail. All important steps in the process from extraction of the brain up to visualization of results will be covered.

3.1. Sample Acquisition and Preparation

1. A good experiment always starts with the construction of a solid experimental design plan covering each step in the process. The illustrations in this chapter are from an experiment that aimed to answer the following questions:
   (a) Can treatment in the Stabilizor system prevent the loss of pCREB (Ser133) phosphorylation shown by outside-in staining in nonstabilized tissue?
   (b) Is pCREB (Ser133) stabile at room temperature that have been processed in the Stabilizor system?

   In order to answer these questions, we collected a series of brains with different incubation time prior to and after stabilization and prior to formalin submerging.
2. Prepare all material required during sample extraction, stabilization, and fixation. Label Maintainor Tissue cards and cups of formalin prior to starting sampling.

3. For IHC, the Stabilizor instrument should be set to run in Auto Structure Preserve mode (see Note 1).
4. When prepared for sampling, make a dry run of the sampling procedure (see Note 2).
5. Sample one mice at the time, calling out which mice is being taken and how it will be processed.
6. Preferentially, kill mice without the use of anesthesia, which have been shown to affect phosphorylation levels (12) (see Note 3).
7. At the time of death, start the *postmortem* clock on the Stabilizor T1 instrument (see Note 4).
8. Extract the intact brain from the mouse head and place the brain in on open Maintainor Tissue card, preferentially with the base of the brain down (see Note 5).
9. Position the brain in the center of the Maintainor card and close the card before pressing the START button on the touch screen.
10. When stabilization is complete, open the card and place the whole brain into formalin solution (see Note 6).
11. Incubate brains between 24 and 48 h in the formalin solution before proceeding to dehydration and paraffin imbedding.
12. Dehydrate samples (see Note 7).
13. Prior to paraffin embedding, cut the brain so that sections will be cut through the center of the brain (Fig. 4); this enables visualization of the formalin-penetration gradient toward the center of the brain (see Note 8).
14. Stain newly exposed surfaces using Tissue Marking dye (see Note 9).

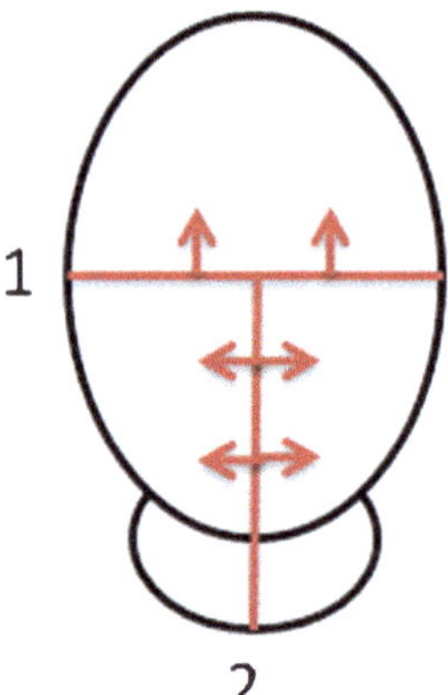

Fig. 4. Schematic drawing of how mouse brains were cut prior to paraffin embedding. Brains were incubated whole in formalin and cut along the *red lines* 1 and 2. Newly exposed surfaces were stained and embedded with the stained surfaces facing upwards in the block. Sections were then cut as indicated by the *red arrows*. In this way, the whole formalin penetration gradient is visual and phosphorylation changes due to slow formalin fixation can be visualized.

15. Embed tissue pieces in paraffin with the stain, newly cut surface facing towards the surface where sections will be cut from (see Note 10).

3.2. Immuno-histochemistry

1. Treat Thermo Microscope slides with Biobond. Follow the manufacturers′ instruction (see Note 11).
2. Cut 4 μm sections from the paraffin block into water. As each section is cut, transfer it to a precoated slide (see Note 12).
3. Dehydrate the slides for 30 min at 58°C (see Note 13).
4. To start IHC procedure, dewax and rehydrate slides (see Note 14).
5. Antigen retrieval is performed in a microwave for 10 min at 750 W followed by 15 min at 350 W 1× TE-buffer at pH 9.
6. Dilute pCREB(Ser133) antibody in Ab diluent S0809 buffer at a 1:250 dilution.
7. IHC staining is done with Envision Real Detection kit, Peroxidase/DAB, Rabbit/Mouse(K5007) using a DAKO Autostainer Plus (see Note 15).
8. As the last step in the autostainer, the slide is counterstained using Mayers hematoxylin (see Note 16).
9. Mount slides using Pertex Mounting medium.
10. Capture images of stained slides at 40× magnification using a ScanScope® XT digital slide scanner (see Note 17).
11. View and select subsections with Aperio's ImageScope viewer.

4. Notes

1. This mode uses less cavity vacuum and pressure in order to better preserve morphology.
2. The dry run consists of going through each step of the experiment making sure all involved personnel know what and how things are going to happen as well as when they are to take action. Also, check that all needed equipment is present and ready to be used. This step is necessary to ensure a swift and flawless sampling with minimal time between killing the animal and stabilization in the Stabilizor system.
3. Cervical dislocation or a scientific Guillotine are examples of preferred killing techniques.
4. *Postmortem* time is calculated preferentially from the moment of death, until the Maintainor Tissue card enters the Stabilizor instrument. Recording and keeping this time as short as possible and standardized between samples is important for high-quality samples.

5. There are several ways of extracting a mouse brain. My preferred way is to use a small scissor and on the removed head, first make a cut in the pelt from the neck and forward. The pelt is then pealed down on each side exposing the cranium. The scissor is then inserted through the base of the skull and a cut is made on the left and front side of the skull. Take care to make several small cuts to avoid cutting into the brain. Using the scissor blades, the top of the skull can now be pride away, toward the right, to expose the brain. Using a spatula, the brain can be carefully scooped out onto an open Maintainer Tissue, preferentially already placed in the Stabilizor instrument.
6. To show outside-in staining for nonstabilized brains, it is important that a suitably large piece is fixed whole in formalin. The optimal size will depend on the rate of change for the phosphorylations of interest. If desired, the stabilized sample can be stored in the Maintainor card, how long should be determined based on need and stability of phosphorylations of interest. pCREB(Ser133) have been shown to be stable for at least 24 h at room temperature after stabilization.
7. We use an automated system, Sakura Tissue Tek VIP E150 Tissue Processor, with the following protocol:
   (a) 20 min incubation in 10% formalin solution.
   (b) 5 min incubation in water.
   (c) 1 h incubation in 70% ethanol.
   (d) 2 × 1 h incubation in 95% ethanol.
   (e) 2 × 1 h incubation in 100% ethanol.
   (f) 3 × 1 h incubation in xylene.
   (g) 1 h incubation in molten paraffin at 60°C.
   (h) 1.3 h incubation in molten paraffin at 60°C.
   (i) 2 h incubation in molten paraffin at 60°C.

   To our knowledge, the use of the Sakura Tissue Tek VIP E150 Tissue processor instrument or specified protocol is not required as long as an established dehydration protocol is used.
8. This is preferentially done by first cutting the brain coronally (ear to ear) through the thickest part. The hind brain is then cut saggitally (front to back) between the two halves.
9. Use different dye color for stabilized and nonstabilized tissue. Marking of cut surfaces will both aid in positioning the pieces in the paraffin block and make it possible to separate pieces with different treatments in a multitissue block.
10. To facilitate optimization of antigen retrieval and antibody dilution, construct at least one block containing pieces of both stabilized and nonstabilized brain, preferably a one half

of the hind brain due to the visually interesting morphological features in the cerebellum.

11. For Stabilizor-treated brain, Biobond coating is necessary. It may not be so critical for other tissue types, but this should be verified experimentally. Avoid treating more slides than what are needed in the near future. Adhesion tends to be slightly less when sections are applied to coated slides that have been stored for long periods of time.
12. Stabilized tissue tends to be more brittle and it will be more difficult to transfer the sections to glass if they have been floating in the water for a period of time.
13. Slides with mounted sections can be stored in a dry place until use.
14. DAKO Autostainer Plus can be used with the following protocol:
    (a) 3 × 5 min incubation in 100% xylene
    (b) 2 × 5 min incubation in 100% ethanol
    (c) 2 × 5 min incubation in 95% ethanol
    (d) 2 × 3 min incubation in water
15. The following protocol can be used by the autostainer:
    (a) 5 min incubation with Endogenous Enzyme Block, 0.5% peroxide solution, to block endogenous peroxidases.
    (b) 30 min incubation with primary antibody.
    (c) 10 min incubation with Envision enhancer polymer coupled to HRP.
    (d) 12 min incubation with DAB to develop signal.
16. The following protocol is used for hematoxylin staining:
    (a) The slide is covered with hematoxylin solution and incubated for 5 min.
    (b) The slide is carefully rinsed with distilled water to remove any trace of unbound hematoxylin solution.
    (c) To increase staining, dip slides 15 times in a 1:5 dilution of saturated lithium carbonate solution (bluing).
    (d) Rinse in water
    (e) 2 min incubation in 95% ethanol
    (f) 2 min incubation in 100% ethanol
    (g) 2 min incubation in 100% xylene
17. Although quite expensive, the use of digital whole slide images are a great complement to ordinary microscope viewing and facilitates sharing and computer-based quantification of staining.

Acknowledgments

The author is thankful to Inga Hansson for skilfully carrying out IHC procedures and Charlotte Emlind-Vahlu for much appreciated advice on the manuscript.

References

1. Cohen, P. (2000) The regulation of protein function by multisite phosphorylation – a 25 year update. *Trends Biochem. Sci.* **25**, 596–601.
2. Voshol, H., Ehrat, M., Traenkle, J., Bertrand, E. and van Oostrum, J. (2009) Antibody-based proteomics-analysis of signaling networks using reverse protein arrays. *FEBS J.* **276**, 6871–6879.
3. Hunter, T. (1987) A thousand and one protein kinases. *Cell* **50**, 823–829.
4. Li Jianlin, G.T.D., Yuan Peixiong, Manji Husseini, K. and Chen Guang (2003) Postmortem interval effects on the phosphorylation of signaling proteins. *Neuropsychopharmacology* **28**, 1017–1025.
5. True, L.D. (2008) Quality control in molecular immunohistochemistry. *Histochem. Cell Biol.* **130**, 473–480.
6. Espina, V., Edmiston, K.H., Heiby, M., Pierobon, M., Sciro, M., Merritt, B., et al. (2008) A portrait of tissue phosphoprotein stability in the clinical tissue procurement process. *Mol. Cell. Proteomics* 7, 1998–2018.
7. Espina, V., Mueller, C., Edmiston, K., Sciro, M., Petricoin, E.F. and Liotta, L.A. (2009) Tissue is alive: new technologies are needed to address the problems of protein biomarker pre-analytical variability. *Proteomics Clin. Appl.* **3**, 874–882.
8. Fox, C.H., Johnson, F.B., Whiting, J. and Roller, P.P. (1985) Formaldehyde fixation. *J. Histochem. Cytochem.* **33**, 845–853.
9. Mandell, J.W. (2003) Phosphorylation state-specific antibodies: applications in investigative and diagnostic pathology. *Am. J. Pathol.* **163**, 1687–1698.
10. Mandell, J.W. (2008) Immunohistochemical assessment of protein phosphorylation state: the dream and the reality. *Histochem. Cell Biol.* **130**, 465–471.
11. Svensson, M., Borén, M., Sköld, K., Fälth, M., Sjögren, B., Andersson, M., et al. (2009) Heat stabilization of the tissue proteome: a new technology for improved proteomics. *J. Proteome Res.* **8**, 974–981.
12. Li, X., Friedman, B.A., Roh M.-S., and Richard, J.S. (2005) Anesthesia and post-mortem interval profoundly influence the regulatory serine phosphorylation of glycogen synthase kinase-3 in mouse brain. *J. Neurochem.* **92**, 701–704.

Chapter 6

An Enhanced Antigen-Retrieval Protocol for Immunohistochemical Staining of Formalin-Fixed, Paraffin-Embedded Tissues

Sergei I. Syrbu and Michael B. Cohen

Abstract

Formalin is the most commonly used fixative for light microscopy because of its preservation of morphological details. A major adverse effect of formalin fixation is formation of cross-linkages between epitopes (amino acid residues) and unrelated proteins by formaldehyde groups. The great majority of monoclonal and polyclonal antibodies used for immunohistochemical (IHC) staining of formalin-fixed, paraffin-embedded (FFPE) tissues necessitate unmasking antigens for antigen retrieval. There are currently two major antigen-retrieval procedures based on treatment of deparaffinized tissue sections with heat or, less commonly, with enzymatic digestion. The use of various antigen-retrieval solutions and heating sources does not allow standardization of IHC staining and minimalization of interlaboratory discrepancies. We developed a novel modified antigen-retrieval protocol for reversing the effect of formalin fixation. The key feature of this protocol is treatment of deparaffinized tissue sections at reduced constant heat (97°C in a water bath) for 40 min in 25 mM Tris–HCl (pH 8.5), 1 mM EDTA, and 0.05% SDS (Tris–EDTA–SDS) buffer. Sections are then immunostained with primary and secondary antibodies conjugated with polymer-labeled Horse Radish Peroxidase. Compared to conventional antigen-retrieval procedures, this protocol more efficiently reverses the effect of formalin fixation of a wide variety of cellular antigens and in most instances decreases the use of primary antibody by 2–40 times, resulting in cost savings. Moreover, this protocol eliminates the need for using different antigen-retrieval methods in the laboratory, which reduces both time and labor for medical technologists.

Key words: Antibodies, Antigen, Retrieval, Formalin fixation, Immunohistochemistry

1. Introduction

Formalin is the standard tissue fixative used in histopathology and research laboratories, although the exact molecular mechanism of tissue fixation is not entirely understood (1–5). The major

Alexander E. Kalyuzhny (ed.), *Signal Transduction Immunohistochemistry: Methods and Protocols*, Methods in Molecular Biology, vol. 717, DOI 10.1007/978-1-61779-024-9_6,

adverse effect of formalin fixation is the cross-linkage of proteins, which masks antigenic epitopes recognized by antibodies used for immunohistochemical (IHC) staining of formalin-fixed, paraffin embedded (FFPE) tissues (3–5). In the past two decades, significant progress has been made in reversing the formalin-induced protein cross-linking. Current antigen-retrieval procedures are based on treatment of deparaffinized tissue sections with heat, including microwaves, autoclaves, steamers, boilers, or combinations of heat and pressure (pressure cookers), and enzymatic digestion (6–14). For each primary antibody, laboratories and manufacturers choose which antigen-retrieval method and antibody concentration is optimal and gives the most reproducible staining. Proteolytic digestion is rarely used in laboratories because it fails to yield reproducible and satisfactory immunostaining for many antigens. The effectiveness of heat-induced antigen retrieval is markedly influenced by pH and ionic strength of the antigen-retrieval buffers. Further, the use of various antigen-retrieval solutions and heating sources does not allow standardization and minimalization of interlaboratory discrepancies. In one study (15) assessing the validation of tissue microarray for IHC staining, using the same antibodies with different antigen-retrieval procedures, the antibody dilution differed up to 20 times (1:20–1:400) among the institutions participating in the study. The need for standardization of the IHC remains a critical issue and there have been many attempts described in literature (16–18). Namimatsu et al. (17) recently described a universal antigen-retrieval method of reversing the effect of formalin fixation with citratonic anhydride and heat. Although it appears that the citratonic anhydride effectively reverses the formalin's effect on tissues, the concentrations of primary antibodies were the same as those used with other antigen-retrieval methods. Moreover, citratonic anhydride is a toxic compound and the retrieval must be performed under a hood to prevent contact with skin and eyes.

We describe here an enhanced antigen-retrieval protocol for formalin-fixed, paraffin-embedded tissues and their IHC staining with a large variety of antibodies. The procedure is based on treatment of deparaffinized tissue sections at uniform and constant heat at 97°C in Tris–EDTA–SDS buffer at pH = 8.5 for 40 min. Compared to conventionally used antigen-retrieval procedures, this standardized protocol more efficiently reverses the effect of formalin fixation of a wide variety of cellular antigens and in most instances decreases the use of primary antibody by 2–40 times.

2. Materials

1. 10× antigen-retrieval buffer: 250 mM Tris–HCl (Tris (hydroxymethyl)amino-methane), pH 8.5, 10 mM EDTA (Ethylenediamine Tetraacetate), 0.5% (w/v) sodium dodecyl sulfate (SDS).
2. Anti-mouse and anti-rabbit antibodies conjugated with polymer-labeled Horse Radish Peroxidase (EnVision + System-HRP from DakoCytomation; Carpinteria, CA).
3. Diaminobenzidine (DAB) chromogen substrate and DAB enhancer (DakoCytomation, Carpinteria, CA).
4. Hydrogen Peroxyde, 30%.
5. Washing buffer (DakoCytomation, Carpinteria, CA).
6. Ethyl Alcohol, 95 and 100%.
7. Xylenes.
8. Harris Hematoxylin.
9. 0.003% Ammonium Hydroxide solution (make fresh daily).
10. Microscope Slides, Superfrost/Plus.
11. Mounting media "Mount-Quick" (Daido Sangyo Co., LTD, Japan).
12. Water bath (Isotemp 205, Fisher Scientific).

3. Methods

3.1. Antigen-Retrieval Protocol

1. Mount 4-μm thick sections of formalin-fixed and paraffin-embedded human tissue specimens onto "plus" slide and dry overnight at 27°C (see Note 1).
2. Prepare the antigen-retrieval buffer by diluting 25 ml of 10× Tris–EDTA–SDS buffer with 225 ml distilled water (see Note 2).
3. Deparaffinize sections in three changes of xylene, 5 min each.
4. Rehydrate in two changes of 100% alcohol and one change in 95% alcohol for 3 min each.
5. Rinse the sections for 3 min in distilled water and block endogenous peroxidase activity with 1% hydrogen peroxide in distilled water for 5 min. Rinse the sections in distilled water (see Note 3).
6. Preheat the water bath containing a covered staining dish with 250 ml Tris–EDTA–SDS buffer until the temperature reaches 97 ± 1°C (see Note 4).

7. Immerse the slides in the staining dish and incubate for 40 min.
8. Turn off the water bath and remove the staining dish to room temperature. Allow the slides to cool for 10–20 min at room temperature.
9. Rinse the sections in washing buffer for 3 min.

3.2. IHC Staining

1. In a humidified chamber, incubate sections with primary antibodies for 30 min. Antibody dilutions are listed in Table 1 (see Notes 5–7).
2. Rinse sections in washing buffer twice for 3 min and incubate sections with appropriate secondary antibodies. These secondary antibodies are prediluted by manufacturer.
3. Rinse sections in washing buffer twice for 3 min and incubate sections with DAB chromogen substrate for 5 min. Rinse with distilled water (see Note 8).
4. Incubate sections for 3 min with DAB enhancer and rinse for 3 min in distilled water.
5. Counterstain with filtered Harris Hematoxylin for 3 min. Rinse well in distilled water (see Notes 7 and 9).
6. Dip for 5 s in 0.003% Ammonium Hydroxide solution and rinse well in distilled water again (see Note 10).
7. Dehydrate sections in graded ethanol (70, 95, and 100%) and xylene (three exchanges) for 1 min each.
8. Coverslip slides manually using mounting media or with an automatic Coverslipper.

4. Notes

1. If using a microwave or pressure cooker instead of a water bath, bake the slides for 1–2 h in an electric oven at 60°C to prevent tissue loss from slides during the heating process. Use only 25 mM Tris (pH 8.5) and 1 mM EDTA buffer; do not use SDS since there will be excessive bubbling with tissue drying in microwaves or complete loss of tissue antigenicity in pressure cookers. See also Note 5.
2. 10× Tris–EDTA–SDS buffer can be stored at room temperature for more than a year if filtered through a 0.45-μm Millipore filter.
3. Make daily fresh 1% hydrogen peroxide in distilled water from stock reagent (30% hydrogen peroxide).
4. Use a precise water bath (97 ± 1°C) to avoid boiling and prevent excessive bubbling due to the presence of SDS. See also Note 1.

Table 1
Comparative study of IHC staining of human tissues

Antigen	Company	Clone/source	PC, citrate	MW, citrate	PK, 5 min	WB, TES
A1AT	Dako	Rabbit	–	1:2,750	–	1:2,750
AFP	Dako	Rabbit	1:3,500	–	–	1:7,000
ALK1	Dako	ALK1	1:50	–	–	1:500
Bcl-2	Dako	124	1:25	–	–	1:500
Bcl-6	Santa Cruz	Rabbit	1:25	–	–	1:250
BOB.1	Santa Cruz	Rabbit	1:500	–	–	1:1,000
Calretinin	Invitrogen	Rabbit	–	1:200	–	1:1,000
CD1A	Dako	010	–	1:50	–	1:500
CD3	Dako	Rabbit	1:100	–	–	1:200
CD5	Dako	SP19	1:10	–	–	1:10
CD8	Dako	C8/144B	1:20	–	–	1:800
CD10	Dako	56C6	1:20	–	–	1:100
CD15	UIHC	1D2	Neat	–	–	1:50
CD21	Dako	1F8	–	–	1:10	1:400
CD23	Novocastra	SP23	1:20	–	–	1:300
CD30	Dako	Rabbit	1:100	–	–	1:200
CD31	Dako	JC70A	1:20	–	–	1:200
CD43	Dako	DF71	–	1:200	–	1:200
CD45	Dako	2B11 + PD7/25	–	1:100	–	1:2,000
CD45RO	Dako	UCHL1	–	1:50	–	1:500
CD79A	Dako	JCB117	1:300	–	–	1:600
CD99	Covance	O13	–	1:200	–	1:200
CD117	Dako	Rabbit	–	1:600	–	1:4,000
c-erbB-2	Novocastra	CB11	1:300	–	–	1:1,000
Chromogranin	Dako	Rabbit	–	1:200	–	1:1,000
CK5/6	Dako	D5/16B4	1:25	–	–	1:25
CK7	Dako	OV-TL	–	–	1:50	1:500
CK8	Dako	C51	–	–	1:50	1:500
CK20	Dako	Ks20.8	–	–	1:50	1:500
CycD1	Dako	SP4	1:50	–	–	1:1,000
ER	Dako	1D5	1:100	–	–	1:400

(continued)

Table 1 (continued)

Antigen	Company	Clone/source	PC, citrate	MW, citrate	PK, 5 min	WB, TES
Gastrin	Dako	Rabbit, 1:1,000	–	–	–	1:5,000
GH	Covance	Rabbit, 1:5	–	–	–	1:25
HBSG	Dako	3E7	–	1:20	–	1:100
Melanosome	Dako	HMB45	–	–	1:100	1:200
HSV2	Dako	Rabbit, 1:4,000	–	–	–	1:4,000
Inhibin	Dako	R1	1:10	–	–	1:50
Kappa	UIHC	2D2	ND	ND	ND	1:300
K903	Dako	34βE12	1:10	–	–	1:100
Lambda	UIHC	4D11	ND	ND	ND	1:300
Melan A	Dako	A103	1:20	–	–	1:100
Ki-67	Dako	MIB-1	1:100	–	–	1:500
MUM1/IRF4	Dako	MUM1p	1:50	–	–	1:1,000
Myogenin	Dako	F5D	1:300	–	–	1:300
Pax5	BD Trans	24	1:200	–	–	1:400
P53	Dako	DO-7	–	1:100	–	1:500
Progesteron	Dako	PgR636	1:200	–	–	1:1,000
Prolactin	Signet	Rabbit, neat	–	–	–	1:5
PU.1	BD Phamagen	G148-74	1:50	–	–	1:250
Synaptophysin	Dako	Rabbit	1:50	–	–	1:250
TDT	Supertech	Rabbit	1:20	–	–	1:100
TSH	Signet	Rabbit, 1:2	–	–	–	1:10
TTF	Dako	8G7G3/1	1:20	–	–	1:200
WT	Dako	6F-H2	1:10	–	–	1:100

Antibody source and antigen-retrieval protocols used
mAb mouse monoclonal antibody; *Poly* rabbit polyclonal antibody; *PC* pressure cooker (citrate buffer pH 6.0); *MW* microwave (citrate buffer pH 6.0); *PK* protease K; *WB/TES* water bath/Tris–EDTA–SDS; *UIHC* antibodies developed at the University of Iowa Hospitals and Clinics; *ND* not done

5. When a high concentration of antibodies (>10 μg/ml) is required for staining and/or staining is very weak, consider using only 25 mM Tris (pH 8.5) and 1 mM EDTA without 0.05% SDS.

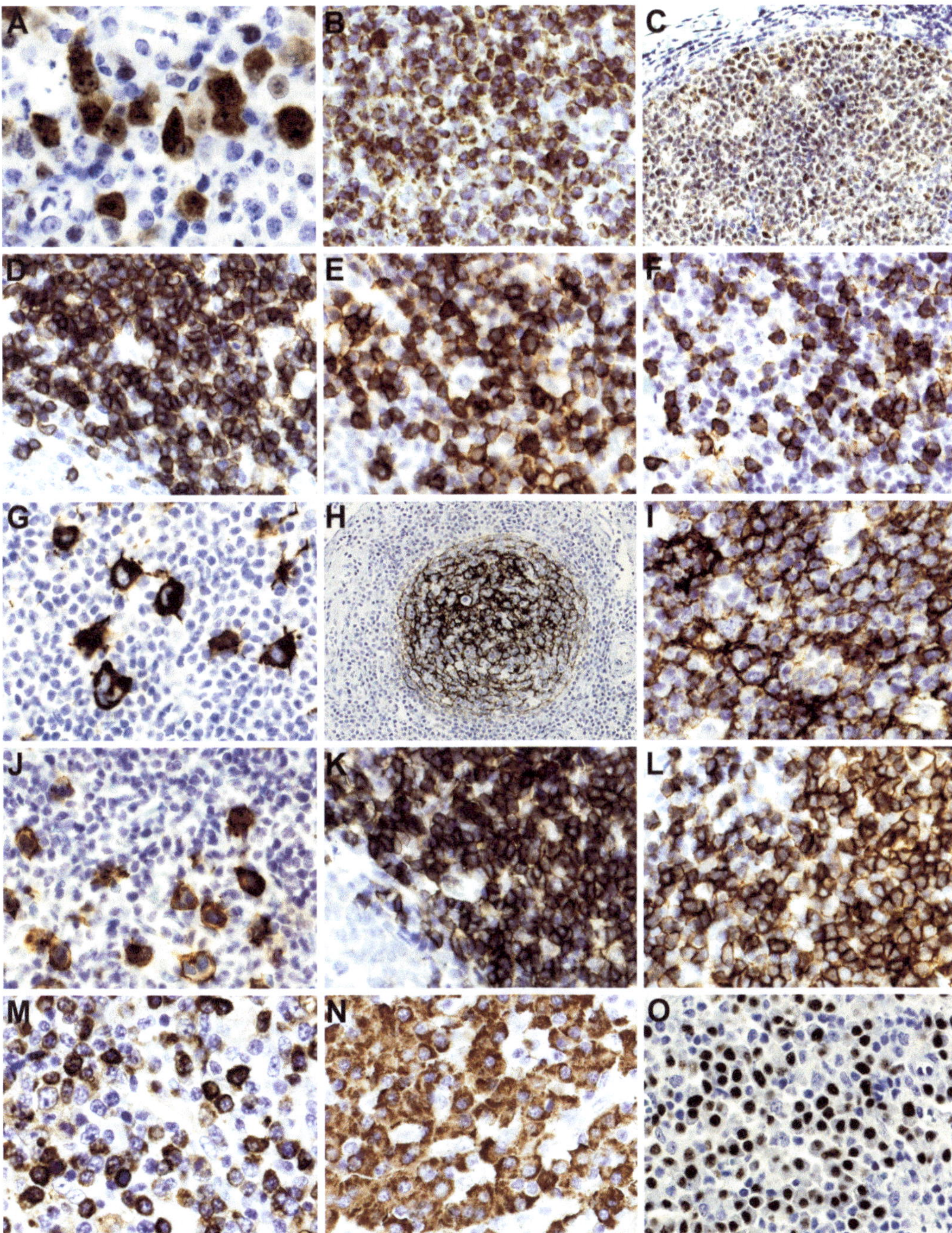

Fig. 1. Representative immunohistochemical (IHC) staining for ALK-1 (**a**, anaplastic large-cell lymphoma), bcl-2 (**b**, tonsil), bcl-6 (**c**, tonsil), CD3 (**d**, tonsil), CD43 (**e**, tonsil), CD8 (**f**, tonsil), CD15 (**g**, Hodgkin lymphoma), CD21 (**h**, lymph node), CD23 (**i**, small lymphocytic lymphoma), CD30 (**j**, Hodgkin lymphoma), CD43 (**k**, lymph node), CD45RO (**l**, lymph node), kappa (**m**, lymph node), lambda (**n**, plasma-cell myeloma), and MUM1 (**o**, diffuse large B-cell lymphoma) using Tris–EDTA–SDS buffer. The antibody dilutions are listed in Table 1. Original magnification 40× A-G and I-O; 20× for H.

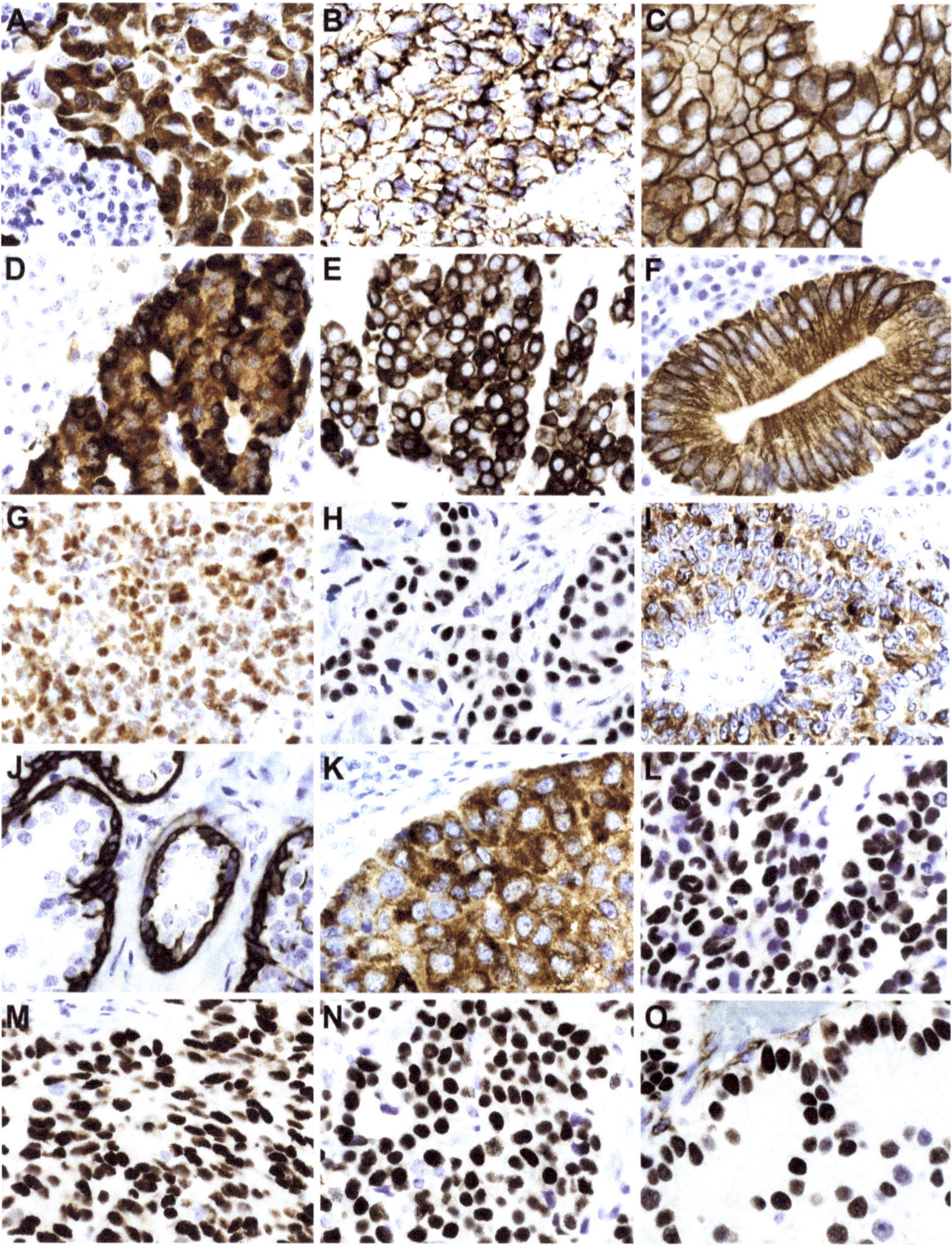

Fig. 2. Representative IHC staining for calretinin (**a**, mesothelioma), CD117 (**b**, GIST tumor), c-erb (**c**, breast carcinoma), chromogranin (**d**, pancreas), cytokeratin 7 (**e**, adenocarcinoma), cytokeratin 20 (**f**, colon), cyclin D1 (**g**, mantle-cell lymphoma), estrogen (**h**, breast), inhibin (**i**, ovarian neoplasm), K903 (**j**, prostate), MART 1 (**k**, melanoma), myogenin (**l**, rabdomyosarcoma), p53 (**m**, diffuse large B-cell lymphoma), progesteron (**n**, breast), and TTF (**o**, thyroid carcinoma) using Tris–EDTA–SDS buffer. The antibody dilutions are listed in Table 1. Original magnification 40×.

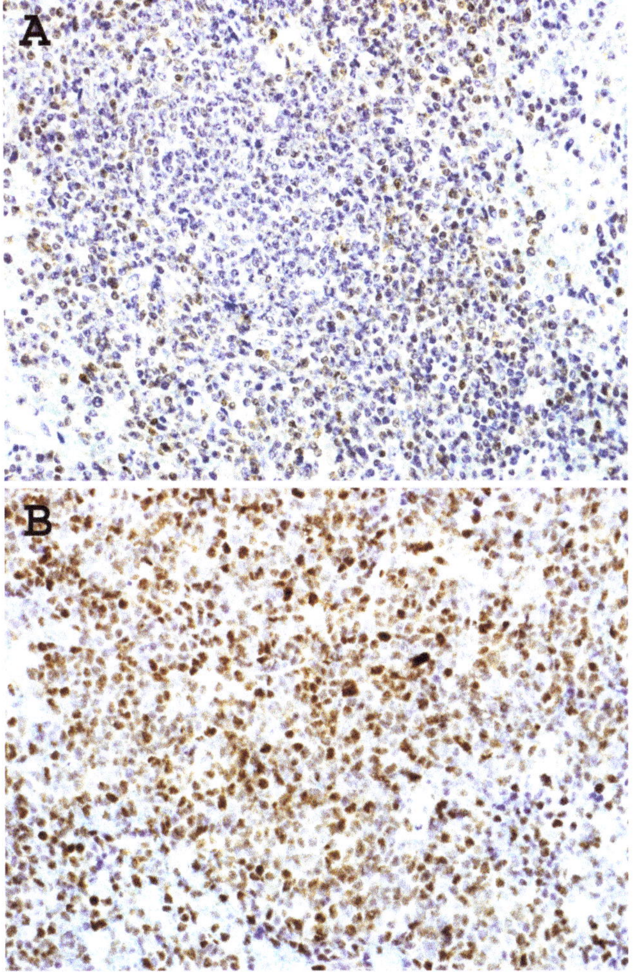

Fig. 3. Comparative IHC staining of a lymph node with mantle-cell lymphoma for cyclin D1 using citrate buffer (pH 6.0) heated in a pressure cooker (**a**, anti-Cylin D1 antibodies at 1:50) or Tris–EDTA–SDS buffer and water bath (**b**, anti-cylin D1 antibodies at 1:800). Original magnification 40×.

6. When there is no staining, consider using another antibody (another clone or polyclonal) directed to different antigenic epitopes. If there is only one commercial antibody, use 25 mM Tris (pH 8.5) and 1 mM EDTA and heat at a higher temperature (for example in a pressure cooker or microwave).

7. We tested 54 different monoclonal and polyclonal antibodies (Table 1, Figs. 1–3) used for IHC staining of a variety of FFPE human-tissue specimens. Using the heat-induced antigen-retrieval method of formalin-fixed specimens in Tris–EDTA–SDS buffer showed enhanced immunostaining at lower concentrations of primary antibody (2–40 times) than after conventional antigen-retrieval protocols.

8. For markers (antigens) with low cellular density, the incubation time with DAB chromogen substrate may be prolonged up to 30 min.
9. For weak nuclear antigens, counterstaining time with Harris Hematoxylin may be decreased from 3 min to 10–30 s for better visualization of such antigens.
10. Sections' treatment with Ammonium Hydroxide is optional.

References

1. Fox, C.H., Johnson, F.B., Whiting, J., and Roller, P.P. (1985) Formaldehyde fixation. *J. Histochem. Cytochem.* **33**, 845–953.
2. Puchtler, H., and Meloan, S.N. (1985) On the chemistry of formaldehyde fixation ant its effects on immunohistochemical reaction. *Histochemistry* **82**, 201–204.
3. Shi, S.-R., Gu, J., Turrens, J., et al (2000) Development of the antigen retrieval technique: philosophical and theoretical bases. In: Shi, S.-R., Gu, J., and Taylor, C., eds. Antigen Retrieval Techniques: Immunohistochemistry & Molecular Morphology. Natick: Eaton Publishing, 17–40.
4. Sompuram, S.R., Vani, K., Messana, E., and Bogen S.A. (2004) A molecular mechanism of formalin fixation and antigen retrieval. *Am. J. Clin. Pathol.* **121**, 120–129.
5. Bogen, S.A., Vani, K., and Sompuram, S.R. (2009) Molecular mechanisms of antigen retrieval: antigen retrieval reverses steric interference caused by formalin-induced cross-links. *Biotech. Histochem.* **84**, 207–215.
6. Shin, R.W., Iwaki, T., Kitamoto, T., and Tateishi, J. (1991) Hydrated autoclave pretreatment enhances TAU immunoreactivity in formalin-fixed normal and Alzheimer's disease brain tissues. *Lab. Invest.* **64**, 693–702.
7. Suurmeijer, A.J.H., and Boon, M.E. (1993) Notes on the application of microwaves for antigen retrieval in paraffin and plastic tissue sections. *Eur. J. Morphol.* **31**, 144–150.
8. Gown, A.M., de Wever, N., and Battifora, H. (1993) Microwave-based antigenic unmasking. A revolutionary new technique for routine immunohistochemistry. *Appl. Immunohistochem.* **1**, 256–266.
9. Leong, A.S.Y., and Milios, J. (1993) An assessment of the efficacy of the microwave antigen-retrieval procedures on a range of tissue antigens. *Appl. Immunohistochem.* **1**, 267–227.
10. Bankfalvi, A., Navabi H., Bier, B., Bocker, W., Jasani, B., and Schmid, K.W. (1994) Wet autoclave pretreatment for antigen retrieval in diagnostic immunohistochemistry. *J. Pathol.* **174**, 223–228.
11. Kawai, K., Serizawa, A., Hamana, T., and Tsutsumi, Y. (1994) Heat-induced antigen retrieval of proliferating cell nuclear antigen and p53 protein in formalin-fixed, paraffin-embedded sections. *Pathol. Int.* **44**, 759–764.
12. Norton, A.J., Jordan, S., and Yeomans, P. (1994) Brief, high-temperature heat denaturation (pressure cooking): a simple and effective method of antigen retrieval for routinely processed tissues. *J. Pathol.* **173**, 371–379.
13. Shi, S.-R., Imam, S.A., Young, L., Cote, R.J., and Taylor, C.R. (1995) Antigen retrieval immunohistochemistry under the influence of pH using monoclonal antibodies. *J. Histochem. Cytochem.* **43**, 193–201.
14. Taylor, C.R., Shi, S.-R., Chen, C., Young, L., Yang, C., and Cote, R.J. (1996) Comparative study of antigen retrieval heating methods: microwave, microwave and pressure cooker, autoclave, and steamer. *Biotech. Histochem.* **71**, 263–270.
15. Zu, Y., Steinberg, S.M., Campo, E., et al (2005) Validation of tissue microarray immunohistochemistry staining and interpretation in diffuse large B-cell lymphoma. *Leuk. Lymphoma* **46**, 693–701.
16. Taylor, C.R., and Cote, R.J. (2005) Immunomicroscopy: a diagnostic tool for the surgical Pathologists. 3rd ed. Philadelphia, Elsevier Saunders, 1–45.
17. Namimatsu, S., Ghazizadeh, M., and Sugisaki, Y. (2005) Reversing the effect of formalin fixation with citratonic anhydride and heat: a universal retrieval method. *J. Histochem. Cytochem.* **53**, 3–11.
18. Shi, S.-R., Liu, C., and Taylor, C.R. (2007) Standardization of immunohistochemistry for formalin-fixed, paraffin-embedded tissue sections based on the antigen-retrieval technique: from experiments to hypothesis. *J. Histochem. Cytochem.* **55**, 105–109.

Part III

Imaging Techniques and High-Throughput Data Analysis

Chapter 7

Imaging Techniques in Signal Transduction IHC

Jerry Sedgewick

Abstract

Augmentation of digital images is almost always a necessity in order to obtain a reproduction that matches the appearance of the original. However, that augmentation can mislead if it is done incorrectly and not within reasonable limits. When procedures are in place for ensuring that originals are archived, and image manipulation steps are reported, scientists not only follow good laboratory practices, but also avoid ethical issues associated with postprocessing and protect their labs from any future allegations of scientific misconduct. Also, when procedures are in place for correct acquisition of images, the extent of postprocessing is minimized or eliminated. These procedures include color balancing (for brighfield images), keeping tonal values within the dynamic range of the detector, frame averaging to eliminate noise (typically in fluorescence imaging), use of the highest bit depth when a choice is available, flatfield correction, and archiving of the image in a nonlossy format (not JPEG).

When postprocessing is necessary, the commonly used applications for correction include Photoshop, and ImageJ, but a free program (GIMP) can also be used. Corrections to images include scaling the bit depth to higher and lower ranges, removing color casts from brightfield images, setting brightness and contrast, reducing color noise, reducing "grainy" noise, conversion of pure colors to grayscale, conversion of grayscale to colors typically used in fluorescence imaging, correction of uneven illumination and flatfield correction, blending color images (fluorescence), and extending the depth of focus. These corrections are explained in step-by-step procedures in the chapter that follows.

Key words: Gamma, Black-and-white limits, Flatfield, Oversaturation, Bit depth, Color fringing, Frame averaging, Lossy compression, Photoshop, GIMP, ImageJ

1. Introduction

With the exception of images in which tonal or color gradations provide experimental evidence (such as electrophoretic samples in which the increasing brightness or darkness of lanes indicates the increased presence of a substance), the augmentation

Alexander E. Kalyuzhny (ed.), *Signal Transduction Immunohistochemistry: Methods and Protocols*, Methods in Molecular Biology, vol. 717, DOI 10.1007/978-1-61779-024-9_7,

of digital images is almost always a necessity. Much of the need for augmentation arises for the following reasons:

- Problems when acquiring the images with the camera, imaging system, lenses, illumination patterns, and specimen itself.
- Low light to the sample and consequent noise.
- Inability to obtain sharp focus at all depths.
- Necessity to conform the dynamic range of tones and colors to fit the output (e.g., hardcopy such as posters and deskjet printouts; devices such as printing presses, laptop projectors, and computer monitors; and software applications such as Microsoft Word and Acrobat) in order to obtain a reproduction that matches the appearance of the original.

Some of the conditions that create the necessity for augmentation are a result of the inexperienced or uninformed operator, but other conditions are simply unavoidable. In the former instance, the use of image processing would be to correct or cover up a mistake, and this method for solving the problem could be perceived as one that borders on scientific misconduct. It would be better if the image was acquired again using imaging systems and software correctly.

In the latter instance – where unavoidable imaging conditions or outputs result in an image that does not represent what was once seen by eye – the use of imaging software to correct and conform images to outputs is essential. Otherwise, the reproduced image does not appear identically (or near-identically) to the original. To maintain image integrity, image augmentation must be done.

1.1. Image Integrity

When images are manipulated, procedures should be implemented to make sure that the following consistently takes place:

- "Raw" images (the acquired images) are saved in a high-resolution format (not as Joint Photographers Expert Group (JPEG), if possible) and archived.
- Images are not manipulated so that features within the image are removed.
- No feature from other images is included, nor is any feature introduced to falsify findings.
- All changes to images are not only documented, but mentioned in writing when including images for publication.

When these procedures are in place, scientists not only follow good laboratory practices, but avoid ethical issues associated with postprocessing, and protect their labs from any future allegations of scientific misconduct. Allegations of scientific misconduct to the Office of Research Integrity as a result of image manipulation or "doctoring" have risen substantially since the year 2000.

When it comes to rules regarding what kinds of manipulation of images are allowed or not allowed, each publication's guidelines need to be reviewed before submission. Generally, some image manipulation is allowed within reasonable limits. However, as mentioned earlier, electrophoretic images and those images in which the brightness or darkness of features reveals an experimental finding are not altered (although, for the latter, the reduction of noise levels is often allowed and it is mentioned in the methods portion of the manuscript). In this article, methods will be shown both to keep a record of changes to images, and to prevent alteration of the original or raw image.

1.2. Applications for the Manipulation of Images

Several image manipulation programs can be used, but the application with the greatest number of functions for this purpose is within Photoshop (1, 2). Photoshop can easily be used to keep a record of changes by making a separate layer for each change, or, in more recent versions of Photoshop (CS2 and greater), changes can be logged to a text file while working on an image.

That, however, does not limit scientists, for other applications contain most of the functionality required. Two mentioned in this article are the free programs ImageJ and GIMP (GNU Image Manipulation Program). Downloads for both of these can be found by using a Google search, or, at this writing, by going to http://www.gimp.org for GIMP and http://rsbweb.nih.gov/ij/download.html for ImageJ.

While the reduction of noise can be accomplished with GIMP, Photoshop, and ImageJ, the exploration of programs specifically dedicated to the reduction of noise is encouraged. Among these are Neat Image, Noise Ninja, Topaz De-Noise, and Noiseware. These programs use processing techniques to remove noise but retain detail, and these are often more effective than GIMP, Photoshop, and ImageJ. Ultimately, these programs will have to be tested on each lab's images.

1.3. What Corrections will be Shown

Because the scope of possible corrections to images is large, essential corrections and manipulations of images will be discussed in this article. But before corrections to an image can be discussed, the way in which the image was acquired must be examined first. If images are not acquired correctly, details are likely to be lost, colors incorrectly interpreted, and signal compromised. When imaging correctly and intelligently at acquisition, potential problems can be avoided. More importantly, resolution can be maintained or improved.

Here are the means presented to improve image quality and integrity at acquisition:

- Oversaturation: keep images within dynamic range of instrument to avoid saturation.
- Bit depth: Use the highest bit depth available.

- Noise: Use techniques to reduce noise and improve discrimination of detail.
- Illumination: Avoid uneven illumination, when possible.
- Color balance: Use means to obtain accurate color interpretation when imaging with a color camera.
- Good Laboratory Practices (GLPs): Write down parameters used when imaging.
- Saving: Save images without loss of data: avoid saving in the JPEG format.

Once these methods for acquiring images are followed, then subsequent corrections in an application may not have to be done, or will not have to be done to as great a degree. In any case, the kind of correction that is applied depends on whether the image was of a sample stained with a chromophore or a fluorophore. In the former instance, the sample would result in the kind of image that has a light background with a darker sample, or a brightfield image. In the latter, the image would contain brighter features against a dark background, or a subset of darkfield images here referred to as fluorescent images.

2. Materials

Photoshop CS3 (Adobe, Inc., San Jose, CA), ImageJ, version 1.41 (Wayne Rasband, NIH, USA), and GIMP 2.6 (GIMP development team, USA) were used for image processing within the Vista (Microsoft, Inc., Seattle, WA) operating system. Images used in figures were acquired at the Biomedical Image Processing Lab (BIPL) at the University of Minnesota. Images were acquired on any one of the following: a Zeiss Axiovert 2 microscope, equipped with a SPOT RT camera (Diagnostic Instruments, Sterling Heights, MI) on SPOT software (version 4.6); an Olympus IX70 microscope with a DVC 1412M camera on DVC View, version 2.2 (DVC Company, Austin, TX). All images were saved as TIFF (Tagged Information File Format) without compression at a 16-bit depth; a custom built (by the author) Second Harmonic Generation (SHG) multiphoton confocal microscope on an Olympus BX50 microscope using external Hamamatsu photomultiplier tubes as detectors, saved at 12-bits/channel as TIFF files.

Images of dialog boxes and image windows were captured from the screen (screen capture) using the Print Screen key and pasting the clipboard at screen resolution (1,024 × 768 pixels) into a new image at identical resolution. All images were set to 300 dot/in. output resolution without resampling (adding or subtracting pixels). White and black maximum values were set to tonal values of 240 and 20 at 8-bits/channel, grayscale, with Dot Gain of 20% embedded in the image file for press reproduction. Additional image-processing steps are indicated in captions, when these were applied.

3. Methods

The corrections and manipulations mentioned in this article include the following, with an indication of whether the correction (see Note 1) is slated for brightfield or darkfield:

Brightfield/darkfield: Change of mode: bit depth, indexed color

Brightfield: Color and brightness correct

Darkfield: Brightness correct

Brightfield: Reduce color noise

Darkfield: Reduce noise

Darkfield: Change color

Brightfield/darkfield: Correct uneven illumination

Darkfield: Blend channels/images

Brightfield/darkfield: Extended depth of focus

3.1. Acquisition

3.1.1. Saturation

When acquiring images with any kind of device, efforts must be made to keep tones within the dynamic range. The dynamic range of the imaging device encompasses the extent of tones from black to white that can be collected. When tones are outside the dynamic range, pure blacks and whites result from a saturation of the device's detector. These pure black and pure white tones are often referred to as "oversaturated" (or clipped) tones.

Features of the image that contain pure blacks and whites are devoid of details that cannot be retrieved in postprocessing. Furthermore, saturated features can no longer be evaluated by comparing tones to other features in the image, since these tones could be at a tonal value above or below the dynamic range of the instrument. Therefore, it is critical to keep all tones within the dynamic range of the imaging device. In postprocessing, an image can be made brighter or darker, but details cannot be retrieved from where these do not exist (Fig. 1).

Most acquisition devices have a means in the acquisition software for displaying features within an image that are saturated. Often, the image is colorized in some way to show these areas, frequently by using what is called a Look-Up Table (LUT) overlay. In this scenario, gray tonal values are matched to a set number of colors. For example, in Olympus confocal software, a hi/lo LUT can be activated so saturated whites are colorized red, and saturated blacks are colorized blue.

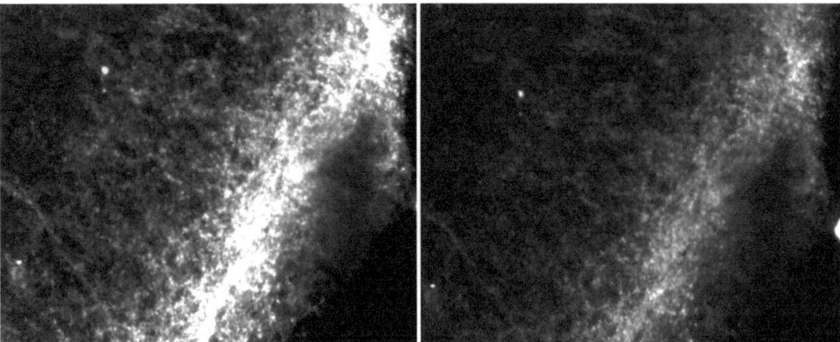

Fig. 1. Image on *left* shows oversaturated areas where bright details are at the maximum tonal value and, thus, are uniform in tone; image on *right* shows same image with bright values within dynamic range of the detector. Images were taken of fluorescently labeled brain sections with a 4× lens, and then cropped to show only a portion of the specimen.

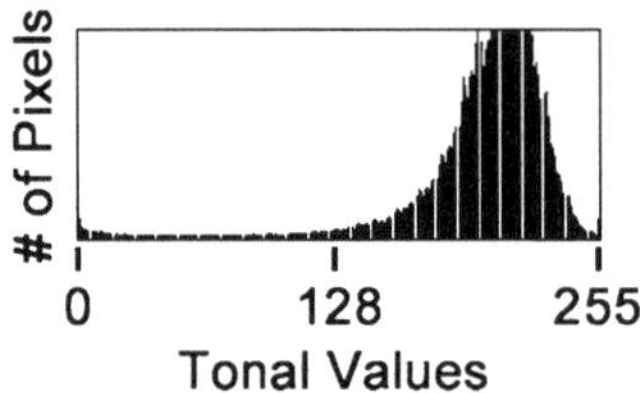

Fig. 2. Histogram display cropped from the Histogram palette in Photoshop CS3. *White, vertical lines* indicate no tonal values at discrete tones as a result of tonal adjustment and consequent rounding errors.

When an overlay is not available, a histogram can be evaluated after the image is taken. The histogram provides the range of tonal values along the *x* axis from black at the *x*,*y* (0) axis to white. The *y* axis graphically indicates the number of pixels occupying the image at each tonal value. A histogram that shows *y* axis pixels at 0 or at the opposite end of the *x* axis would indicate that areas of the image are oversaturated (Fig. 2).

The way the image appears on the monitor can never be used as a means to evaluate the brightness or contrast of an image. Monitors, even when made by the same manufacturer, are subject to differences in how images are displayed. The light in the room (ambient light), the position of the head in relation to the monitor (especially with LCD flat screens), and monitor settings all contribute to variability in the appearance of the image. Overlays and histograms, on the other hand, reveal areas of oversaturation that may not be apparent by eye.

When an image contains oversaturated values, the time in which the detector is exposed to light (exposure) can be adjusted until saturation no longer occurs. Other means for keeping tones within the dynamic range of the imaging device might include:

- Attenuating the power of the light source
- Varying an aperture diameter
- Setting a black level, contrast, or pedestal
- Changing the ISO (for a camera) or gain (e.g., for confocal systems)

Another means (not mentioned above) for adjusting the image to enable lightening or darkening of areas within the dynamic range of an imaging device is called gamma. This is a reference to a calculation that is made in which an exponent used to change numeric tonal values within an image (for purposes of gamma calculations, the tones are given values between 0 and 1). When the gamma value is 1, then tones remain unchanged and grayscale values are linearly related to each other (except when the detector is at the limits of its sensitivity and noise overcomes signal). When this value is greater or less than 1, then tones are changed logarithmically and tones are no longer linearly related. Because evaluation of images in science so often relies on darkness or lightness of related features, the gamma is best kept at 1. However, in instances in which the linear relationship of tones is irrelevant, the gamma can be changed.

3.1.2. Bit Depth (Tonal Depth)

Imaging systems may include a means for choosing the bit depth of the image. The term "bit depth" is used to describe the number of bits (binary digits) used to record the illumination level of a pixel. An 8-bit image is made up of pixels that each contain the possibility of 2 to the power of 8, or 256 illumination levels from black to pure white. A 12-bit image is made up of pixels that each contain the possibility of 2 to the power of 12, or 4,096 illumination levels, and so on. The higher the bit depth, the greater the number of tonal divisions.

The advantage to using a higher bit depth lies in the numbers: when calculations are used when making adjustments to tonal values, rounding errors do not have as great an overall effect on the image. Pixel values do not include decimal places, and so a tonal adjustment that results in an increase of, say, 127.5 would then have to be increased to 128. This can lead to the elimination of discrete tones in a low-bit depth image.

3.1.3. Noise

Notably, in situations in which illumination levels are low – what is more likely when imaging fluorescence – noise creates nonuniform pixel-to-pixel variation in areas where tones should be uniform. The appearance of noise makes an image look "grainy"; what is sometimes called "salt-and-pepper" noise.

When the specimen that is being imaged does not move, noise can be reduced. For many imaging systems, the noise takes

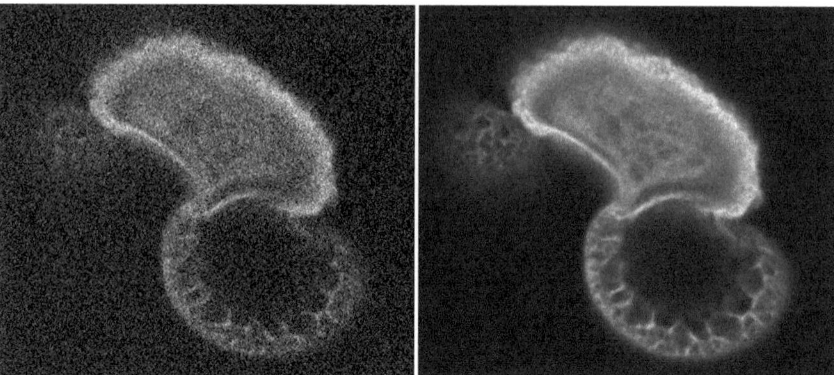

Fig. 3. Image on *left* was taken as a single image; on *right* as an average of eight images. Note the level of noise on the *left* compared to the *right*. Image is of autofluorescing pollen grain, acquired via a 20× lens on a custom built Second Harmonic Generation (SHG) confocal microscope, cropped to show detail.

on a random pattern for each image, so several images can be taken and then averaged with each other to reduce the noise. This option is often available in acquisition software for scientific imaging systems, referred to as Frame Averaging. A means for reducing noise may also be available as a Kalman averaging filter. In either case, the extent of the reduction of noise depends on how many images are averaged, with that number increasing exponentially for a similar factor of improvement. Thus, the averaging of two images achieves the same factor of improvement as four images, and so on (Fig. 3).

3.1.4. Focus

Some imaging systems include a means for optically sectioning specimens labeled with fluorescent dyes, such as confocal instruments. These can incrementally section specimens so that several planes of focus can be subsequently combined so that all are in focus. Most imaging instruments do not provide this capability, but new options in software now provide methods for accomplishing focus at several depths. In Photoshop and ImageJ, the option is called Extended Depth of Field. For ImageJ, it is a free plug-in (http://bigwww.epfl.ch/demo/edf).

When acquiring images, it is useful to take several pictures at incremental levels of focus. In postprocessing, these can be put into an image stack for a resulting in-focus image at all image depths.

3.2. Correcting Uneven Illumination

Providing even illumination across the width and length (the image field) of the specimen improves image appearance and the ability to obtain true color from center of the image to the edges. When images are measured for tonal levels, and when different areas of more than one image are compared, even illumination is a requirement. Otherwise, measurements will not be accurate because one area of the image will be more illuminated than another (3).

To create even illumination on a microscope with brightfield images, Kohler illumination must be set. For information on how this is done, refer to the documentation provided by Florida State University: http://micro.magnet.fsu.edu/primer/anatomy/kohlerjava.html. For fluorescently labeled specimens, check documentation for aligning light sources, or contact the appropriate local sales representatives.

Even under the best of conditions when aligning lamps and lenses to obtain even illumination, uneven illumination is most likely an inevitability. The illumination problem can either be corrected in the imaging system's acquisition software, or in post-processing. Check the acquisition software for flatfield correction (or shading correction or blank field correction); each reference of an image that was taken with the illumination source on and the specimen removed so that only the light pattern is taken. Often, the background is subtracted as well to remove dead pixels (individual pixels that are unnaturally bright as a result of detector defects). "Background" is an image that was taken with the illumination turned off. Double check this information.

If the software has a means to correct uneven illumination, then use the correction when acquiring images. Each lens and zoom (if available) will require that you go through the process of creating a flatfield and background image because the illumination pattern will change with magnification. The flatfield/background correction only needs to be done once at the beginning of a session for each magnification. For any new sessions, flatfield and background images will have to be imaged again.

When flatfield correction is not available in the acquisition software, both the flatfield image and the background image can be acquired and saved for correction in Photoshop/GIMP or ImageJ (see Note 2). For correction, the flatfield images are divided into the specimen images.

3.2.1. Color Balance

When acquiring brightfield images with a color camera, procedures are followed to ensure accurate color interpretation of the specimen. Cameras have an inherent tendency to overemphasize one color, and this creates an overall tinting of the image, what is called a color cast. Color balancing remedies the color cast.

Generally, the camera is color balanced to the spectrum of the light source through a procedure found in the documentation that came with the camera. Some cameras autobalance, but often the results are varied from one session to the next, and so it is best to manually balance for consistent results. This only needs to be done at the beginning of each session.

Cameras may provide the option of setting the color balance to a preset for the light source used. Transillumination on a microscope is generally provided by a tungsten–halogen source.

The light can be attenuated, but a dimmed light will emit with a different spectrum than a light that is full on, and so, for consistency, the light is best kept at full power. If it needs to be attenuated, neutral density filters can be placed in the light path. When the light is full on, then the camera can be set at the tungsten light source. Manual color balancing may still be necessary, but the amount of color shift from session to session will most likely be marginal.

3.2.2. Good Laboratory Practices

The scientific community is increasingly devoting attention to falsified *visual* data (images). As a result, it is becoming even more important to write down procedures and imaging system settings in the event that visual data is challenged. To date, attention is devoted to image manipulation after acquisition, but it is only a matter of time until challenges will also be levied for the means in which images are acquired. Here, too, adjustments in settings can be made to mislead and misinform the scientific community, and so it is crucial to detail steps and procedures when imaging so that documentation can be retrieved to refute challenges.

3.2.3. Saving Images

The way in which images are saved can be the difference between keeping all the visual data and throwing varying degrees of the data away. When the image can be saved in a common format, then TIFF (Tagged Information File Format) is best used, for it is the most widely used format, which retains all the data. If imaging system manufacturers create their own file format, then save in that format so that all the data are retained. Even when that format cannot be read by other software programs, the image is retained because it is the original or raw image. The raw image can be retrieved later if there are any challenges in regard to the image.

A guaranteed way to eliminate data is to save in the JPEG format. This format creates a smaller file size, in the worst case, by grouping pixels into blocks. Often, the strength of image compression can be chosen so that the loss of visual information is indiscernible by eye, but, in any case, the file size is made smaller by eliminating visual information (Fig. 4). For that reason, a JPEG compression is called a lossy compression.

Another means for saving is less obvious. Sometimes, image acquisition software saves an image by simply copying the image off the monitor, and then it is put into a commonly used format, such as the BMP (Bitmap) format. This is done automatically by the software. Images often intrinsically contain far more pixels than what are used to display the image on the screen, and so a so-called "screen shot" will almost always have less pixel resolution. Be sure to completely understand the formats used with your imaging system so that images are saved with no loss of visual information.

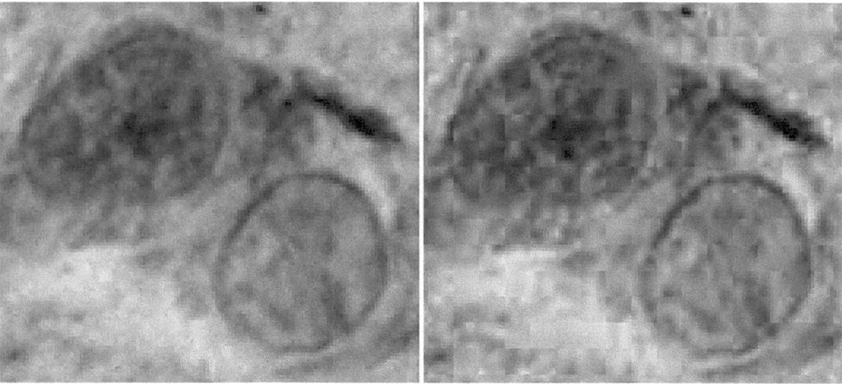

Fig. 4. Image on *left* was saved with low Joint Photographers Expert Group (JPEG) compression; on *right* with high compression. Note reorganization of image information into visible blocks.

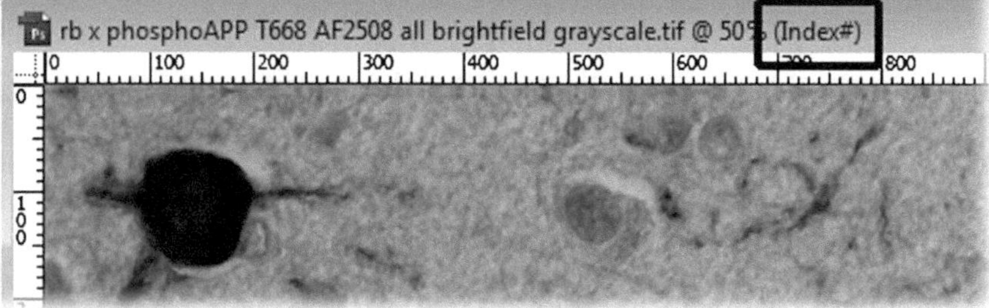

Fig. 5. *Top* of image window in Photoshop showing text to indicate an Indexed Color image.

3.3. Typical Procedure for Processing Brightfield and Darkfield Images

3.3.1. Changing Mode

The mode (or type) of images produced in science include those images made up of gray tones (grayscale), color images comprising the three primaries for light (red, green, and blue), and images that have been assigned colors based on other image information, such as the gradient of gray tones, called Indexed Color images. It may be necessary to change the mode/type of image in order to use all the functions available in the image-processing software, or because the image is being prepared for purposes of obtaining image measurements. Indexed color images typically create the greatest loss of functionality. These can be identified in the image window in Photoshop and GIMP (Fig. 5), or when checking the image type in ImageJ on discovering that functionality is limited (see Note 3). In ImageJ, an Indexed Color image is referred to as 8-bit Color, found in the menu under Image > Type, or in the image window simply as 8-bit.

To make a change in the mode, go to the following menu options and then change the image mode to the preferred mode. Indexed color images are typically changed to RGB Color, unless grayscale is desired:

Photoshop and GIMP: Image > Mode

ImageJ: Image > Type

3.3.2. Color to Grayscale

Note that a mode change from a color image to grayscale may not result in an image that adequately shows contrast in features containing dyes, and when the image is of a fluorescently labeled sample, brightness may be sacrificed. That is because preset percentages of the primary colors are generically used when making the conversion to grayscale. The following functions can be used in order to either visually determine percentages of the colors used when interpolating grayscale from color, or to select the predominate color to match the predominate fluorescent color.

In Photoshop and GIMP, the Channel Mixer function provides a means to interpolate grayscale from color. In Photoshop, select Image > Adjust > Channel Mixer. In GIMP, select Colors > Components > Channel Mixer. In the Channel Mixer box, check the Monotone checkbox and adjust the Red, Green, and Blue channels (components) visually until the image contains a matched contrast appearance as the original.

In brightfield, knowing *complementary* (opposite) colors helps: red darkens blue and green (e.g., trichrome), green darkens orange and red (e.g., hematoxylin and eosin), and blue darkens yellow and brown. Note that blue can often be set at a negative value to increase contrast, and that the Channel Mixer settings can be saved and applied to related images.

For fluorescence, choose 100% of the *similar* color. Thus, 100% red would be the correct setting for a fluorescent label that emits in the orange to red wavelengths, such as rhodamine.

In ImageJ, a channel mixer feature is not available in the standard package. However, for fluorescent images in the standard red, green, and blue colors, the channels can be split from the image, and the appropriate channel chosen:

In ImageJ, go to Image > Color > Split Channels. Choose the channel with brightest fluorescent labeling.

However, images from scientific cameras may be saved as color in a single channel (red, green, or blue). For these images, the color image must be made into a composite (three-channel) image. An additional step is necessary:

1. In ImageJ, go to Image > Color > Make Composite.
2. Then, Image > Color > Split Channels. Choose the channel with the brightest fluorescent labeling.

3.3.3. Bit Depth

Images acquired in 8-bit pose no problem when opening and visualizing images in ImageJ, Photoshop, and GIMP. However, 12- and 16-bit files can be problematic. Images saved in 12-bit do not open as 12-bit images in Photoshop, because the program accepts 8-, 16-, and, in more recent versions, 32-bit images only. So, images destined for Photoshop must be saved as 16-bit, even if acquired on a 12-bit camera. Camera manufacturers either scale

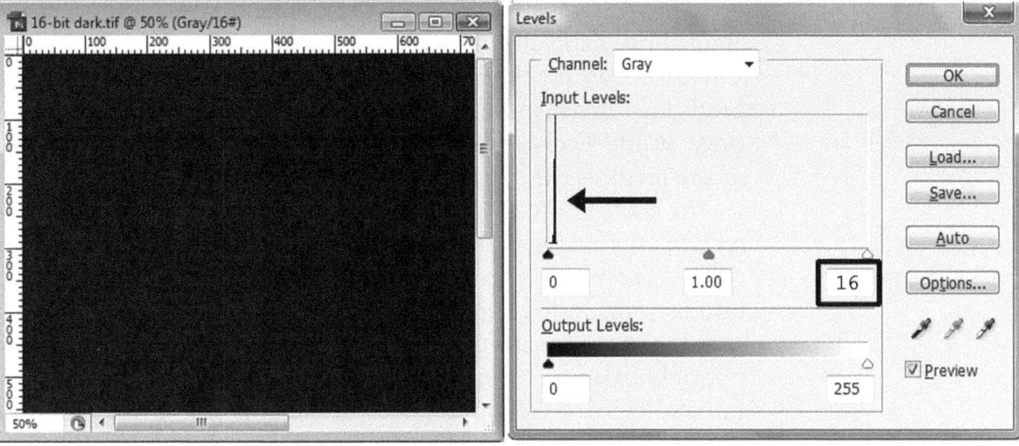

Fig. 6. Image on *left* is saved at 16-bits with 12-bits of image information (four empty bits). To scale 12-bit image to 16-bits, the number 16 is typed into the input white slider box in the Levels dialog box (*right*). Note how histogram is only evident in the expected 1/16th of the histogram (*arrow*) at *right*.

their 12-bit image to 16-bit through a simple multiplication of each pixels' tonal value, or the 12-bit file is saved with the original tonal values, resulting in 4-bits with tonal values at 0. The latter is often more true than the former because manufacturers wish to keep original values untouched.

The result of 4-bits at 0 (four "empty" bits) is an image that appears dark in Photoshop, and one that shows tonal values only at the extreme left end of the histogram display. To scale these values to 16-bit and restore brightness, pixels are multiplied incrementally (from brightest to darkest) by a value of 16. This is simply done in Levels within Photoshop (see Note 4):

1. Under Image > Adjustments, choose Levels. In the Levels dialog box, type 16 into the white input box (Fig. 6).

   12- or 16-Bit images do not open in GIMP as of this writing. Both bit depths open in ImageJ, and so no additional adjustments need to be made.

   Note that when images are scaled to 16-bit (0–65,535 including 0), the readout in Photoshop's Info box shows 15-bit values (0–32,768). The intrinsic values, however, are the full 16-bit range.

3.4. Brightfield

3.4.1. Correcting Color and Brightness

Color correction in postprocessing relies on white, gray, and black parts of the image as reference areas (4). Because white, gray, and black contain equal parts of the three primary colors that comprise an image derived from light, these tones can be queried by software and then the red, green, and blue values can be matched. The percent difference applied to these tones can then be applied equally to every pixel in the image.

The software can "choose" these areas automatically, but the algorithms generally do not work well with scientific images. Thus, it is best for the user to choose these areas of the image. For brightfield images, the most convenient tone to choose is white since nearly every image includes some white areas where no image features exist (background areas).

In both Photoshop (5, 6) and GIMP (automatic color correction in ImageJ is not available in the standard version), the color correction also corrects for brightness when using the function recommended here: the Levels adjustment. When the white area is chosen within the image (by clicking on it), the software not only balances color by matching red, green, and blue colors, but also increases all three values to the uppermost tonal value of the image's dynamic range (determined by the bit depth). The same percentage of increase is applied to all other values from the whitest value incrementally to the lowest tonal value.

Given this scenario, it is important to choose a white area of the image that is brightest; otherwise, brighter tonal values in other parts of the image will max out at the limits of the dynamic range (saturate). Brightfield images from microscopes tend to be unevenly illuminated, even when Kohler illumination is applied on the microscope, and often a brighter center spot results. That brightest area is not always evident by eye. The image can be thresholded to find the brightest part of the image before using the Levels function (Fig. 7), or an iterative process can be used in which white areas are clicked on by using the mouse and the overall effect is evaluated by eye until a satisfactory color adjustment is made. In either approach, the overall color cast presented by the camera is removed. After color correction, differently colored areas within what should be the background white can result, especially with some cameras, and when widely varied uneven illumination patterns exist.

3.4.2. Color Sampler Tool and the Info Box

When using Photoshop and GIMP, markers can be placed on the image to provide locations in which tonal values can be read out. The markers do not print: these appear only on the computer screen.

In Photoshop, the markers are placed using the Color Sampler Tool located beneath the eyedropper tool. Click and hold on the eyedropper tool to reveal the Color Sampler Tool. Then click on significant parts of the image, depending on what information is necessary. A single pixel can be sampled at the precise marker point, or surrounding pixels can be averaged, depending on how the Sample Size is set. For averaging, select the pixel neighborhood to 5 × 5 or greater from the submenu.

If the Info palette is not opened, it can be activated by selecting Info under Window. The info palette will show the pixel tone readouts at each marker selection. The readouts typically appear as RGB

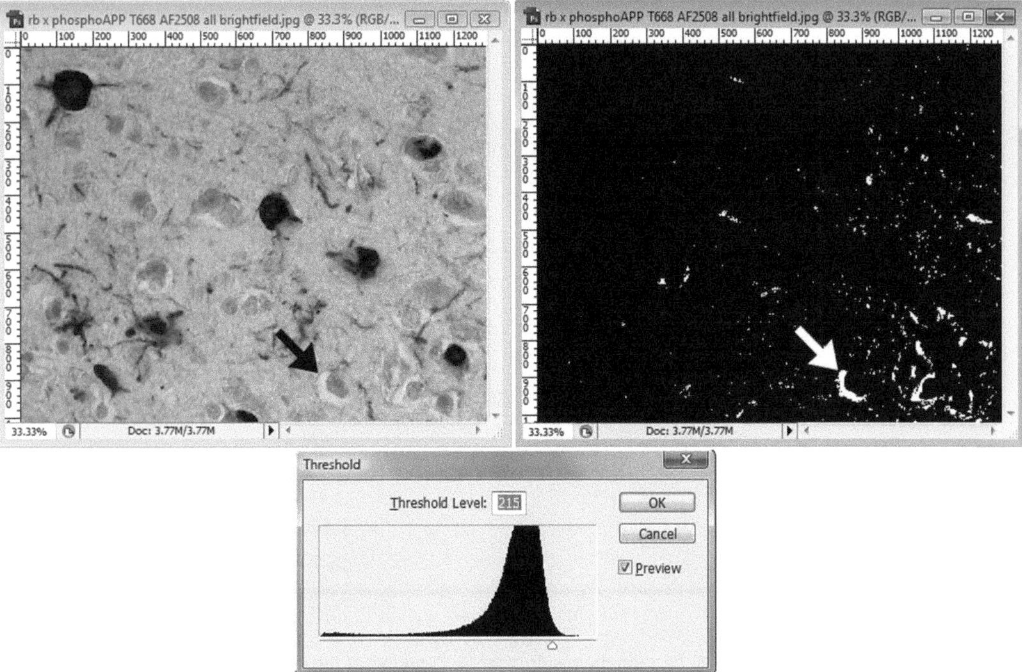

Fig. 7. It can be difficult to visually determine the brightest area on the image at the *left*. Image at *right* has been thresholded: all values are black under a cutoff value set by the slider under the histogram in the Threshold dialog box. *Arrow* points to the brightest area. Image was taken with a 20× lens, cropped and converted to grayscale.

units for color images, or as K: units for grayscale images (K is the percentage of ink that would be deposited on paper were the image to be printed on a press). By clicking on a small eyedropper tool within the Info palette, the readout units can be changed (Fig. 8).

Markers are placed on images before tonal adjustments are done. Except for background areas, markers are placed on significant parts of the image. Detritus and nonspecific features are not considered significant and these are allowed to saturate.

To find the brightest background area without using a thresholding step, simply move the Color Sampler Tool over the background while noting values in the readout of the Info palette. Place a marker where the readout shows the highest values.

In GIMP, the color sampler readout palette is called Sample Points. GIMP has a second readout that shows pixel values at cursor positions called Pointer Information. These windows can be activated by selecting these under Windows > Dockable Dialogs. Sample points can be put in position by holding down the Control key (PC) or the Apple key (Mac), then clicking in the Ruler area surrounding the image, and dragging sample point to the desired position: no tool is available for this function in the GIMP toolbox. If rulers do not show around the image, activate these under View > Show Rulers.

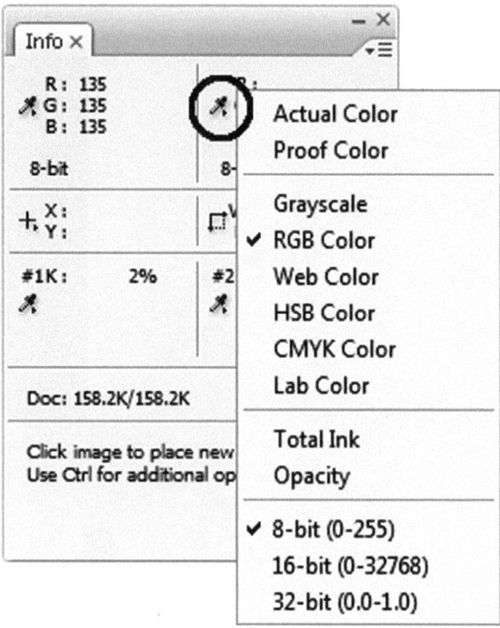

Fig. 8. Clicking the eyedropper tool (*circled*) in the Info palette reveals a drop-down list. From that list, units for readouts can be chosen. For convenience, so that only one tonal range need be recalled, 8-bit color is often chosen.

The number of surrounding pixels desired for averaging pixels around and including sample point (Radius) is available when a dialog box (Color Picker) appears automatically under the toolbox. Within that box, the Pick Mode is available. To avoid changing the foreground and background colors, check Pick Only.

In ImageJ, color can be adjusted under Image > Adjust, then select Color Balance. Each channel can be selected and then, typically, the Maximum levels can be adjusted by eye for each channel to create equal red, green, and blue values in the white areas. A sampler point cannot be placed to track red, green, and blue values while changing levels. Each time a channel is adjusted, click the Apply button. If needing to start again, click Reset. For excessive color shifts, it is difficult to achieve good color balance.

3.4.3. Thresholding to Find Brightest Area

If wishing to threshold the image to find the brightest part, do the following in Photoshop:

1. Image > Adjustment > select Threshold.
2. In the Threshold dialog box, move the slider until only the whitest areas appear.
3. Using the Color Sampler Tool, mark the center of this area.

In GIMP, find the Threshold tool under Colors. In ImageJ, the threshold tool is under Image > Adjust. Set the Threshold drop-down list at the bottom of the dialog box to Black and White.

3.4.4. Color Balancing and Brightness

Color and brightness adjustments are done as follows in Photoshop:

1. First, place Color Sampler Tool markers on the brightest and darkest *significant* parts of the image.
2. Under Adjustment > Image, select Levels.
3. In the lower right of the Levels dialog box, click the white eyedropper tool and then click on the marker that is placed on the brightest part of the image, or, if a marker was not placed, click on what looks like the brightest point by eye. On clicking, the color cast will disappear.

   If the brightest point was not clicked, and that point is located somewhere else on the image, then iteratively click at various points until the color cast disappears and no part of the image gets overly bright (saturated).

   The image may correct in parts and retain color in other parts when illumination is uneven. The image will have to either be corrected for uneven illumination, or re-acquired using flatfield or shading correction in the acquisition software (if available).

   The image may be oversaturated in background areas in all parts of the image. In that case, the image is overexposed and will have to be re-acquired at correct exposures.
4. Expand range of tones, if desiring more contrast, by moving the black slider toward the histogram (right).

   If you have placed a marker on the darkest significant part of the image, be careful to avoid moving slider so that these areas read 0. If reproducing the image to the press, the darkest significant areas or features readout at a minimum of 20 on an 8-bit scale (see below for more information).

   If a marker has not been placed, take readings from darkest significant areas and note readouts in the Info palette. Be sure that the darkest values do not saturate at 0.

3.4.5. More About Setting the Darkest Black Value

Taking readings of the darkest areas is critical because flatscreen monitors often show blacks as darker grays, so the tendency is to increase the darkness until values reach the lowest limit of the dynamic range and details are lost. For reproduction, it is critical to keep the darker tones at a value that could easily be viewed by eye as lacking in contrast. For most reproductions, the darkest significant values should not be less than 20 (on an 8-bit, 0–255 scale) or a high-end printing press, and closer to 25–30 for most press reproduction. Limits of 25–30 for lower limit of dark values also translate well for laptop presentations, laserjet printing, and possibly, poster/inkjet printing.

To lighten blacks, do the following:

In Levels, move the bottomost black slider (output levels) to the right until values increase to the desired level.

In GIMP, the Levels dialog box can be found under Colors. In ImageJ, if wishing to adjust levels incrementally, under Image > Adjust, select Color Balance. In the drop-down list at the bottom of the dialog box, select All. Use the Maximum to increase brightness and the Minimum to increase the darker values. The darker values cannot be lightened using output levels in ImageJ, nor can pixel values at specific locations be tracked as level adjustments are made.

3.4.6. Darkfield: Brightness Correct

For fluorescently labeled samples, the colors are "pure." Pure colors do not contain visible color casts and, therefore, do not require color balancing. However, because differently colored fluorescent labels are shown together, and because detectors have varying sensitivities to wavelength ranges, the brightness level should be made uniform. That is done by increasing the overall brightness to a value just shy of saturation for each fluorescent color.

The tendency in setting brightness for the pure colors used in fluorescent labeling is to increase brightness to the point of saturation so that all details are lost in the brightest features. That is especially true for pure red, blue, and indigo, and not as true for green, orange, yellow, and cyan. Human eyesight, along with poor display from a computer screen in the blue to purple range, both contribute to a diminished ability to perceive brightness in the red to purple to blue range. Thus, it is important to track tonal levels at the brightest significant locations using the Color Sampler Tool and the Info palette.

The second concern when setting brightness levels is to retain visible detail in the darker regions. With fluorescent images, the background is black, and a darker black background creates the perception of greater brightness in the bright features (higher contrast). The tendency is to adjust values so that the background is set to the lowest possible value (0). However, when background is set to tonal values less than 20 (on an 8-bit, 0–255 scale), a loss of visible detail ensues when the image is reproduced to an output. That is especially true when projecting the image on a laptop projector at meetings, and when publishing. Although the image may not look contrasty on a laptop screen, the darker regions will always contain visible details if set at a tonal value of 20 or greater.

The decision about the set point for background values then lies on the intended output for the image. Two images may have to be created: one for display on a screen for the principal investigator, and another for destination to outputs. When the image is destined for viewing on a computer screen, a decision must be made about the level for the background. A suggested rule might be to set the blacks at a brighter level to ensure that details are not lost in the darker areas, even if that setting might result in a background that is not pure black.

A final tendency is to set the brightness levels using the Brightness/Contrast function in Photoshop. In versions of Photoshop previous to CS3, the value for the brightness slider is added to every other pixel in the image. The converse is true for the contrast slider where values are subtracted. In that scenario, the tendency would be add tonal values by adjusting the Brightness slider, and then, because the black levels become too bright, to subtract by adjusting the Contrast slider. The consequence would then be the elimination of tonal values in the image: tones that once comprised the image – visual data – are "thrown away."

To set brightness levels in Photoshop while minimizing loss of visual data (all tonal adjustments inevitably remove tones because of rounding errors), follow these steps:

1. Place markers on the image using the Color Sampler Tool (described earlier). Place on brightest features and on black featureless background region.
2. Under Image > Adjustments, select Levels. Using the white triangle slider, move slider toward the histogram (left) while paying attention to the readouts in the Info palette. Adjust until readouts for the brightest points are shy of saturation (255 for an 8-bit image).
3. Move black slider triangle toward histogram (right) until readouts are set at 20 (or, if meant for computer display, just shy of 0).

   If background levels need to be increased, use the black triangle slider at the bottom of the Levels dialog box and move to the right until readouts for the black level are at 20 (or greater).

To adjust brightness levels in GIMP, under Colors, select Levels and perform the same steps. In ImageJ, to adjust brightness similarly, under Image > Adjust, select Color Balance and move the Maximum or Minimum sliders.

3.4.7. Brightfield: Reduce Color Noise

Color noise in brightfield imaging arises from chromatic aberration and scattering at borders of dark features. The result is unnatural colors at borders of features (other sources of noise can also result from the use of digital cameras, such as noise resulting from heat, but here a differentiation is made between color noise and noise associated with CCD and CMOS chips). This source of noise is often unavoidable and cannot be ameliorated by setting

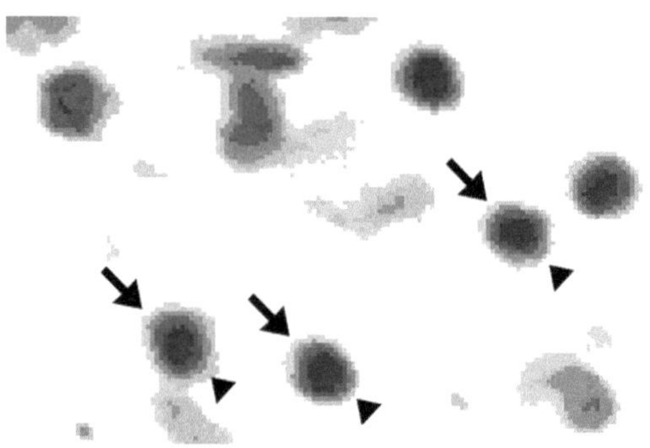

Fig. 9. Lighter values at edges indicate where colors were unnatural in this grayscale image. A color image was posterized by assigning a limited number of tones to emphasize gradients, and then made to grayscale.

Kohler illumination properly on a microscope, or by purchasing a more expensive camera or optic (though better optical coatings and cameras that do not interpolate color will improve the odds). In fact, in terms of scattering, the human visual system also perceives unnatural colors when dark objects are lit from behind. The appearance of unnatural colors at edges is referred to as color fringing (Fig. 9).

Color fringing can be minimized by blurring the color component of an image, but not its lightness component. The lightness component comprises the noncolored details, or simply the perceived differences in tonal values from white to black. While it has already been said that a color image comprises red, green, and blue components, a color image also comprises a color component and a grayscale (lightness) component, and it can be divided in this manner as well. A means for doing that lies in dividing its components to the L*A*B* mode, where the three channels are divided into a Lightness channel, along with A (green–magenta) and B (blue–yellow) color channels. The channel is left untouched, but the color channels are blurred with a Gaussian filter.

While color fringing is rarely perceived in images by microscopists (because fringing is only visible on zooming in), it is often prevalent in those images in which single cells are discrete against a white background, and in instances in which tissue is sparse and there is a greater appearance of edges. Because the reduction of color fringing does not visibly affect either the resolution of the image or the appearance (except that colors may lose some intensity), the removal of color fringing may be considered as a standard postprocessing step. This step is especially important when features are measured for quantitative data because unnatural colors affect the ability for quantitation programs to separate features of interest from surrounding areas.

To reduce color fringing in Photoshop, follow these steps:

1. Under Image > Mode, select LAB Color.
2. Open or reveal the Channels palette (Windows > Channels).
3. In the Channels palette, click on the A channel to select it. Make sure the other channels are unselected (not blue highlighted).
4. Under Filter > Blur, select Gaussian.
5. Set the Gaussian blur to 2–5. The level of blur depends on the extent of color noise: overblurring can cause a color change, underblurring may not remove unnatural colors.
6. Click on the B channel in the Channels palette and repeat Steps 4–5.
7. Click on the Lab (top) layer in the Channels palette to reveal all channels.
8. Under Image > Mode, select RGB Color.

If colors are not satisfactorily removed, revert file to its last saved state (File > Revert) and run steps again, increasing Gaussian blur level.

To reduce color fringing in GIMP, follow these steps:

1. Under Colors > Components, select Decompose.
2. In the next dialog box (Extract Channels), click on the arrowhead from the drop-down list and select LAB. Check Decompose to Layers. A new image window will appear in addition to the original color image.
3. In the Layers palette, three layers appear marked L, A, and B. Click eye icons to turn off L and A and blur B: Under Filters > Blur, select Gaussian Blur.
4. In the Gaussian Blur dialog box, set horizontal and vertical size to 5–10 points. The level of blur depends on the extent of color noise: overblurring can cause a color change, underblurring may not remove unnatural colors.
5. Unclick eye icon for layer B and click on layer A. Repeat Steps 3–4.
6. Click on eye icons so that all icons appear.
7. Under Colors > Components, select Recompose.
8. Select original color image. The color noise correction was applied to this image.

If satisfied with color fringing removal, close LAB color image and save the original color image.

If dissatisfied, close LAB color image and revert original color image (File > Revert) to its last saved state. Run steps again choosing a different Gaussian blur value.

A method for reducing color fringing in ImageJ does not exist in the standard program.

3.4.8. Darkfield: Reduce Noise

Camera systems include several sources for noise. The most visible source of noise is a result of both heat, and amplification of background noise (and signal) from increases in voltage (gain) after microvoltages are read from the detector, and/or when excessive voltage is applied to the detector (when the detector multiplies voltages that result from striking photons, as in photomultiplier tubes). The noise tends to be random, and so an effective way to reduce noise lies in averaging images, as mentioned earlier.

However, when imaging live processes, or when detecting subresolution fluorescently labeled specimens and dim luminescence, frame averaging may not be feasible and the resulting noise is inevitable. For these images, a median filter can be effective if not applied at a strength that is too great. A median filter, unlike a Gaussian, preserves the borders of features while reducing nonuniform variation in pixels in areas where these should be uniform.

Because the median filter is a coarse application of noise reduction, the following steps show how to fine tune the effects of median filtering in Photoshop and GIMP:

1. Duplicate the existing layer in Photoshop: Under Layers, choose Duplicate Layer.
2. Under Filter > Noise, select Median.
3. Adjust slider until noise is reduced and features begin to look artificial: usually a value between 2 and 3.
4. In Layers palette (Windows > Layer), adjust opacity slider until features appear by eye to be slightly blurred.

In GIMP, follow steps above, except that Median is found under Filter > Noise, and it is called Despeckle. Set Radius for same effect, with threshold values either at limits (0 and 255) or choose as desired. In ImageJ, layers cannot be made with the standard program, but the image can be median filtered: under Process > filters, choose Median and set Radius.

3.4.9. Darkfield: Change Color

The reproduction of color on output devices can appear darker and with less contrast than colors appear through the microscope. That is especially true with blue to purple–blue colors. The blue-colored fluorophore DAPI, for example, typically appears dark on the computer screen and darker yet when printed or published.

Output devices (e.g., printing presses), and display devices (e.g., computer screens) display not only a subset of colors that are interpreted by the human visual system (when not color blind), but characteristic shifts in color. The range of colors a device is capable of displaying or outputting is known as its gamut, and in almost every device the colors that do not reproduce correctly are the pure colors. For scientists working with fluorescence, the handful of colors that are used are pure, and so it is expected that most colors will not reproduce faithfully.

Pure color includes 100% of either red, green, or blue. These include violet, blue, cyan, green, yellow, orange, red, and magenta. Of these, the yellow and orange reproduce brightest in publication, with other brightness levels following in this order: green, red, cyan, magenta, blue, and violet. Because of that, cyan is substituted for blue, and orange for red, when applicable (in conventional colocalization experiments red, green, and the combination of red and green – yellow – are used).

Images may have to be made into grayscale images before recolorizing, if in color already. That must be done before converting from one pure color to another. If images are converted directly to grayscale (by changing the mode), there is a loss of brightness, unless the predominant color is green:

Converting from color to grayscale in Photoshop:

1. Under Image > Adjustments, select Channel Mixer.
2. Click the Monochrome check box and then adjust sliders according to the following table. Note that these are suggested settings from author's experience:

Color	Red Channel	Green Channel	Blue Channel
Red–blue	100	0	0
Magenta	100	0	0
Violet	0	0	100
Blue	0	0	100
Cyan	0	0	100
Blue–Green (Aqua)	0	100	0
Green–Blue	0	100	0
Green	0	100	0
Green–Yellow	0	100	0
Yellow	100	0	0
Orange	100	0	0
Red	100	0	0

Follow the same steps in GIMP. Find the Channel Mixer dialog box under Color > Components: select Channel Mixer.

This function works differently in ImageJ, depending on whether the image is 8-bit color or RGB Color. If it is 8-bit color (check under Image > Type), then change the LUT to Gray from the list: Under Image > Lookup Tables, select Gray. If the image is RGB Color, then only red, green, or blue colors will produce gray values that replicate color values. The RGB Color image can be split into its components (Image > Color > Split Channels) and the two undesired images can be closed.

Set colors in Photoshop in the following way:

1. Under Image > Adjustments, select Levels.
2. In the Levels dialog box, using the Output slider (slider at bottom of the Levels box), set white triangle slider (on right) according to the following table:

Color	Red Channel	Green Channel	Blue Channel
Red–blue	255	0	128
Magenta	255	0	255
Violet	128	0	255
Blue	0	0	255
Cyan	0	128	255
Blue–green (Aqua)	0	255	255
Green–blue	0	255	128
Green	0	255	0
Green–yellow	128	255	0
Yellow	255	255	0
Orange	255	128	0
Red	255	0	0

In GIMP, follow the same steps except that Levels is found under Colors (Colors > Levels).

In ImageJ, if the image is grayscale or 8-bit color, colors can be assigned by a Lookup Table in red, green, blue, cyan hot, yellow hot, magenta, and orange hot. Under Image, choose Lookup Table and then choose from list. Note that combinations of colors can be chosen so that the brighter fluorescence is a different color than background fluorescence. These combinations can add vibrancy to images.

3.4.10. Brightfield/Darkfield: Correct Uneven Illumination

Nearly every image taken by a microscope contains some degree of uneven illumination. Much of this is the result of an uneven illumination source, but some can arise from the image itself: fluorescently labeled features, when grouped together, can also create brighter surrounding areas.

Densitometric or intensity measurements, when measured at varying spatial locations, require correction of uneven illumination. That is done when acquiring the image, as mentioned earlier. Images destined for measurements and for stitching to other images (such as when several fields are taken and then connected together to make a large image) also require correction. If that correction does not take place when acquiring the image because the function is not available, then the illumination must be corrected in postprocessing. Representative images always appear at higher quality when uneven illumination is corrected.

If it is necessary to correct for uneven illumination in postprocessing, then it is best to save a flatfield image for this purpose. That can be done, as described earlier, by taking an image with the specimen removed of the illumination source. The image should be exposed so that all areas of the flatfield image are under the dynamic range upper limit of the imaging system (under 255 for 80-bit, 4,095 for 12-bit, and 32,768 for a 16-bit image or lower). A flatfield image needs to be taken for each magnification (and zoom), if more than one objective/zoom is used during the imaging session. A method is described below for using Photoshop or GIMP to correct uneven illumination when a flatfield image is available.

When the sample itself causes uneven illumination, or when the flatfield image is not available, the image itself can serve to eliminate flatfield illumination problems. Use Photoshop, GIMP, or ImageJ to correct for uneven illumination in this instance using the Flatfield correction using image method.

When a flatfield image is available, in Photoshop

1. Open flatfield image and place Color Sampler marker on brightest point (usually near the center). Use thresholding method (mentioned earlier) if unsure.
2. Under Image > Adjustments, select Levels.
3. Move white triangle slider to the left to increase brightness until Info palette readout indicates a value for the uppermost limit of the image (255 for 8-bit, 4,095 for 12-bit, and 32,768 for 16-bit).
4. Save the flatfield image.
5. Select the flatfield image (Select > Select All), and Copy (Edit > Copy).
6. Open the specimen image to which the flatfield will be applied.
7. Paste (Edit > Paste) onto the specimen image. This will create a layered image with the flatfield image on top, and the specimen image below in the background layer.
8. Under Image > Adjustments, select Invert. This inverts the grayscale values of the flatfield image.
9. In the Layers palette (Window > Layers), click on the layer mode drop-down arrow and select Hard Light from the list (this multiplies values above a pixel tone of 128 from the flatfield image).

10. Use the Color Sampler Tool to place markers on the background areas of the image from center points to edges. In an evenly illuminated image, these should read out at identical values.
11. Change the Opacity slider in the Flatfield image layer while keeping an eye on the Info Box. Adjust until all Color Sampler markers are close to the same amount.

    If it is impossible to adjust the opacity slider and get values that are within 2–4 points of each other, then placement of the color sampler markers may be on a feature that cannot be visualized, or the edge values are too different from the center values. Move markers into different positions, and mark edges closer to the center and attempt again.

    Sometimes, the illumination is so uneven that this method can only reduce uneven illumination by a degree that cannot be accomplished with a single flatfield image. This step can be followed by the method for correcting uneven illumination with the sample itself.
12. Move Color Sampler Markers to the brightest and darkest significant areas of the image.
13. Restore any loss of contrast: Under Image > Adjustments, select Levels. Adjust the white triangle slider (left) and the black triangle slider (right) to values mentioned in the "Adjust Brightness" steps earlier.
14. In GIMP, use the same steps, except that the sampler tool is applied in a different way (mentioned earlier), and Levels is found under Color in the menu (Color > Levels).

ImageJ is more straightforward because it contains a means for dividing one image with another:

1. Open the flatfield and specimen image.
2. Under Process, select Image Calculator.
3. In the Image Calculator dialog box, set the Operand and Images 1 and 2.
4. Be sure to check 32-bit result!
5. Set the bit depth to less tones, either 8- or 16-bit: under Image > Type, select desired bit depth.

The image may have to be adjusted in Levels or Color Balance to restore the tonal levels before correcting uneven illumination (unless the image is intended for densitometry).

Using image itself to correct uneven illumination

1. Open specimen image.
2. Under Layer, select Duplicate Layer to make a duplicated layer above the original image. Duplicate again if desiring to keep original image untouched.

3. Under Image > Adjustments, select Invert.
4. Under Filter > Blur, select Gaussian Blur.
5. Set value for Gaussian blur by eye, often between 50 and 100. Set so that all features are blurred to the degree that bright and dark areas remain, but nothing in the image is recognizable.
6. If image is in color, under Image > Adjustments, select Desaturate. For brightfield images in which the color is unnatural in the background, this step can be skip1.ped: the inverted color image can help to correct unnatural colors.
7. Follow Steps 9–13 in the previous method for the remaining steps.

For GIMP, follow the same steps, except that for Step 3, under Colors, choose Invert. For Step 6, under Colors, choose Desaturate.

In ImageJ, the steps are as follows for grayscale and those color images in which a grayscale result is acceptable:

1. Under Image, select Duplicate Image to create a second image.
2. Under Process > Filters, choose Gaussian Blur.
3. Set Gaussian blur to 40–80, or when image blurs enough so that uneven illumination pattern is seen without image details.
4. Under Process, select Image Calculator.
5. In the Image Calculator dialog box, set the Operand and Images 1 and 2. Be sure to check 32-bit Result.
6. Set the bit depth to less tones, either 8- or 16-bit: under Image > Type, select desired bit depth.

To keep the image in color, do the following:

1. Split the channels (Image > Color > Split Channels).
2. Duplicate the channel that contains enough gray levels to Gaussian blur for a flatfield image.
3. Divide the red, green, and blue channels into the Gaussian blurred image.
4. Recombine the images: Under Image > Color, select Merge Channels. In the Merge Channels dialog box, select the relevant images for the red, green, and blue channels. For grayscale, indicate None.

3.4.11. Darkfield: Blend Images

The various wavelength ranges and associated colors of separate images, saved as channels on a confocal, can be merged or blended so that equal amounts of the colors appear. This is easily done in Photoshop, GIMP, and ImageJ using methods that follow.

Photoshop and GIMP allow for brightness adjustments while images are merged so that the effects can be seen as the adjustments are made.

In Photoshop and GIMP:

1. Open the green, yellow, or orange colored image first, since these are perceived as the brightest. Open additional colorized images one by one.
2. For each additional image, select image (Select > Select All), copy image (Edit > Copy).
3. Paste (Edit > Paste) additional image on the first image to create a layer above the green, yellow, or orange colorized image layer.
4. For each additional layer, click the layer mode drop-down arrowhead in the Layers palette and select Lighten or Screen. Lighten allows values greater than 128 to appear when upper layer values are less than 128 (the centermost tone). Screen is equivalent to projecting more than one image onto a screen at equal intensities.
5. It is very likely that the color of one layer will overpower the other layers, and so these will have to be adjusted by the eye so that background values are neutral. Click on the layer that overpowers other layers:

If it is an additional layer, *decrease* brightness by reducing the Opacity slider in the Levels palette.

If it is the bottom layer, *increase* brightness of additional layers (using Brightness method mentioned earlier and causing some values, possibly, to oversaturate): make minimal adjustments!

In ImageJ, colocalization can be done as follows:

1. Open the red, green, and blue images, or any two.
2. Under Image > Color, select Merge Channels.
3. Choose relevant channels in the Merge Channels dialog box.

3.5. Extended Depth of Focus

More that one image can be taken at several different depths to obtain all the focal planes. These several different images can then be recombined and processed so that only the in-focus parts of each image show. In that way, a highly magnified image with several different planes that are out of focus can be made into an in-focus image. The generic term for this image correction is Extended depth of focus.

Several recent software programs, as of this writing, offer the ability to create extended depth of focus for brightfield images, although the technique can also be used for fluorescence-labeled images that are from standard microscopes. The best of these use a means for determining out-of-focus areas by comparing several images and then locally removing blur, such as what is done with neighborhood-based, deconvolved images. Other programs mask out unfocused areas to then only reveal what is in focus. The latter is how Photoshop accomplishes extended focus as of version CS4; it is not implemented as of this writing in GIMP, and the ImageJ plug in varies in its efficacy so it is not discussed.

At least three images need to be taken in order to have enough images for software to work. Start with the top of the image and find the first parts of features that are in focus, and then continue focusing downward, taking a picture each time new parts of features appear in focus. Erring on the side of too many pictures can only aid in the final result. Save the pictures for each series of images to its own folder.

What follows is the rest of the steps in Photoshop:

1. Under File > Scripts, select Extended Focus.
2. Find the folder with the series of images. Check "Blend Images" box. Click OK and allow Photoshop time to interpret images.
3. Under Layer, select Flatten Image.

4. Notes

1. *History of Corrections Made.* No additional steps were included in postprocessing methods to either save corrections that were made or to record how the corrections were made. For the latter, more than one means exists for keeping a record of postprocessing corrections. In more recent versions of Photoshop, all steps taken can be recorded to a log file. This is found in Edit > Preferences > General and it is called History Log. By choosing both Metadata and Text in the dialog box, along with Detailed from the Edit Log Items drop down, every step is recorded in great detail. The downside to the Log, however, is that the user's name is not entered and it is easy to keep the Log running through several sessions if it is not de-activated. In other words, the History Log runs until it is unchecked.

   For saving corrections, since the CS3 version, images can be made into Smart Objects (Layer > Smart Objects > Convert to Smart Object). When that is done, tonal adjustments can only be made on layers by choosing adjustments from the bottom of the Layers palette (called Adjustment Layers). Filters, such as the Gaussian filter, are applied to the background image, and each filter setting is recorded.

   Adjustment Layers have been available since Photoshop 6, but it is easy to forget that an adjustment layer is active when applying another correction. For this reason, it is convenient to make a habit of duplicating layers (Layer > Duplicate) and then applying a correction to duplicate layers, thereby preserving the original layer. Layers can be labeled with the correction made, and later deleted, if necessary. Be sure to save the file with records of corrections in the Photoshop format, preserving the layers and the original.

Having spoken about the importance of preserving the original image, this chapter did not adequately address how critical it is to save the original. In that spirit, it is best to also duplicate your image before making corrections (Image > Duplicate). Even better, save all images to CDs or DVDs on discs that are not rewritable so that these are universally read.

2. *ImageJ and GIMP.* The use of ImageJ and GIMP can be frustrating until some familiarity is gained. The greatest annoyance in both is that the programs only occupy part of the computer screen without blocking out other applications. On top of that, each image window becomes its own object, and in GIMP, each image window also contains the menu items.

   In both programs, it is least confusing to minimize all other applications when these are active. In GIMP, it is useful to expand each new image window to fill the computer screen to avoid using the menu of the image window that lies behind.

3. *Macintosh Versus PC.* For Photoshop, some differences in placement of menu items exist. In this article, references were only to PC placements. Some items found in the Edit menu on a PC are found in the Photoshop menu on a Macintosh.

4. *Changing the Minimum and Maximum Values in Levels.* To keep tones within the printing press range, many commercial photographers set the white and black eyedropper tools in Levels to 240 and 20, respectively. This can also be done for research, if desired. However, researchers also project images at meetings and view images onscreen. Thus, in this book chapter, the white and black limits are not emphasized.

References

1. Blatner, D. and Fraser, B. (2004) Real World Photoshop CS. Peachpit, Berkeley, CA.
2. Kelby, S. (2003) The Photoshop CS Book for Digital Photographers. New Riders, Thousand Oaks, CA.
3. Leong, F. J. W.-M., Brady, M. and O'D McGee, J. (2003) Correction of uneven illumination (vignetting) in digital microscopy images. J. Clin. Pathol. 2003; 56(8): 619–621.
4. Margulis, D. (2002) Professional Photoshop: The Classic Guide to Color Correction, 4th Edition. Wiley, San Francisco, CA.
5. Sedgewick, J. (2008) Scientific Imaging with Photoshop: Methods, Measurement and Output. Peachpit, Berkeley, CA.
6. Sedgewick, J. (2002) Quick Photoshop for Research: A Guide to Digital Imaging for Photoshop 4x, 5x, 6x, & 7x. Kluwer Academic/Plenum, New York.

Chapter 8

Practical Considerations of Image Analysis and Quantification of Signal Transduction IHC Staining

Michael Grunkin, Jakob Raundahl, and Niels T. Foged

Abstract

The dramatic increase in computer processing power in combination with the availability of high-quality digital cameras during the last 10 years has fertilized the grounds for quantitative microscopy based on digital image analysis. With the present introduction of robust scanners for whole slide imaging in both research and routine, the benefits of automation and objectivity in the analysis of tissue sections will be even more obvious. For in situ studies of signal transduction, the combination of tissue microarrays, immunohistochemistry, digital imaging, and quantitative image analysis will be central operations. However, immunohistochemistry is a multistep procedure including a lot of technical pitfalls leading to intra- and interlaboratory variability of its outcome. The resulting variations in staining intensity and disruption of original morphology are an extra challenge for the image analysis software, which therefore preferably should be dedicated to the detection and quantification of histomorphometrical end points.

Key words: Immunohistochemistry, Signal transduction, Image analysis, Tissue microarray, Whole slide imaging, Proximity ligation assay, Cancer

1. Introduction

In contrast to the average concentration determined by immunoassays of antigenic proteins extracted from tissue samples, immunohistochemistry (IHC) allows determination of local amounts. The combination of spatial information and quantity is of particular relevance to assays for signal transduction (ST) factors, which typically are differentially expressed in activated tissue regions and subcellular compartments.

Alexander E. Kalyuzhny (ed.), *Signal Transduction Immunohistochemistry: Methods and Protocols*, Methods in Molecular Biology, vol. 717, DOI 10.1007/978-1-61779-024-9_8,

Still, quantitative studies of ST factors in situ have until recently been severely restricted by the general, practical limitations in classical manual microscopy of tissue sections stained by IHC:

- Labor-intensive microscopy
- Unstandardized and/or monotonous inspection resulting in high intraobserver variation
- Subjective signal detection and adjudication of intensity resulting in high interobserver variation
- Discrete scoring categories
- Limited documentation

Conversion to computer-assisted analysis of digital images of IHC-stained tissue sections presents a promising way to reduce these problems, and has shown important benefits to the local quantification of ST factors by:

- Automation and standardization of data collection
- Objective measurements of quantities and spatial parameters with continuous scales
- Documentation allowing review, control, and recalculation

The present chapter describes general procedures recommended for image-based analysis of IHC-stained ST factors in tissue sections, and presents examples of materials, methods, and protocols used for their quantification in tumor tissues. Though the examples are all from the field of breast cancer, where automated quantitative IHC-analysis has already had an important impact in both research and diagnosis, the principles are generally applicable for studies of ST factors.

2. Materials

Efficiency in quantitative analysis of IHC-stained tissue sections requires automation and coordination of the digital imaging, data management, and image analysis processes. Still, the equipments involved are typically physically separated:

- Digital imaging by microscope-based systems or whole slide imaging scanners
- Data management (images, metadata, and results) by central or local server
- Image analysis by local computers including laptops

As for other research-oriented applications of IHC, the in situ detection and quantification of ST factors will typically involve tissue microarrays (TMAs, see Note 1) (1). In this case, the software controlling image acquisition, data management, and analysis must do this in a core-specific manner.

2.1. Digital Imaging (See Chapter 7)

The digital images of IHC-stained tissue sections can be acquired by microscope-based systems or by whole slide imaging scanners.

Microscope-based imaging traditionally has been a labor-intensive procedure aiming mainly at depicting a subjectively selected, small, representative, and/or characteristic tissue region of interest for publication purposes. However, motorization and software-control of the microscope and its associated equipment can convert it into a highly automated imaging station with a substantial throughput for what concerns number of slides, number of tissue samples, and tissue surface area (see Note 2).

The quality of whole slide scanners has improved remarkably during the last 5 years (2), and within the next decade, it is quite likely that majority of research microscopes used for IHC imaging will be substituted by scanners (see Note 3).

2.2. Data Management

Histomorphometrical research involving digital images and image analysis can easily generate terabytes of data with complex relationships, thereby creating a requirement for local or externally hosted server-capacity. Beyond the obvious need for server-efficiency, it is crucial to use imaging and quantification software, which allow data management to be fully integrated in the workflow and ascertain data integrity. For IHC images, the software must also provide tools for the operator's simple and efficient final review of the data achieved by highly automated quantitative analysis, and it should be easy to export the computed end-points to other programs, e.g., for advanced statistical analysis.

2.3. Image Analysis

In order not to occupy computers integrated in the imaging system, the image analysis and data management preferably should be controlled by separate workstations and if required a dedicated server.

3. Methods

Before turning to the examples of methods for quantitative analysis of digital images of IHC-stained tissue sections (Subheadings 3.4 and 3.5), a few central concepts are described in Subheadings 3.1–3.3 and the associated notes.

3.1. General Concepts of Digital Images

Digital images consist of pixels (short for Picture Elements). Each pixel is associated with one or more intensity values, typically contributing to a grayscale or full color image.

Thus, each pixel is defined by a position and a number of intensity values that are sometimes referred to as "features" of the image (see Note 4). In standard color images, the features are Red (R), Green (G), and Blue (B), which is the spectral information associated with each individual pixel.

3.2. Image Processing and Analysis

In the following discussion, we will distinguish between image processing and image analysis.

Image processing is concerned with the transformation of an image from one form to another, which is in some way more convenient than the original. Image processing is often used for facilitating subsequent image segmentation, and in that context referred to as a preprocessing step. Typically, the goals of preprocessing are to enhance image structures of relevance to the application and to suppress the noise.

Image analysis, on the other hand, is typically concerned with the extraction of quantitative information from images. Extraction of useful quantitative information from images is very often critically dependent on the ability to combine spectral, spatial, morphological, contextual, and relational information.

Here, we describe both traditional and novel approaches that will allow scientists to address a very broad range of typical challenges encountered in quantitative microscopy.

3.3. Image Analysis Protocols

In order to set up image analysis protocols, a number of steps are typically required:

1. *Preprocessing:* This is a set of operations allowing for enhancement of relevant image structures for use in image segmentation, combining spectral, local spatial, and local morphological information (see Note 5).
2. *Definition of taxonomy:* In Visiomorph™, a label tool can be used for naming relevant structural content to be recognized, and a "teach-by-example" feature allows the user to train the system by digitally "painting" examples of the defined structures in the image.
3. *Segmentation:* Selection and definition of a decision rule combining the defined image features and training data, based on which images are segmented into their relevant components (see Note 6).
4. *Postprocessing:* This allows for a further refinement of the segmentation result based on morphological, contextual, and relational information (see Note 7).
5. *Output:* Quantification of end-points is based on the segmented image and may also include the original feature image, the derived feature images, and combinations of other end-points (see Note 8).

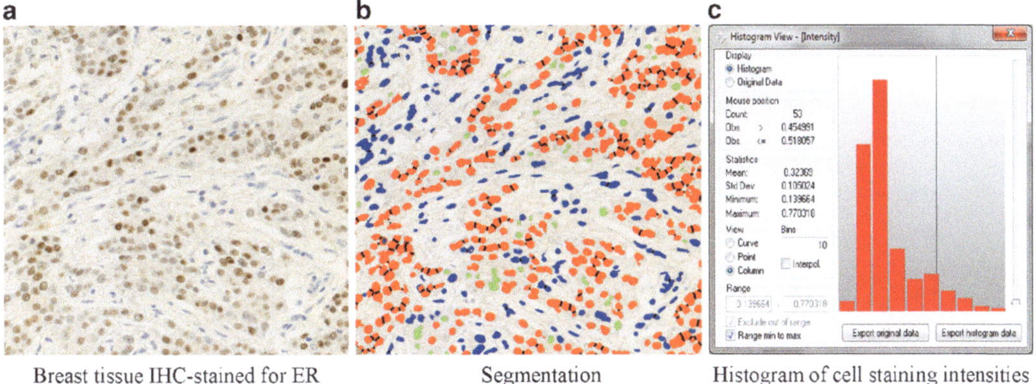

Fig. 1. In this example, a breast tissue section was stained for estrogen receptor (ER) by IHC (**a**). In the actual field of view, 75% of the cells had positively stained nuclei according to the segmented image (**b**). However, their average staining intensity was just 30% according to the derived feature image and the segmentation, and the variation in staining intensity was demonstrated by the histogram showing the frequency of cells with various intensity levels (**c**).

Once an image analysis protocol has been defined for a given application, large volumes of image data can be processed with minimal user-interaction allowing highly standardized output.

Figure 1 shows an example of central steps in a typical image analysis protocol. Though the intensity measure for the nucleus stain is on a continuous scale (here 0–100%), the output can be presented in a histogram format, thereby grouping and enumerating the cells belonging to various distinct staining intensity levels. The scorings can be even further categorized by converting the output to scoring methods such as the Allred score, where a discrete scoring category (range 0–8) represents the estimated proportion and intensity of positively stained tumor cells within the region of interest (3, 4).

Though the scoring by discrete categories is mainly a remnant from the subjective adjudication of IHC stainings by classical microscopy, it is still widely used also for digital images not only when adjudicated by the operator on a computer monitor, but also when analyzed automatically and objectively by image analysis protocols, which actually produces outputs on a continuous scale. As automated image analysis of digital images becomes more established in microscopy, the acceptance of results on a continuous scale is likely to become widely accepted and beneficial for improved data mining.

3.4. Examples of Analysis of IHC Images

Figure 2a–d shows four common IHC staining types: p53, ER, HER2, and Ki-67. From an image analysis point of view, there are just two really different types of staining quantification in this set: nuclear staining as seen for p53, ER, and Ki67, and membrane staining as seen for HER2. The further analysis described in Fig. 2e–j is using p53 and HER2 as examples.

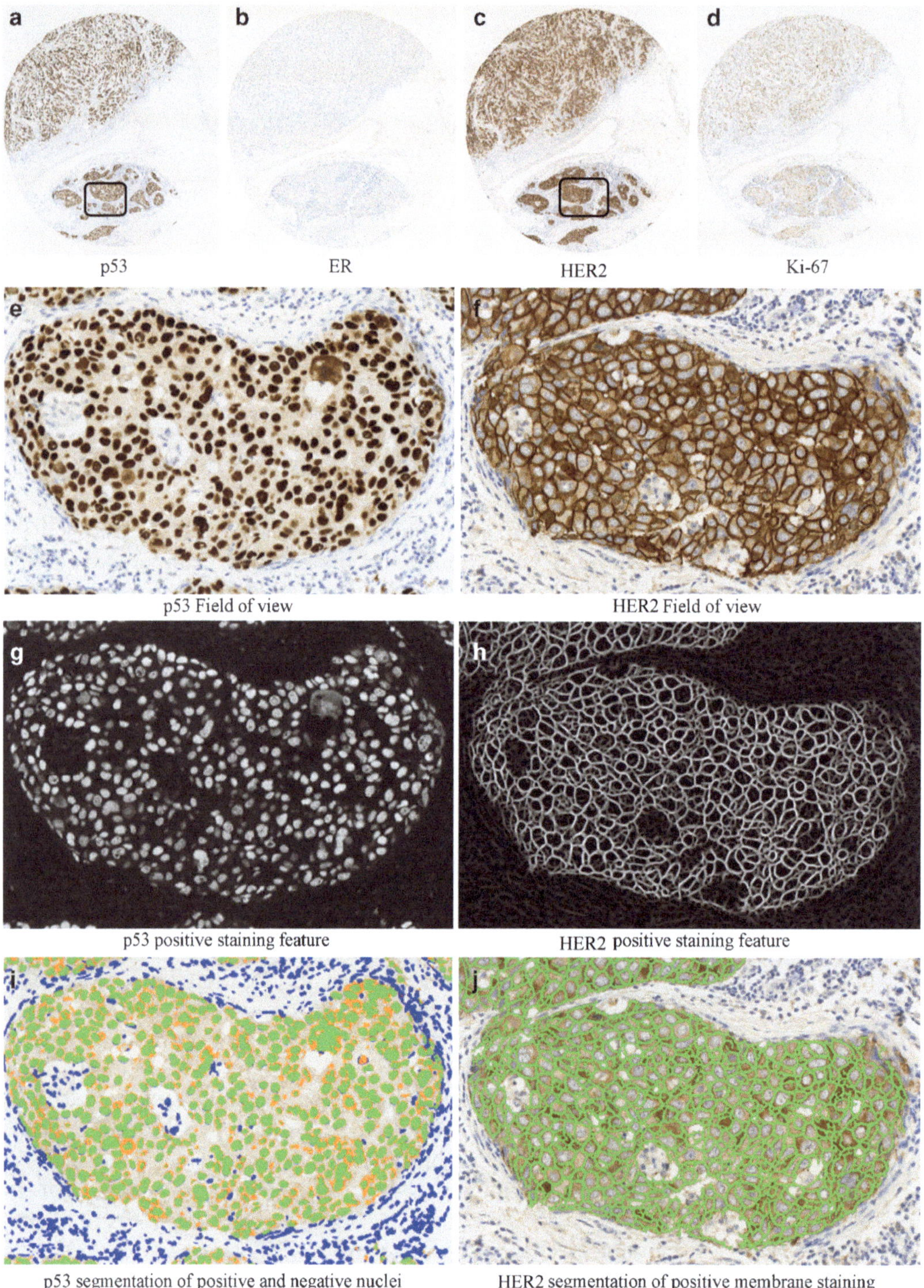

Fig. 2. Examples of IHC-staining for p53 (**a**), ER (**b**), HER2 (**c**), and Ki67 (**d**) in serially sectioned TMA core of breast cancer tissue. The quantitative analysis of the nuclear staining for p53 (**e**, **g**, and **i**) was based directly on spectral information, whereas the analysis of the membrane staining for HER2 (**f**, **h**, and **j**) was based on a combination of spectral and morphological information by inclusion of a polynomial local linear filter in Visiomorph™.

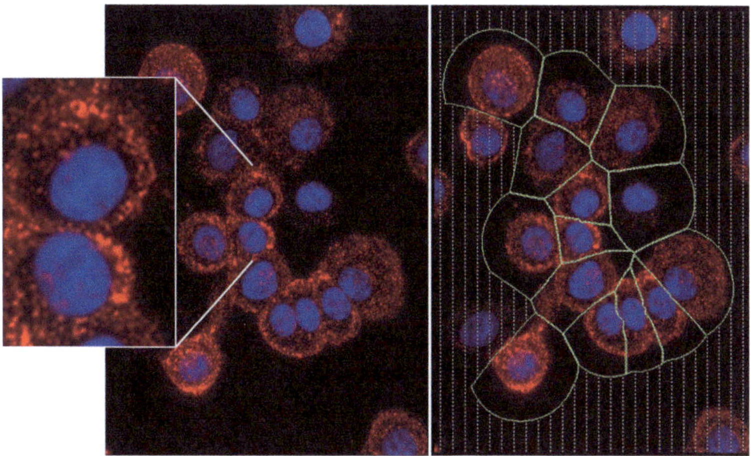

Fig. 3. Automated separation of cells exhibiting cytoplasmic PLA-signals. The watershed algorithm of the Visiomorph™ software used the DAPI-stained nuclei, and a user-defined maximum orthogonal distance to the closest nucleus, to define the pseudocytoplasmic region of each cell. This allowed automated estimation of the load of PLA-signals per cell (Original image kindly provided by Dr. Mats Gullberg, Olink Bioscience, Uppsala, Sweden).

3.5. Protocol Example: In Situ PLA

The proximity ligation assay technology, in situ PLA is an IHC technique highly suitable for studying the transient molecular interactions in ST pathways (5, 6). Its application of two bifunctional PLA-probes for binding to pairs of target-specific antibodies allows the in situ detection and quantification of e.g., receptor dimerization, receptor-specific phosphorylation, receptor–ligand interaction, and in a modified form of PLA, even the specific binding to DNA of transcription factors. Typically, the in situ PLA results in bright fluorescent spots each derived from a single molecular event, which can be automatically quantified by image analysis based on either counting of low to moderate numbers of signals, or intensity measures of wide dynamic ranges of signals. It is also possible to use chromogenic substrates for development of PLA signals for bright field visualization and detection. In order to associate the quantities to individual cells and subcellular locations, an automated separation of objects is required. This can be achieved by a postprocessing step including a watershed algorithm based on the nuclear DAPI-stain as shown in Fig. 3.

4. Notes

1. Of relevance to IHC of ST factors, the capacity of the imaging systems can be increased for what concerns tissue sample numbers by using TMAs or multitissue blocks rather than having just a single tissue section per slide.

2. Typically, a microscope-based imaging system will include:
   - Microscope (motorized: z-axis, objectives, filters, light source, etc.)
   - Digital camera (cooled, color/monochrome)
   - Stage and optional slide loader with bar code reader (motorized: $xy(z)$-axis)
   - Computer and software for controlling image acquisition through coordination of these devices
3. Typically, a whole slide scanner-based imaging system will include:
   - Scanning device (line or tile scanning principle)
   - Computer and software for whole slide image acquisition and for image subsampling (high resolution whole slide images are normally digitally subsampled before image analysis)
4. The term "image feature" can be a little confusing, but it is standard in statistical decision theory, which is an important discipline in the context of image segmentation and analysis.
5. Preprocessing is based on spectral and spatial/morphological information.

   The spectral information is typically conveyed by color models (see Table 1) offered as standard in Visiomorph™ as well as in other commercially available image analysis packages. Image analysis software with capabilities for manipulating color components are highly useful for the more demanding preprocessing tasks, which may be required for achieving a good separation of image structures that are stained differently in a given application.

   Spectral information is rarely sufficient for identifying the structures of interest in an image. Typically, spatial/morphological information, either alone or in combination with spectral information, is required in order to obtain sufficiently good image segmentation results.

   Filtering is often used for enhancing image structures or suppressing noise. Visiomorph™ provides a range of well-known basic filters that are also available in most other commercially available software packages due to their general usefulness and computational simplicity. The basic filters included are listed in Table 2, along with their typical application.

   The polynomial filters (see Table 3) are a novel class of filters implemented in the Visiomorph™ system, and have proved to be very powerful tools for enhancing the frequently occurring structural features in digital images of tissue sections, such as "blobs" (e.g., cells and particles) or linear structures

Table 1
Color models

RGB	*Description*: The classical spectral representation of a color image is Red, Green, and Blue, where each color band has a given certain intensity at each given pixel location *Comments*: For many practical image analysis tasks, use of the raw input color bands does not provide sufficiently robust segmentation results across multiple images where either acquisition conditions or staining intensities vary considerably
rgb (Chromaticity)	*Description*: Color chromaticities are sometimes referred to as tristimulus values or trichromatic coefficients. The chromaticities provide information about the content of red, green, and blue relative to the total image intensity for each pixel *Comments*: Color chromaticities are often very useful as input features for segmentation algorithms. Of particular relevance to chromogenic IHC is the usefulness of chromaticity for quantifying the intensities of red/brown and blue staining
HSI (Hue, saturation, intensity)	*Description*: Sometimes useful for segmentation purposes in the sense that it separates image intensity (I) from color hue (H) and color saturation (S) *Comments*: The inherent discontinuity of the Hue feature makes is very sensitive and difficult to interpret. The Saturation can be used as a measure of staining intensity
Color contrasts	*Description*: Color contrasts (Red-Green, Red-Blue, and Green-Blue, and Green-Blue) are useful for enhancing areas in an image with a particular staining, while at the same time reducing the impact of non-specific variations in image intensity *Comments*: For whole slide imaging, intensity gradients across the image are rarely a problem, but for microscope systems they are a frequently observed issue

Table 2
Basic filters

Mean filter	Smoothing and noise removal. Does not preserve edges
Standard deviation filter	Used for enhancing major changes in intensity (edges), and sometimes for identifying regions in images with high or low variability in intensity (texture)
Minimum filter	Used as basic building block for constructing grayscale morphological operations (erosion)
Maximum filter	Used as basic building block for constructing grayscale morphological operations (dilation)
Median filter	Smoothing and noise removal. The median filter is edge-preserving, but slower than the mean filter

Table 3
Polynomial filters

Polynomial local linear	Enhances linear structures of a certain width. As an example, this filter is very useful for identifying positive stained membranes
Polynomial blobs	Enhances "blob"-shaped objects of a certain size. This filter can sometimes assist in the detection of cells and subcellular structures such as nuclei or nucleoli, and it is especially useful for detecting roundish signals in e.g., digital images of tissue sections stained by PLA, CISH, or FISH

(e.g., blood vessels and membranes). The size of the filter is defined to match the individual application.

6. Image segmentation refers to the process of partitioning a digital image into multiple segments (sets of pixels). The goal of segmentation is to simplify and/or change the representation of an image. Image segmentation is typically used to locate objects and boundaries (lines, curves, etc.) in images. More precisely, image segmentation is the process of assigning a label to every pixel in an image such that pixels with the same label share certain visual characteristics. The result of image segmentation is a set of segments that collectively cover the entire image, or a set of contours extracted from the image. Each of the pixels in a region is similar with respect to some characteristic or computed property, such as color, intensity, or texture. Adjacent regions are significantly different with respect to the same characteristic(s).

   Most image analysis software packages have some type of image segmentation capability, such as thresholding and phase analysis that are simple and intuitive methods for segmenting an image. In practice, however, a more robust approach is often required in order to obtain useful results across larger batches of images and different studies. Table 4 includes a list of segmentation methods, and their respective merits and shortcomings are described.

7. Postprocessing. Even when image segmentation works well, it does not necessarily represent the end-result that allows for a meaningful segmentation. The segmented image represents a set of different objects (i.e., groups of pixels), of which we may only wish to characterize objects of a certain type, size, shape, and perhaps even objects in a certain context or relation to other objects. Postprocessing is a step, which is applied

Table 4
Segmentation methods

Thresholding	*Description*: Perhaps the most common segmentation method. The method typically defines a threshold for a given *feature*, and assigns one class to all pixels with a feature value above or equal to that value, and another class for the rest *Comment*: The most obvious shortcoming is that thresholding is very sensitive to the day-to-day variations in image quality caused by small changes to acquisition parameters or staining properties. This limits the applicability of this method to automate analysis across studies and even batches of images from the same study. Sometimes, it is possible to reduce the influence of such variability by introducing robust preprocessing steps
Bayes classifier	*Description*: The Bayes classifier has a number of properties that make it very useful for a wide range of image segmentation applications. It is a multivariate statistical classifier capable of working on any number of input features, allowing a combination of spectral and spatial/morphological features in the segmentation rules, making this a very powerful tool for image analysis *Comments*: Eliminates the need to define entire intensity range for each of the image classes of interest. Fast to train
K-means clustering	*Description*: As above, but can sometimes handle situations where (e.g., staining) intensity varies between images/sections *Comment*: Works best where the image classes are well separated in color and/or other features. Fast to train but classification is much slower than the Bayesian

Table 5
Postprocessing steps

Change object(s) into other object(s) based on	Size, shape, context (surrounding objects/labels and extent of surrounding objects/labels), relation (nearest objects/labels)
Change object(s) morphology using	Erosion, dilation, open, close (binary morphological operations) Skeletonize (and identify end-points/branch points) Fill holes Separate objects and/or space between objects

to the segmented image, and provides an opportunity to incorporate important application knowledge, through the use of morphological, contextual, and relational information. In Visiomorph™, it is possible to define multiple postprocessing steps, where each step can be defined in a high-level (abstraction) language. Some of these are listed in Table 5.

8. The output can typically be planimetric based on the segmented image and/or intensity-based when measured on original or derived features.

The planimetric outputs are often:

- Number
- Perimeter
- Interface length

The feature outputs are often:

- Mean, median, mode
- Standard deviation, entropy
- Max, Min

Also combined output of the above can be achieved including e.g., sum, difference, multiple, or ratio.

References

1. Kononen, J., Bubendorf, L., Kallioniemi, A., Bärlund, M., Schraml, P., Leighton, S., et al. (1998) Tissue microarrays for high-throughput molecular profiling of tumor specimens. *Nat. Med.* **4**, 844–847.
2. Rojo, M.G., Bueno, G., and Slodkowska, J. (2009) Review of imaging solutions for integrated quantitative immunohistochemistry in the Pathology daily practice. *Folia Histochem. Cytobiol.* **47**, 349–354.
3. Allred, D.C., Clark, G.M., Elledge, R., Fuqua, S.A., Brown, R.W., Chamness, G.C., et al. (1993) Association of p53 protein expression with tumor cell proliferation rate and clinical outcome in node-negative breast cancer. *J. Natl. Cancer Inst.* **85**, 200–206.
4. Harvey, J.M., Clark, G.M., Osborne, C.K., and Allred D.C. (1999) Estrogen receptor status by immunohistochemistry is superior to the ligand-binding assay for predicting response to adjuvant endocrine therapy in breast cancer. *J. Clin. Oncol.* **17**, 1474–1481.
5. Fredriksson, S., Gullberg, M., Jarvius, J., Olsson, C., Pietras, K., Gústafsdóttir, S.M., Ostman, A., Landegren, U. (2002) Protein detection using proximity-dependent DNA ligation assays. Nature Biotechnol. **20**, 473–477.
6. Söderberg, O., Gullberg, M., Jarvius, M., Ridderstråle, K., Leuchowius, K.J., Jarvius J., et al. (2006) Direct observation of individual endogenous protein complexes in situ by proximity ligation. *Nat. Methods* **3**, 995–1000.

Chapter 9

Flow Cytometric Analysis of Cell Signaling Proteins

Maria A. Suni and Vernon C. Maino

Abstract

In recent years, techniques that combine the use of phospho-specific antibodies and multiparameter flow cytometry have been developed for the detection of protein phosphorylation at the single cell level. Flow cytometry is uniquely suited for this type of analysis, as it can measure functional and phenotypic markers in the context of complex cell populations. Phosphorylation can be assessed simultaneously in multiple cell subsets, and due to the small sample sizes required, and the rapid analyses of large numbers of cells in this approach, rare cell analysis is possible without the ex vivo expansion of cells.

In this chapter, we detail flow cytometric protocols for the detection of intracellular phospho-proteins in samples derived from whole blood and peripheral blood mononuclear cell preparations. These protocols define steps for cell activation, fixation, permeabilization, and staining by phospho-specific and phenotyping antibodies. We discuss technical difficulties inherent to this technique and suggest solutions to commonly encountered problems. Additionally, we show examples of phospho-protein detection in lymphocyte subsets, dendritic cells, and monocytes activated with various stimuli, including mitogens, cytokines, and superantigens. Finally, we highlight a potential clinical trial application for this flow cytometric assay as a platform for pharmacodynamic monitoring of kinase inhibitors.

Key words: Phosphorylation, Cell signaling, Phospho-specific antibodies, Intracellular staining, Multiparameter flow cytometry, Whole blood, PBMC

1. Introduction

The transient phosphorylation of intracellular signaling proteins is critical to the regulation of most aspects of cellular activity. Phosphorylation is controlled by an assortment of kinases and phosphatases, which propagate signaling events along many different signal transduction cascades, including the ones that regulate cell differentiation, growth, proliferation, and apoptosis, and those that are involved in responses to cytokines, chemokines, and external stress factors (1). Phosphorylation events

Alexander E. Kalyuzhny (ed.), *Signal Transduction Immunohistochemistry: Methods and Protocols*,
Methods in Molecular Biology, vol. 717, DOI 10.1007/978-1-61779-024-9_9,

are rapid and reversible, with the phosphorylated state most often corresponding to the activated state of the protein. Consequently, the measurement of phosphorylation reveals information about the selective activation of signaling cascades by specific stimuli, the kinetics of signaling, and the downstream targets involved (2).

In recent years, the increased availability of antibodies that recognize phospho-specific protein epitopes in intracellular environments has led to the development of new flow cytometric techniques for cell signaling analysis (2–7). These assays detect signaling events in individual cells and can offer several advantages over the traditionally used bulk assays such as Western blotting and ELISA (8, 9).

Flow cytometry allows single-cell analysis within heterogeneous primary cell populations such as whole blood or peripheral blood mononuclear cells (PBMCs). This method can define phosphorylation events in minor populations without concern that the effect will be diluted by nonresponding cells. Also, by virtue of its multiparameter capability, flow cytometry enables the simultaneous identification of multiple cell populations and the analysis of their phosphorylation states, which results in an information-rich data set from a single staining tube. The required sample sizes for this technique are relatively small, which allows the analysis of rare cell subsets, such as dendritic cells (DCs), without the need for cell sorting and ex vivo expansion.

The protocol for the detection of intracellular phospho-proteins by flow cytometry is fast and flexible. In its most simplified form, the technique involves treating a cell sample briefly with a stimulus to induce phosphorylation, and then fixing the cells with a cross-linking agent, typically formaldehyde. Cells are subsequently permeabilized and then stained with antibodies against phospho-proteins and cell surface antigens. Multiparameter flow cytometry is employed to measure phosphorylation events within the cell populations of interest (10).

Permeabilization reagents are generally alcohol-based (ethanol, methanol) or contain detergents (e.g., saponin, Tween-20, Triton X-100) (10, 11). Both types of permeabilization reagents have distinct advantages and limitations. For example, while buffers containing nonionic detergents like saponin or Tween-20 maintain staining by most cell surface antibodies, they do not support optimal detection of some phospho-proteins, such as the nuclear signal transducer and activator of transcription (Stat) proteins, with currently available antibodies. Conversely, permeabilization reagents with high methanol concentrations are compatible with a wide variety of nuclear and cytoplasmic phospho-proteins, but can compromise the detection of many cell surface markers by antibodies directed against native epitopes (10). Careful consideration should therefore be given not only to the choice of permeabilization buffers, but also to the selection of cell surface antibody clones, as they can differ significantly in their abilities to stain cells under various permeabilization conditions. The selection of phospho-specific antibodies that bind with high affinity and specificity in fixed and permeabilized cells is equally important, as many antibodies used in Western blotting do not perform well in flow cytometry. Commercial phospho-antibodies optimized for flow cytometry applications have become increasingly available; however, it is important to consider performance requirements and the fixation and permeabilization conditions when selecting panels of reagents for this application.

In this chapter, we describe protocols for phospho-protein detection in whole blood and PBMC samples, using one of the two permeabilization buffers that have yielded robust phospho-specific staining in our laboratory (Buffer III methanol-based and Buffer IV, which consists of a mixture of detergents). Although these permeabilization conditions present some challenges with cell surface staining, it should be noted that with adjustments to the staining sequences in the protocol, it is possible to accomplish good labeling of most cell surface markers, while obtaining optimal intracellular phospho-protein detection. The wider dynamic ranges obtained with these permeabilization buffers enable semiquantitative phospho-protein analysis.

2. Materials

2.1. Cell Activation

1. Refer to Table 1 for commonly used cell activators and their suppliers, storage conditions, final concentrations, and examples of phospho-protein targets (also see Note 1).
2. 1× Phosphate buffered saline (PBS) (Ca^{2+} and Mg^{2+} free). Store at room temperature. Warm an aliquot to 37°C to resuspend PBMC for activation.
3. Complete RPMI medium (cRPMI): Supplement RPMI-1640 (Sigma, St. Louis, MO) with 10% heat-inactivated fetal bovine serum and 1% penicillin/streptomycin (Gibco, Rockville, MD). Store at 4°C. Warm to 37°C to wash cryopreserved PBMC.
4. Kinase inhibitor, Imatinib mesylate (American Custom Chemicals Corporation, San Diego, CA) is dissolved at 20 mM in DMSO and stored as single-use aliquots at –20°C. Dilute 1:10 in sterile 1× PBS on the day of use.

2.2. Cell Fixation, Permeabilization, and Staining

1. 5× BD™ Phosflow Lyse/Fix Buffer (used with whole blood; BD Biosciences, San Diego, CA): Prepare working stock by diluting 1:5 in deionized water. Warm to 37°C before use.

Table 1
Cell activators and their sources, stock concentrations, storage temperatures, final concentrations, and phoshorylation targets

Activating ligand	Vendor	Stock concentration	Storage (°C)	Final concentration	Phospho-protein
Human rIFNα	PBL Biomedical Labs, Piscataway, NJ	1×10^6 IU/ml in PBS	−80	1,000 IU/ml	Stat1
Human rIL-2	R&D Systems, Minneapolis, MN	1×10^6 IU/ml in PBS	−80	1,000 IU/ml	Stat5
Human rIL-4	BD Biosciences, San Diego, CA	0.1 mg/ml in PBS	−80	100 ng/ml	Stat6
Phorbol 12-myristate 13-acetate (PMA)	Sigma, St. Louis, MO	0.1 mg/ml in DMSO	−20	200 ng/ml	ERK1/2, p38 MAPK
Staphylococcal enterotoxin B (SEB)	Sigma	1 mg/ml in PBS	+4	10 μg/ml	ERK1/2, p38 MAPK
Lipopolysaccharide (LPS)	Sigma	0.1 mg/ml in PBS	−20	200 ng/ml	ERK1/2
ODN2006 Type B CpG oligonucleotide	InvivoGen, San Diego, CA	1 mg/ml in di water	−20	6 μM	NFκB
CD3 (clone UCHT1)	BD Biosciences	1 mg/ml in PBS	+4	5 μg/ml	ERK1/2
CD28+CD49d costimulus	BD Biosciences	0.1 mg/ml in PBS	+4	1 μg/ml	ERK1/2

2. 32% Paraformaldehyde (PFA) solution: Store at room temperature.
3. BD Phosflow Perm Buffer III (BD Biosciences): Cool to −20°C before use, or 10× BD Phosflow Perm Buffer IV (BD Biosciences): Prepare working stock by diluting 1:10 in 1× PBS.
4. Wash buffer: Add 1% bovine serum albumin (Sigma) and 0.1% sodium azide to 1× PBS. Store at 4°C.
5. Several phospho-specific and phenotyping antibodies are listed in Table 2 (all from BD Biosciences). Store the antibodies at 4°C, protected from light.
6. 1% PFA solution: Dilute 32% PFA 1:32 in 1× PBS. Store at 4°C.

Table 2
Fluorescent-conjugated phospho-specific and cell surface antibodies used in the experiments shown

Monoclonal antibody	Clone
Alexa Fluor® 647 anti-ERK1/2 (pT202/pY204)	20A
Alexa Fluor® 647 anti-p38 MAPK (pT180/pY182)	36/p38
Alexa Fluor® 647 and PE anti-Stat1 (pY701)	4a
Alexa Fluor® 647 anti-Stat5 (pY694)	47
Alexa Fluor® 647 anti-Stat6 (pY641)	18/p-Stat6
PE anti-NFκB (pS529)	K10-895.12.50
CD3 PE and PerCP	SK7
CD4 FITC	SK3
BD Horizon™ V450 CD11c	B-ly6
Alexa® Fluor 488 CD14	M5E2
CD16 PE	3G8
Alexa® Fluor 488 CD19	HIB19
CD56 PE	B159
CD123 PerCP-Cy5.5	7G3
HLA-DR PE-Cy7	L243
Lin-1 FITC	Cocktail

3. Methods

3.1. Whole Blood Activation and Staining

Most flow cytometric analyses of phosphorylation involve an activation step, although phosphorylated signaling proteins can be constitutively expressed in some cell types (12). Stimuli for activation can be broadly reactive (e.g., PMA) or selective (e.g., Toll-like receptor-active compounds). Several commonly used stimuli are listed in Table 1. Figure 1 illustrates the general scheme for activation, fixation, permeabilization, and staining of samples using this methodology.

Whole blood should be collected into heparin or EDTA blood collection tubes (BD Vacutainer®, Franklin Lakes, NJ, or equivalent), stored at room temperature, and used within 8 h after draw for best cellular responses (see Note 2).

1. For each test and control condition, aliquot 0.2–2 ml of whole blood into a 15- or 50-ml conical polypropylene tube. The blood volume will depend on the number of staining tubes planned for each condition and the number of events to be acquired. Typically, 200 μl of whole blood is adequate for each staining tube (see Note 3).
2. Add the stimulus at optimal dose and incubate the samples in a 37°C water bath or a humidified CO_2 incubator for the required time (see Note 4).

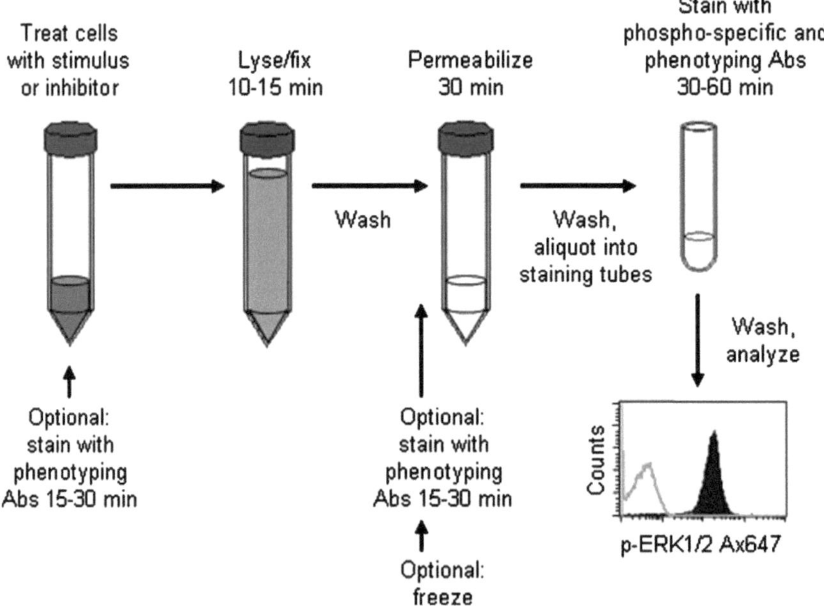

Fig. 1. Flowchart of the phospho-protein flow cytometric assay. Whole blood or PBMC are treated with various stimuli (e.g., mitogens, cytokines, superantigens) or inhibitors (e.g., kinase inhibitors) and samples are then processed as described in the text. After staining with phospho-specific and phenotyping antibodies (Abs), the samples are analyzed on a flow cytometer. Note the variations of the basic method to enable cryopreservation and to enhance cell surface staining.

3. Lyse erythrocytes and fix leukocytes with an appropriate lysing and fixing solution, such as Phosflow Lyse/Fix Buffer. Add at least nine volumes of warm 1× Lyse/Fix Buffer to each tube and mix well by vortexing. Incubate 10–15 min at 37°C (see Notes 5 and 6).
4. Centrifuge tubes at 500×*g* for 10 min at room temperature. Aspirate supernatant and vortex the tubes to loosen pellets. Add 10 ml of PBS to each tube and centrifuge at 500×*g* for 10 min (see Note 7).
5. Aspirate supernatant to obtain as dry a cell pellet as possible without disturbing it. Then, vortex the tubes to loosen the pellets (see Notes 8 and 9).
6. Permeabilize the cells with an appropriate permeabilization solution, such as Phosflow Perm Buffer III or IV. Add 1–2 ml of cold Perm Buffer III to each tube while vortexing gently. Incubate on ice for 30 min. Alternatively, add 1–10 ml of Perm Buffer IV to each tube while vortexing gently. Incubate at room temperature for 30 min (see Notes 10 and 11).
7. Add 5–10 ml of wash buffer to the tubes and centrifuge at 500×*g* for 10 min at room temperature. Decant supernatant and loosen the pellets. Repeat the wash step (see Note 12).
8. Resuspend the cells in wash buffer and aliquot 100 μl of cells into 12×75 mm staining tubes. Add the appropriate amounts of fluorescent-conjugated phospho-specific and phenotyping antibodies and incubate at room temperature for 30–60 min, protected from light (see Notes 13 and 14).
9. Add 3 ml of wash buffer to each tube and centrifuge for 10 min at 500×*g*. Decant supernatant and loosen the pellets.
10. Add 200–400 μl of wash buffer to each tube, vortex, and store at 4°C, protected from light. Analyze, preferably within 4 h. For overnight storage of the samples, resuspend the cells in 1% PFA solution and store at 4°C, protected from light. Examples of phospho-protein staining in whole blood samples are shown in Figs. 2–4.

3.2. PBMC Activation and Staining

PBMC can be isolated from blood using Ficoll-Paque™ solution (GE Healthcare Bio-Sciences, Uppsala, Sweden) following the manufacturer's instructions. Alternatively, PBMC can be harvested by centrifugation in Cell Preparation Tubes (CPT™, BD Vacutainer). Wash the cells once in PBS (300×*g*, 10 min) and resuspend in warm PBS at 2–5×10^6 PBMC/ml. Remove any clumps with a pipette tip or use a cell strainer. Proceed to step 1 below (see Note 15).

Cryopreserved PBMC should be thawed at 37°C, and washed using warm cRPMI (13). Resuspend cells in cRPMI in a 15- or a 50-ml tube at a final concentration of 2–5×10^6 PBMC/ml.

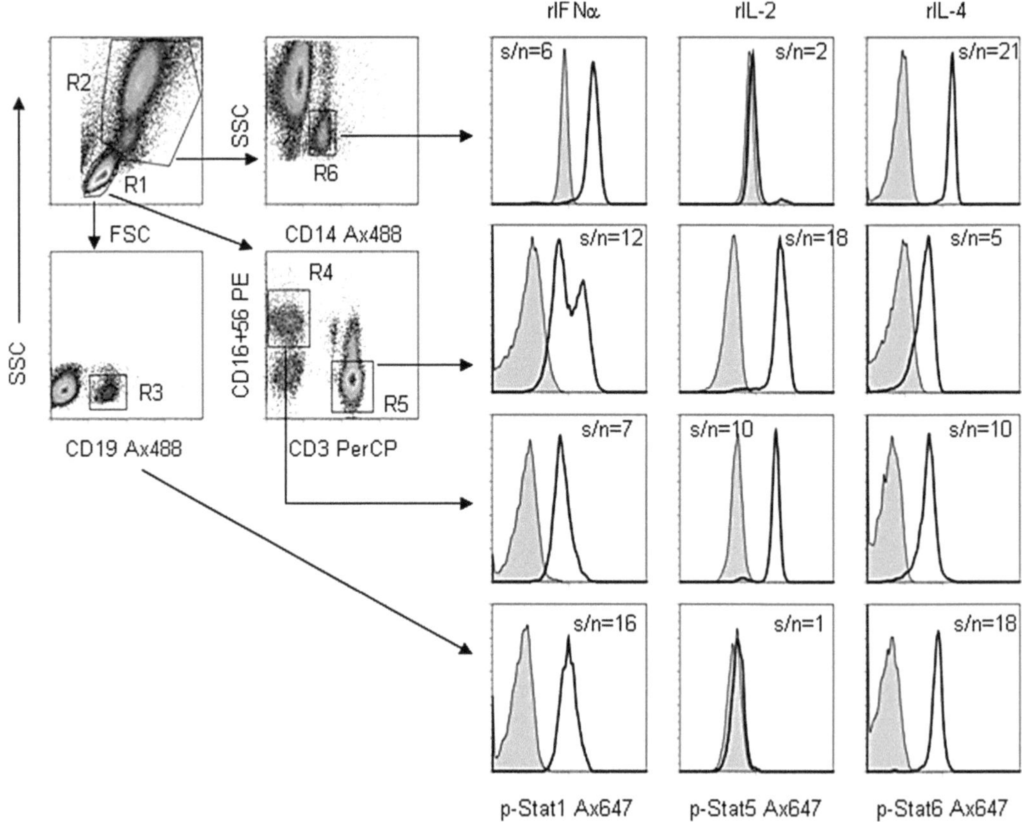

Fig. 2. Examples of cytokine-induced phospho-Stat expression in lymphocyte subsets and monocytes. Whole blood samples were simultaneously stained with fluorescent antibodies against CD14, CD19, CD16, and CD56, and activated with human recombinant IFNα or IL-2 or IL-4 for 15 min. The blood samples were lysed, fixed, and permeabilized (Perm Buffer IV) according to the protocol and then stained with fluorescent-labeled antibodies to CD3 and p-Stat1 or p-Stat5 or p-Stat6. CD3 PerCP staining was performed post-permeabilization, because PerCP can be damaged by Perm Buffer IV treatment. The data analysis was performed using FlowJo analysis software (Tree Star, Inc., Ashland, OR). The T, B, NK cell, and monocyte populations were gated as depicted by the *arrows* and the phospho-Stat signaling profiles were visualized in these populations by using histogram overlays (cytokine-treated *open histogram*, untreated *shaded histogram*). Note the differences in the patterns and levels of induction of the phospho-Stats by different cytokines. Signal/noise (s/n) = MFI stimulated/MFI unstimulated.

Rest the cells for 3–4 h in a 37°C humidified CO_2 incubator before activation. Pellet the cells by centrifugation and resuspend in warm PBS. Remove clumps, if present (see Note 16).

1. For each test or control condition, add 0.5–4 ml of cell suspension to a 15- or 50-ml conical polypropylene tube (see Note 17). The cell number and volume will depend on the number of staining tubes planned for each condition, and the number of events to be acquired. For example, in assays that assess phosphorylation events in rare cell subsets such as DCs, at least 1×10^6 PBMC per staining tube is recommended.
2. Add the stimulus at optimal dose and incubate the samples in a 37°C water bath or a humidified CO_2 incubator for the required time (see Note 4).

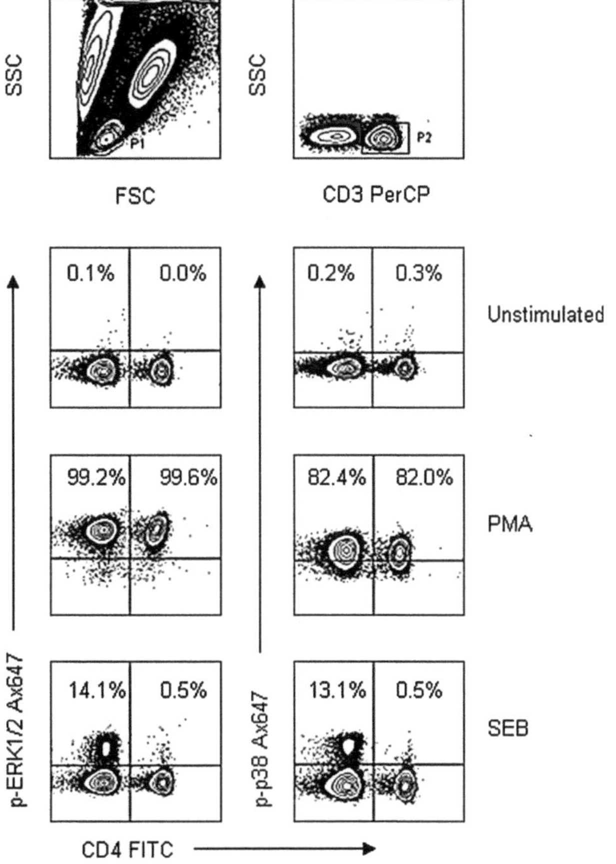

Fig. 3. Examples of mitogen- and superantigen-induced phospho-ERK1/2 and phospho-p38 expression in T cells. Whole blood samples were activated with PMA or SEB for 15 min, after which they were lysed, fixed, and permeabilized (Perm Buffer III) according to the protocol. The samples were subsequently stained with fluorescent-conjugated antibodies against CD4, CD3, and p-ERK1/2 or p-p38. The data analysis was performed using BD FACSDiva™ software. The percent positive values in the *upper left* and *upper right quadrants* of the contour plots depict the frequencies of CD3+CD4− and CD3+CD4+ cells expressing p-ERK1/2 or p-p38. As shown in the contour plots, SEB induced p-ERK1/2 and p-p38 expression predominantly in CD4− T cells, while a low frequency of CD4+ T cells (0.5%) was also positive for these phospho-proteins. In cases like this, two-dimensional contour plots allow better visualization of small changes in cell frequencies and/or fluorescence intensities than one-dimensional histograms.

3. Fix the cells by adding 32% PFA solution to the cells to a final concentration of 4%. Vortex and incubate for 10–15 min at 37°C (see Note 5).
4. Add 10 ml of PBS to each tube and centrifuge at 500 × *g* for 10 min at room temperature (see Note 7).
5. Aspirate supernatant to obtain as dry a cell pellet as possible without disturbing it. Then, vortex the tubes to loosen the pellets (see Notes 8 and 9).
6. Permeabilize the cells using an appropriate permeabilization solution, such as Phosflow Perm Buffer III or IV. Add 1–2 ml

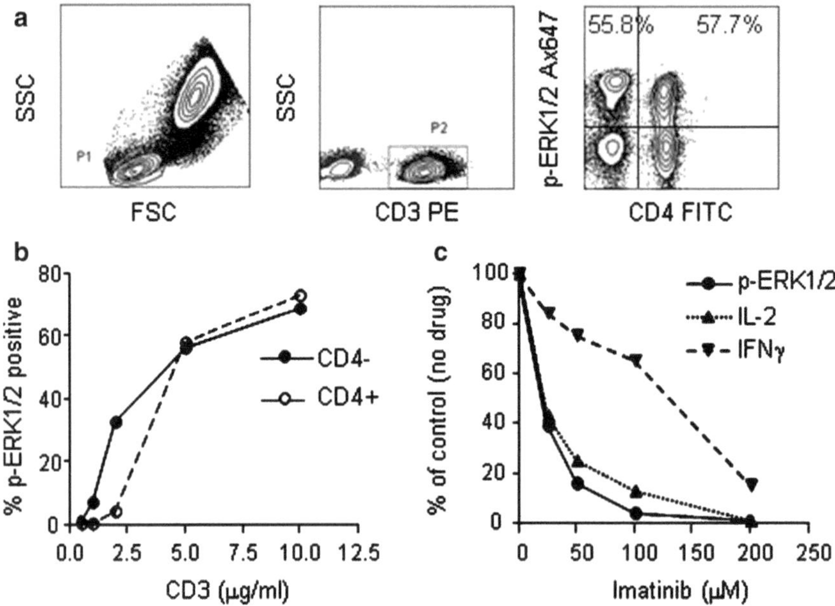

Fig. 4. Measuring T cell dose responses to a stimulus or an inhibitor using the phospho-protein flow assay. (**a**) Whole blood was stimulated with CD3+CD28+CD49d antibodies for 5 min. The samples were lysed, fixed, and permeabilized (Perm Buffer IV) according to the protocol, and stained with CD4 FITC, CD3 PE, and anti-p-ERK1/2 A×647. (**b**) An example of a dose response of T cells to purified CD3 antibody in the presence of CD28+CD49d costimulus. Note that in this example, the CD4+ T cells have a slightly higher threshold for p-ERK1/2 induction than the CD4– T cells. (**c**) An example of a drug inhibition assay. Whole blood was pretreated with the kinase inhibitor imatinib (25–200 μM) for 45 min and activated with CD3+CD28+CD49d antibodies for 5 min. The effect of imatinib on intracellular cytokine (IL-2, IFNγ) expression was assessed in parallel, after a 5-h culture, following the standard intracellular cytokine assay protocol (15, 16). The frequencies of CD3+ T cells expressing p-ERK1/2, IL-2, or IFNγ were determined and the drug effect was calculated as: % of control = activation response with imatinib/activation response without imatinib × 100%. As shown in the graph, imatinib inhibited p-ERK1/2 expression in activated T cells in a dose-dependant manner (IC50 = 20 μM). A very similar dose response curve was obtained for IL-2 (IC50 = 20 μM), while IFNγ expression was less affected in this donor (IC50 = 130 μM). This example illustrates the utility of flow cytometry in studying the relationship between protein phosphorylation and downstream cytokine expression following in vitro stimulation. It further illustrates the potential of the phospho-protein flow assay as a rapid technique for monitoring signal transduction inhibitors, and as a useful pharmacodynamic assay to assess drug sensitivity (17, 18).

of chilled Perm Buffer III to each tube while vortexing gently. Incubate on ice for 30 min. Alternatively, add 1–5 ml of Perm Buffer IV to each tube while vortexing gently. Incubate for 30 min at room temperature (see Note 11).

7. Add 5–10 ml of wash buffer to the tubes and centrifuge at 500 × *g* for 10 min at room temperature. Decant supernatants and loosen the pellets. Repeat the wash step (see Note 12).
8. Resuspend the cells in wash buffer and aliquot 100 μl/tube into 12 × 75 mm staining tubes. Add the appropriate amounts of fluorescent-conjugated phospho-specific and phenotyping antibodies and incubate for 30–60 min at room temperature, protected from light (see Note 18).

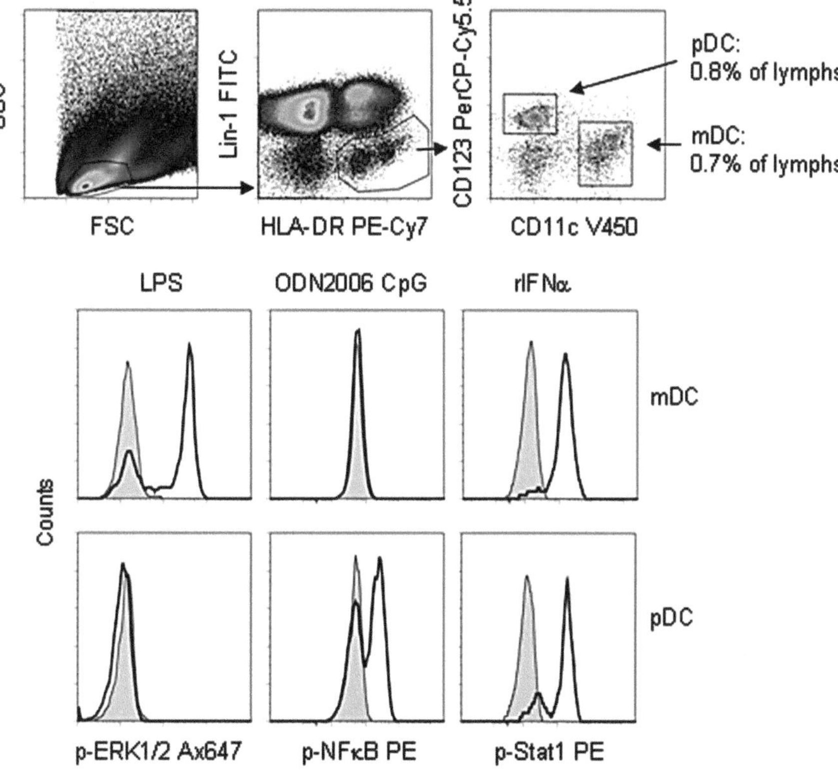

Fig. 5. An example of phospho-protein analysis in rare cell populations: Toll-like receptor (TLR)-specific or cytokine-mediated signaling in DC subsets. PBMC were first stained with fluorescent-conjugated Lin-1 (CD3, CD14, CD16, CD19, CD20, CD56) and CD11c antibodies for 20 min and then activated with LPS (ligand for TLR4 on mDC), ODN2006 CpG (ligand for TLR9 on pDC), or human recombinant IFNα (binds the IFNα receptor) for 20 min. The cells were fixed and permeabilized (Perm Buffer IV) according to the protocol and stained with fluorescent antibodies against HLA-DR, CD123, p-ERK1/2, and either p-NFκB (nuclear factor κB) or p-Stat1. The data analysis was performed using FlowJo analysis software. The myeloid DCs were identified by their Lin-1-/HLA-DR+/CD11c+ phenotype, while the plasmacytoid DCs were identified by their Lin-1-/HLA-DR+/CD123+ phenotype. The histogram overlays (*open histogram* stimulated, *shaded histogram* unstimulated) show the TLR4-specific p-ERK1/2 expression in mDC and TLR9-specific p-NFκB expression in pDC. IFNα-induced p-Stat1 expression was detected in both mDC and pDC.

9. Add 3 ml of wash buffer to the tubes and centrifuge at 500 × *g* for 10 min at room temperature. Decant supernatants and loosen the pellets.
10. Add 200–400 μl of wash buffer to each tube, vortex, and store at 4°C, protected from light. Analyze, preferably within 4 h. For overnight storage of the samples, resuspend the cells in 1% PFA solution and store at 4°C, protected from light. An example of phospho-protein detection in activated PBMC is shown in Fig. 5

3.3. Acquisition and Analysis

1. Follow instrument setup and quality control procedures appropriate for your flow cytometer.
2. Refer to your cytometer software reference manual for a detailed protocol for compensation setup. Use cells or

compensation control beads (e.g., BD CompBead anti-mouse Ig,κ and negative control, BD Biosciences; or equivalent) as appropriate to set the photomultiplier tube voltages and compensation values (14).

3. Acquire enough events for each sample to obtain statistically meaningful data for all populations of interest.
4. Identify and gate the cell populations of interest. For example, for the analysis of phosphorylation events in T cells, first draw a gate around small lymphocytes, followed by a gate around the CD3+ population (Fig. 2). Then, display data for each phospho-marker within this population using histogram overlays (Fig. 2) or contour plots (Fig. 3). Commonly, the activation profile of each phospho-marker is overlaid with the profile of its untreated (unstimulated) control (Figs. 2 and 5). The signaling responses are typically expressed as signal-to-noise ratios or fold changes above untreated, by dividing the mean fluorescent intensity (MFI) of the activated histogram profile by that of the untreated profile. Alternatively, the activation responses can be expressed as percent positive cells above the untreated control (Fig. 3).

3.4. Assay Controls

As is true for all functional assays, the inclusion of positive controls to each experiment is important, because the lack of positive staining by a phospho-antibody could have several causes unrelated to the lack of phosphorylation. For example, a phospho-epitope might be inaccessible to antibody binding due to inadequate fixation and/or permeabilization, while inadequate washing, or staining with an antibody conjugate with free dye present, might lead to high background staining that can mask the differences between signals from stimulated and unstimulated samples. Since phosphorylated extracellular signal-regulated kinase (p-ERK1/2) staining in PMA-activated lymphocytes is generally robust, it can serve as a useful positive stimulation control in these assays. Additionally, commercially available lyophilized control cells have recently become available. These stimulated and unstimulated control cells (BD Biosciences) can be used as positive and negative controls for staining by several different antibodies against phospho-proteins induced by various stimuli.

4. Notes

1. The cell activators should be frozen in single use aliquots. After preparing working stocks in sterile PBS, keep the stimuli on ice until use.
2. For whole blood stimulation conditions that require Ca^{2+} (e.g., superantigen stimulation), blood should be drawn into

sodium heparin blood collection tubes, as EDTA and citrate chelate calcium. For PMA stimulation, blood collected into EDTA is preferred because of improved cell recoveries and scatter characteristics.

3. Typically, 0.2–1 ml blood is dispensed into a 15-ml, and 1–2 ml blood is dispensed into a 50-ml tube.
4. The optimum stimulation kinetics vary and should therefore be determined for each stimulus and phospho-protein.
5. In order to preserve phosphorylation states, it is important to fix the cells immediately post-activation.
6. In rare cases, 1:10 dilution of blood with the Lyse/Fix Buffer can result in incomplete erythrocyte lysis. This can be prevented by diluting blood 1:20 instead of 1:10 with the Lyse/Fix Buffer.
7. Fixed cells are more buoyant than live cells and are therefore more difficult to pellet. In order to minimize cell loss, cells are centrifuged at higher g-forces and for longer times.
8. Decanting the supernatant is not recommended here, since it can leave a residual volume that will dilute the permeabilization buffer. This could compromise the staining of some phospho-proteins.
9. Fixed samples can be stored at –80°C for at least 1 month. For frozen storage, resuspend the cells in 1 ml of PBS, cap the tubes tightly, and place in a freezer.
10. For optimal phospho-Stat staining intensities, increase Perm Buffer IV volumes with increasing blood volumes, e.g., when treating cells from 2 ml of whole blood, use at least 10 ml of Perm Buffer IV.
11. Perm Buffer IV can also be used at 0.5× concentration (dilute stock 1:20 in PBS) for the detection of most phospho-proteins. Although this milder permeabilization condition can result in reduced signal-to-noise ratios of phospho-Stats, it significantly improves staining intensities of most phenotyping antibodies, while also improving cell recoveries.
12. Washing must be thorough in order to completely remove the perm buffer. Residual perm buffer can negatively impact the staining of PE and PerCP conjugates.
13. Some cell surface epitopes are not maintained under these fixation and permeabilization conditions. In these cases, surface staining can be performed on live cells for 15–30 min, either before or during activation (before or during step 2) or after fixation, before permeabilization (after step 5).
14. Optimal titers for some cell surface antibodies are different on live and fixed/permeabilized cells. Titration of such antibodies under the relevant fixation and permeabilization conditions

can reduce nonspecific background staining and increase signal-to-noise ratios.

15. Incubating the PBMC at 37°C for 1 h before stimulation might improve signals for some phospho-proteins.
16. The resting of thawed PBMC in X-VIVO 15 medium (Lonza, Walkersville, MD) instead of cRPMI can improve the recovery of DCs.
17. Typically, 0.5–1 ml of cell suspension is dispensed to a 15-ml tube and 2–4 ml is dispensed to a 50-ml tube, although using these volumes is not critical.
18. Cell surface staining can also be performed before activation (before step 2) or after fixation (after step 5). If cell surface staining is performed before activation, cells are resuspended in 300–500 μl of PBS and stained for 15–30 min at room temperature. Cells are then either pelleted by centrifugation (300 × *g*, 10 min) and resuspended in warm PBS, or simply diluted to their final concentration with warm PBS and activated as instructed in step 2 in Subheading 3.2.

Acknowledgments

The authors thank Drs. Smita Ghanekar and Guo-Jian Gao for helpful discussions and for reviewing this manuscript.

References

1. Cohen, P. (2000) The regulation of protein function by multisite phosphorylation–a 25 year update, *Trends Biochem. Sci.* *25*, 596–601.
2. Krutzik, P. O., Irish, J. M., Nolan, G. P., and Perez, O. D. (2004) Analysis of protein phosphorylation and cellular signaling events by flow cytometry: techniques and clinical applications, *Clin. Immunol.* *110*, 206–221.
3. Fleisher, T. A., Dorman, S. E., Anderson, J. A., Vail, M., Brown, M. R., and Holland, S. M. (1999) Detection of intracellular phosphorylated STAT-1 by flow cytometry, *Clin. Immunol.* *90*, 425–430.
4. Montag, D. T., and Lotze, M. T. (2006) Successful simultaneous measurement of cell membrane and cytokine induced phosphorylation pathways [CIPP] in human peripheral blood mononuclear cells, *J. Immunol. Methods* *313*, 48–60.
5. Krutzik, P. O., and Nolan, G. P. (2003) Intracellular phospho-protein staining techniques for flow cytometry: monitoring single cell signaling events, *Cytometry A* *55*, 61–70.
6. Varker, K. A., Kondadasula, S. V., Go, M. R., Lesinski, G. B., Ghosh-Berkebile, R., Lehman, A., et al. (2006) Multiparametric flow cytometric analysis of signal transducer and activator of transcription 5 phosphorylation in immune cell subsets in vitro and following interleukin-2 immunotherapy, *Clin. Cancer Res.* *12*, 5850–5858.
7. Irish, J. M., Hovland, R., Krutzik, P. O., Perez, O. D., Bruserud, O., Gjertsen, B. T., et al. (2004) Single cell profiling of potentiated phospho-protein networks in cancer cells, *Cell* *118*, 217–228.
8. Montag, D. T., and Lotze, M. T. (2006) Rapid flow cytometric measurement of cytokine-induced phosphorylation pathways [CIPP] in human peripheral blood leukocytes, *Clin. Immunol.* *121*, 215–226.
9. Haas, A., Weckbecker, G., and Welzenbach, K. (2008) Intracellular phospho-flow cytometry

reveals novel insights into TCR proximal signaling events. A comparison with Western blot, *Cytometry A 73*, 799–807.

10. Perez, O. D., Mitchell, D., Campos, R., Gao, G. J., Li, L., and Nolan, G. P. (2005) Multiparameter analysis of intracellular phosphoepitopes in immunophenotyped cell populations by flow cytometry, *Curr. Protoc. Cytom. Chapter 6*, Unit 6.20.
11. Chow, S., Hedley, D., Grom, P., Magari, R., Jacobberger, J. W., and Shankey, T. V. (2005) Whole blood fixation and permeabilization protocol with red blood cell lysis for flow cytometry of intracellular phosphorylated epitopes in leukocyte subpopulations, *Cytometry A 67*, 4–17.
12. Desplat, V., Lagarde, V., Belloc, F., Chollet, C., Leguay, T., Pasquet, J. M., et al. (2004) Rapid detection of phosphotyrosine proteins by flow cytometric analysis in Bcr-Abl-positive cells, *Cytometry A 62*, 35–45.
13. Disis, M. L., dela Rosa, C., Goodell, V., Kuan, L. Y., Chang, J. C., Kuus-Reichel, K., et al. (2006) Maximizing the retention of antigen specific lymphocyte function after cryopreservation, *J. Immunol. Methods* ***308***, 13–18.
14. Maecker, H. T., and Trotter, J. (2006) Flow cytometry controls, instrument setup, and the determination of positivity, *Cytometry A* ***69***, 1037–1042.
15. Maino, V. C., and Picker, L. J. (1998) Identification of functional subsets by flow cytometry: intracellular detection of cytokine expression, *Cytometry* ***34***, 207–215.
16. Maecker, H. T. (2004) Cytokine flow cytometry, *Methods Mol. Biol.* ***263***, 95–108.
17. Chow, S., Patel, H., and Hedley, D. W. (2001) Measurement of MAP kinase activation by flow cytometry using phospho-specific antibodies to MEK and ERK: potential for pharmacodynamic monitoring of signal transduction inhibitors, *Cytometry* ***46***, 72–78.
18. Tong, F. K., Chow, S., and Hedley, D. (2006) Pharmacodynamic monitoring of BAY 43-9006 (Sorafenib) in phase I clinical trials involving solid tumor and AML/MDS patients, using flow cytometry to monitor activation of the ERK pathway in peripheral blood cells, *Cytometry B Clin. Cytom. 70*, 107–114.

Chapter 10

CytoSys: A Tool for Extracting Cell-Cycle-Related Expression Dynamics from Static Data

Jayant Avva, Michael C. Weis, Radina P. Soebiyanto, James W. Jacobberger, and Sree N. Sreenath

Abstract

Computational models of biological processes are important building blocks in Systems Biology studies. Calibration and validation are two important steps for moving a mathematical model to a computational model. While calibration refers to finding numerical value of the coefficients such as rate constants in a mathematical model, validation refers to verifying that the calibrated model behaves the same as the biological system under previously unseen conditions such as environmental changes (e.g., drug treatment) or mutations. In lieu of direct measurements of rate constants, modeling of the molecular mechanisms that govern biological behaviors may be able to use dynamic expression profiles of reactant biomolecules for calibration. For validation, similar data, obtained under new conditions, are probably better than direct measurements of rate constants. In any case, direct measurement of rate constants is almost always impractical and difficult or impossible. Here, we show a computer-assisted methodology to extract embedded dynamic profiles of cell-cycle proteins from statically sampled, multivariate cytometry data guided by heuristics assembled from canonical cell-cycle knowledge. The methodology is implemented using standard "list mode" cytometry data-processing software followed by CytoSys – a software tool with an easy-to-use graphical interface. We demonstrate the use of CytoSys with a case study of exponentially growing, human erythroleukemia cells and extract the dynamic expression profiles of cyclin A for calibrating an existing deterministic mathematical model of the cell cycle.

Key words: Model calibration, ODE, In silico, Dynamic time and expression profiles, Flow cytometry, Cell cycle

1. Introduction

The aim of systems biology is to know biological systems with the precision and predictability of engineering. Large data sets and computational models are the tools of systems biology. To understand the complex properties of cellular biochemical control

Alexander E. Kalyuzhny (ed.), *Signal Transduction Immunohistochemistry: Methods and Protocols*, Methods in Molecular Biology, vol. 717, DOI 10.1007/978-1-61779-024-9_10,

systems, larger networks are modeled as interactions between components or subsystems (1). One subcellular system that connects all others is the signaling network. Segments of this network have been a favorite within systems biology and have been the frequent subject of mathematical modeling (2, 3). In developing these models, investigators must deal with estimating parameters that are rarely available from the literature. These parameters include reaction rate constants, initial values of biochemical concentrations, Hill coefficients, etc. (4). The current best solution to this problem is to obtain measurements of the molecules in the pathway under study from timed, synchronized samples, largely using antibody-based assays such as Western blots (5), ELISA (6), and flow cytometry (7).

Regulation of the cell cycle is an attractive biological process for mathematic modeling of biochemical networks because it is well studied at the biochemical and genetic level; it is a closed loop with a demarked beginning and end, and it is easily manipulated with a plethora of drugs, gene transfer, siRNA, and gene-deletion strategies. Of the several ways in which biomolecules can be measured, cytometry offers significant advantages. Foremost is that asynchronous populations can be interrogated, and, by virtue of that asynchrony, cells will be distributed at each quantitative position along a programmed expression profile. Therefore, cytometric data contain the dynamic expression profile for the interrogated molecules, and thus, the output of the cell's expression program over the lifetime of the program from a known beginning to a known end. The second advantage of modern cytometry is that several parameters can be measured simultaneously, and therefore, with the appropriate choice of parameters, each expression profile can be unambiguously derived from the primary data. Such data sets are a rich source of information with outputs that are exactly equivalent in form to the outputs of dynamical mathematical models of cell-cycle regulation (8–30). The purpose of this chapter is to present a data set that contains a resolvable expression profile for cyclin A, a cell-cycle regulating and regulated protein, and a software tool, CytoSys, which assists in deriving this expression profile. The principles of cytometry, discussed above, are also illustrated.

2. Materials

2.1. Flow Cytometry

For a comprehensive introduction to flow cytometry, see refs. (31–33). For a methodological overview of the technology and science applied to the mammalian cell cycle, see refs. (34, 35).

The major "wet science" methodological concerns relevant to this chapter are preparing a single-cell suspension, fixation, and staining for intracellular antigens (since most proteins of interest within cell-cycle studies are expressed inside the cell). There are several aspects in each of these areas to be concerned with, but the area has been well reviewed (36–39).

Fluorescent antibodies are intermediate reporters; the amount of fluorescence signal is proportional to the abundance of the epitope in the cell, and therefore becomes a proxy for the quantity of the epitope and often the larger molecular context (e.g., the whole protein containing the epitope or even a protein complex). It is important to note that in mathematical modeling of biochemical reactions, use of methods that report relative quantities are an intermediate step. Models will be most meaningful when they are realized in absolute quantities – molecules and molarity. We will report elsewhere an approach toward that goal. The work described here yields expression profiles in relative quantities.

The approach to extracting expression profiles, demonstrated here, was first explored by Jacobberger et al. (40). Cyclin B1 was quantified as a function of DNA content, and the levels expressed in G1, S, and G2 were quantified. In that study, we introduced the idea that the cyclin B1 distributions in G1 and G2, which were not resolved in time by DNA content (as is S phase), could be reduced to a phase-specific kinetic expression profile by fitting a series of Gaussian distributions based on more uniform expression (a single Gaussian) from another, more uniform cell-cycle compartment. The mitotic cells were used to obtain the most narrow log-normal distribution within the cell cycle and the G1 and G2 phase expression profiles were obtained by plotting the centers of a Gaussian series vs. average phase times for typical cells. As stated above, S phase expression was uniquely resolved as a function of DNA content. A similar approach was used and extended by Frisa and Jacobberger (41). In that study, the frequency information within regions set across the bivariate (DNA vs. cyclin B1) distributions was used as a surrogate for time spent within the expression range of each region. Using the frequency information puts the ideas presented in ref. (40) into action. However, fitting distributions manually (e.g., in spreadsheet programs) is tedious and clumsy in practice. To combine region setting and Gaussian fitting in a semi-automated manner, we have developed both a working methodology and a software program to extract expression profiles from multivariate distributions of cell-cycle regulated epitopes. While the measurements made by this approach are relative, they are correlated.

The analysis that we will step through here utilizes data from a sample of exponentially growing K562 cells that were fixed and stained for cyclin B1, cyclin A2, phospho-S10-histone H3 (pHH3), and DNA content as described in ref. (39). Generally, analyses of this type will include:

1. A proliferating population of cells, sampled at one or more times, with or without some treatments, fixed either by formaldehyde/MeOH or formaldehyde/detergent methods (36).
2. Staining for one or more cell-cycle-regulated epitopes. The focus could be on the protein, as in this case, cyclins A2 and B1, or the epitope – e.g., pHH3. The samples do not have to include DNA content, although this facilitates the analysis here by isolating the three major interphase subphases (G1, S, and G2). The element that is necessary for a complete analysis is that the expression should be able to be followed as a closed loop without ambiguity.
3. List mode data with enough total events such that each region defining a data subset is populated with statistically significant data. In making this determination, if cells are not limiting, then each region should contain approximately 100–400 events (so that the coefficient of variation for accurate detection is between 5 and 10%). However, if cells are limiting, then keep in mind that statistical significance is positively affected by the number of cells in the bounding regions, and the target values of 100–400 cells can be substantially lower.
4. Compensation and background, nonspecific antibody-binding controls. For compensation controls, either antibody-binding beads or cells stained singly with each antibody for the probes with spectral overlap problems can be used. Nonspecific binding has to be determined independently using each antibody in indirect assays with and without the specific antibody. These results can then be mapped to the data using multiple conjugated primary antibodies in direct staining procedures. If there are cells that are essentially biologically negative within the population, these are the best nonspecific binding controls.

2.2. Analytical Subsystems

The time profile extraction method is implemented in MATLAB-based graphical user interface (GUI) software, named CytoSys. CytoSys has been designed for users who are knowledgeable in the analysis of cytometry data. Specifically, CytoSys expects data that have been preprocessed and conditioned (explained in Subheading 2). The data flow diagram in Fig. 1 places CytoSys in context. The processes that take place in each of the blocks indicated are: (a) *data capture* using a cytometer and subsequently saved as readable list mode files (FCS format is preferred); (b) *data conditioning*; data parsing (cluster identification), and expression and frequency extraction using listmode processing

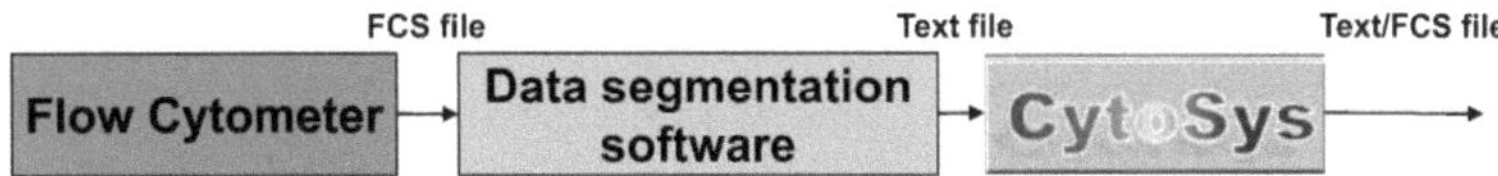

Fig. 1. Data processing diagram: Cytometry data is preprocessed and clustered using a data segmentation program such as WinList™ or FlowJo™ and written into text files, which are input to CytoSys.

software (here, we use WinList™ (Verity Software House)); (c) assembly of expression and frequency data into expression profiles within the context of cell cycle using CytoSys.

As stated before, the frequency of events in any region of the cytometric data space is proportional to the time cells in the population existed in that state space. Therefore, the derivable time profile of measured variables consists of unequally spaced points in time. The frequency axis (hereafter referred to as time axis) is itself relative and is in terms of percentage of cell cycle. CytoSys provides the ability to linearly interpolate between these unequally spaced points to create data that are continuous, i.e., a single curve.

3. Methods

We have developed a methodology to convert statically sampled multivariable cytometry data into time profiles using heuristics guided by literature and an experienced biologist. The germ of the idea behind our methodology here, as mentioned in Subheading 2.1, was initially discussed and performed by hand as in refs. (40, 41). We describe the methodology by example (details in Note 1) in this section, starting with the raw data from flow cytometry (46).

3.1. Data Preprocessing

Multivariate cytometry data must be preprocessed before it can be available for time profile extraction, as follows:

(a) Fluorescence compensation: Subtracting overlapping signal (light) coming from one of two fluorophores. Since this is a common cytometric procedure and can be done either in software (as was done here) or at the instrument, we will not discuss or illustrate this.

(b) Doublet discrimination: This is perhaps unique to cell-cycle analysis because it is easily performed. In this procedure, cell aggregates (such as due to cell-to-cell adhesion) are removed from the analysis by virtue of their DNA content integrated to peak signal value ratios. This is also a very standard procedure.

(c) Subtraction of nonspecific antibody binding: This procedure removes noise generated by nonspecific binding of antibodies to cellular constituents. In the example processed here, cyclin A2 is not expressed in G1 (i.e., expressed below the level of detection). Therefore, we can use the direct relationship of the size (light scatter) of G1 cells to background antibody binding to subtract this component. In WinList, this is set up exactly like spectral compensation except that the subtraction coefficients are set by viewing a plot of the parameter to be subtracted vs. light scatter of G1 cells. An example of the procedure can be viewed (41).

All these procedures can be done with standard listmode processing software.

3.2. Data Segmentation

We start with the data after the preprocessing steps delineated in Subheading 2.1 have been completed. Regions are the most atomic feature of the data set. A *region* is a segmented portion of the data. In Fig. 2a, the data enclosed by gray boundaries are regions.

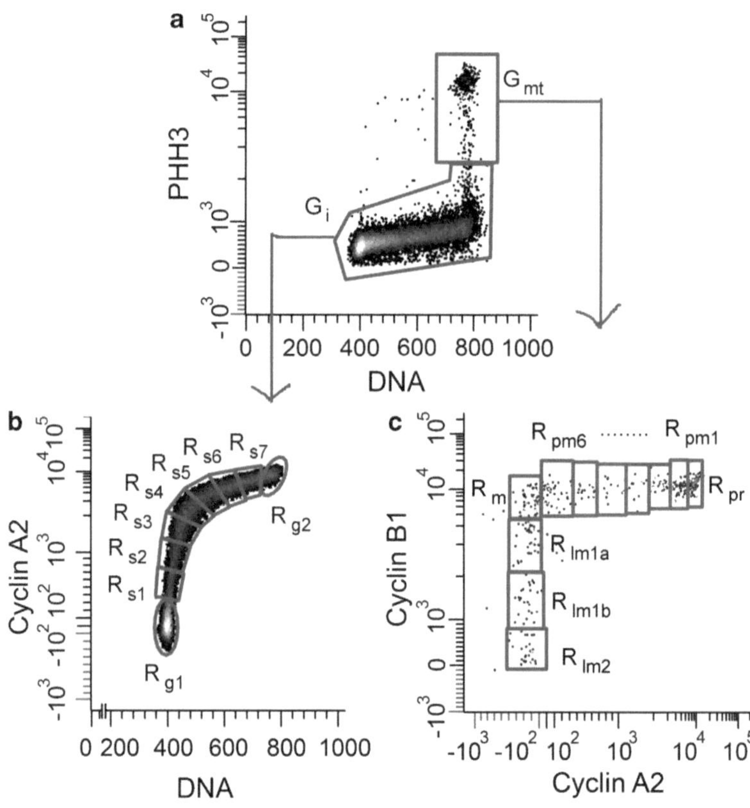

Fig. 2. Data clustering using WinList™. (**a**) Separation of the mitotics from the cells in interphase using the data variation in a PHH3 vs. DNA scatter plot. (**b**) Interphase: The lower oval is G1; The *upper oval* is G2; all the other regions in between are S (**c**) Mitosis: This plot must be read from the upper right to the lower left, following the data clusters. The *rectangular box* on the extreme *right* is prophase.

The goal is to first reduce multidimensional data to meaningful two-parameter histograms that will allow us to move along a multidimensional centroid without ambiguity. Therefore, our first sets of regions are gates. It is common to mix the terms gate and region, but here they are specific. In Fig. 2a, the gates G_{mt} and G_i define mitotic and interphase populations, respectively. In general, gates and regions are set from prior knowledge (as in Fig. 2b), but the general rules for region setting sufficient for any data set are: (1) boundaries are set at the interface of a two-dimensional Gaussian cluster (e.g., R_{g1} and R_{g2} in Fig. 2a); (2) boundaries are set when two-dimensional data change "directions" (e.g., R_{pm6}, R_m, and R_{lm1b} define a change in direction for the relationship between cyclin B1 and cyclin A2 in mitotic cells in Fig. 2c); (3) long stretches of approximately unidirectional data may be divided by an arbitrary number of boundaries for convenience (e.g., R_{s1} through R_{s3} in Fig. 2b). Although the terms segment, bin, gate, cluster, and region are used somewhat interchangeably in the literature, here we use "region" for the lowest level of data organization.

Operationally, we do the following:

1. Using preprocessed data, plot PHH3 vs. DNA content, and set regions (R_i and R_{mt}) as in Fig. 2a, that divide the cell cycle into interphase and mitotic cells (Fig. 2a).
2. Plot cyclin A2 vs. DNA content, gated through R_i, then set regions (R_{g1}, R_{g2}) around the Gaussian clusters that are equivalent to G1 and G2 cells as shown in Fig. 2b.
3. Segment S phase by a series of regions that follow the shape of S phase defined by cyclin A vs. DNA content. Cyclin A is expressed linearly through S, which creates a curve on a log-linear plot. These regions are termed R_{s1}...R_{sn}.
4. Segment M phase by setting a region, R_{pr}, around the Gaussian cluster of double positive cells. The Gaussian nature of the cluster is difficult to see in Fig. 2c. Figure 5 (top) shows a single parameter view that is a better illustration. Create a region, R_m, at the corner of the switch in data direction. Then, create a region, R_{lm2} at the bottom of the cyclin B1 tail. Although difficult to see here, in populations that are enhanced for cytokinetic cells, a distinct Gaussian cluster can be observed. Finally, we create arbitrary numbers of regions that divide the stretches of data between R_{pr} and R_m and R_m and R_{lm2}. The stretch between R_{pr} and R_m are termed R_{pm1}...R_{pmn} and those between R_m and R_{lm2} are termed R_{lm1a}...R_{lm1b}, etc. The naming is arbitrary. These names were chosen because R_{pr} is enriched for prophase cells; R_{pmx}s are enriched for prometaphase; R_ms are enriched for metaphase; R_{lm1x}s are enriched for anaphase and telophase, and R_{lm2} is enriched for telophase and cytokinetic cells.

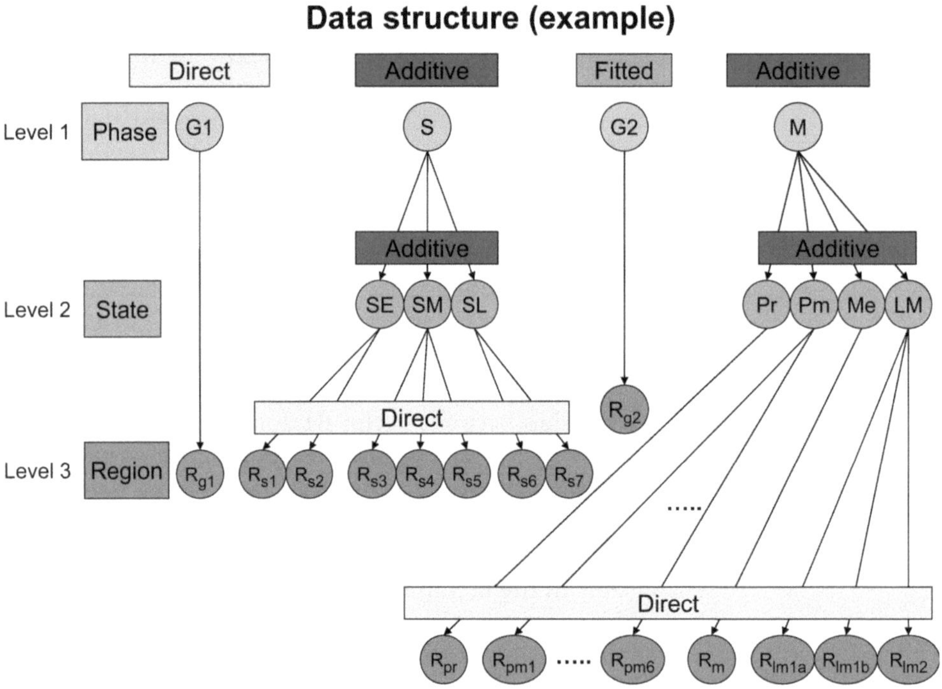

Fig. 3. CytoSys data structure: Data is organized into phases, and each phase into constituent subphases, each subphase into constituent regions. The level at which each phase is processed and the type of processing are indicated in rectangular boxes that indicate whether the level was treated as a consolidated whole (G1) or whether the level was noise corrected using a Gaussian fitting process (G2) or whether we drilled down from the given level to an additional level of detail (S and M).

Segmentation is ripe for automation using image-processing algorithms. We have tried a few commercially available routines, but have not found any that work consistently.

After regions are set, grouping into states or subphases is essential for the extraction of dynamics. Accordingly, we have created a hierarchical data structure as shown in Fig. 3 for organizing information extract from low-level regions and assigning it to higher levels.

3.3. Data Structure

The hierarchy of data within the CytoSys program has three levels. We do not include gates. The lowest level, as stated above, is composed of regions. States are one level above regions, and one or more region will form a state. States are the unit of organization with some biological meaning. In its simplest form, a cell state defined by the level and direction (net synthesis or net degradation) of a biomolecule. A state can be defined by absence of expression as well, and there is the possibility that a state is defined by one region. Thus, the region R_{g1} defines the state G1. When a state is defined by a single region, then regions and states appear equivalent (but see below). The highest organization level is defined by the cell-cycle phases, G1, S, G2, and M. States aggregate to phases. Using Fig. 3, the G1 phase, G1 state, and R_{g1}, i.e., the phase, state, and region appear to be equal. However, the region, R_{g1}, is defined operationally on a specific data set as in Fig. 2. The state, G1, is defined by one or more analytical parameters within the specific analysis – here, the absence of cyclin A2, the lowest levels of phosphorylation of histone H3, and DNA content. The G1 phase includes all of the known biological properties for G1 that can be appropriately inferred for K562 cells. Therefore, they are not equivalent but related.

In cases where the highest level has no distributional information with respect to a particular biomolecule, the expression level of the defining parameter within the phase is constant (this constant value will later be modified – see Notes 2 and 3). The G1 phase for cyclin A2 is such a case. In the data hierarchical arrangement, this processing mode is termed "direct." When the distribution within the highest level is defined by one or more states, defined by multiple regions, we term the processing mode "additive." Finally, when there is distributional information within the highest level that is defined by only one state and only one region, and we can extract that information by multiple Gaussian fitting, as described by Jacobberger et al. (39), we term the processing mode "fitted."

3.4. Primary Information Extraction

The frequency with which we randomly observe cells within a defined cell-cycle compartment (phase, state, or region, as we have defined them here) is proportional to the time spent in that compartment. This is axiomatic, from first principles, and underlies the value of cell-cycle analysis by cytometry. This property of cytometric data is key to our analysis here. Therefore, after regions and hierarchal relationships have been defined and enacted as described in detail below, a set of center measures (mean or median) for each parameter and the frequency with which it occurs within a specific region will be computed by CytoSys.

3.4.1. Data Aggregation or Conditioning

Data aggregation or conditioning is operational and the purpose is to track and work with the segmented data in an ordered but flexible manner. In Subheading 3.2, we described gating and region setting (Fig. 2), and in Subheading 3.3, we described the overall hierarchical structure that we use to organize the primary data. To facilitate further processing, we created the file storage structure shown in Fig. 4. First, a Data Directory is the repository of all data sets. A data set is a directory that contains three subdirectories: PROFILES, RESULTS, and WORK. PROFILES contains list mode data files specific to regions from the segmentation step. It also contains a variable name file with the names of the parameter

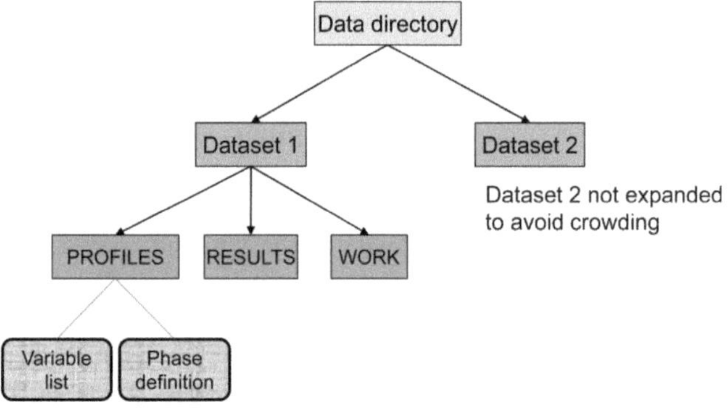

Fig. 4. Folder structure in CytoSys.

variables for which expression profiles will be computed. In this example, PROFILES contains a file, Rg1.txt that has the form:

Event	P1	P2	P3	P4	P5
1	u1	v1	w1	x1	y1
2	u2	v2	w2	x2	y2

where cyclin A2 is P2. Therefore, PROFILES contains a file, "variable_list.txt" with the information, "cyclin A2, 2," denoting that the biological name of the parameter is "cyclin A2" and the value is in the second column. This directory will also contain a file, "phase_definition.txt," which will contain successive lines, one for each high-level phase defined in Fig. 3. Each line contains directives for processing from a top-down perspective. The first term of the line is the phase followed by directives for states, regions, processing type. Thus, in our example, the first line of "phase_definition.txt" is: "G1;[G1];{[Rg1]};[0]:{}:{}." This identifies a G1 phase with a single state, G1, a single associated region file, Rg1.txt, and a directive, "[0]," for "direct processing" as illustrated in Fig. 3. If the directive is "additive," a more complex set of directives are required. For example, the second line is "S;[SE,SM,SL];{[Rs1,Rs2].[Rs3,Rs4,Rs5].[Rs6,Rs7]};[2]:{2.2.2}:{0.0.0.0.0.0.0}" with a definition for an S phase, three states, with a set of associated files that will be processed additively. [2] is the additive processing at the phase level; {2.2.2} directs additive processing at the state level, and the final term, {0…0}, directs processing at the region level. Processing is ordered from left to right, equating to beginning to end in the cell cycle. The purpose of this structure is to retain data in the preprocessed/prestatistic state with a direct mapping (moving through the cell cycle from beginning to end) from the cytometry list mode processing software to CytoSys. The RESULTS and WORK directories are automatically created during CytoSys execution. The results of the analysis using CytoSys can be stored in RESULTS. Every variable created or modified during the entire analysis is automatically stored in the WORK folder during a CytoSys run. This is essentially the workspace, and can be utilized to recreate an earlier analysis.

3.4.2. A Specific Example for Cyclin A2: A Priori Knowledge

It is known that the DNA content is constant in G1, elevated in S, and constant again in G2 and M at twice the level of G1. Indeed, synthesis of DNA defines S and the two gaps between S and M. We do not have that kind of knowledge about the expression of the other parameters in this example analysis, but we can intuitively sketch the expected expression profile from a cytometric plot of cyclin A2 vs. DNA content. Additionally, previously published Western blot data at different time points for synchronized cells provides a general idea of the expected profiles. From previously published work, we expect cyclin A2 levels to be low in G1, rise linearly as a function of DNA content in S, rise in G2, and fall in M. Similarly, we expect cyclin B1 to rise in an exponential fashion from some point in G1 through G2 and decrease dramatically in M, and finally, we expect pHH3 to increase like a housekeeping function in interphase but rise dramatically in M and decrease at the end of mitosis.

1. G1 phase: To process data in G1 phase to extract time profile, we adopt the following steps.
   Step: The G1 list file, Rg1.txt is processed using the Variable list: "cyclin A2,2" and the Phase definition: G1;[G1];{[Rg1]}; [0]:{2}:{0}. CytoSys then calculates the number of events in Rg1.txt and the mean or median value for the expression of cyclin A2 within that region, state, and phase, since they are equivalent in terms of cyclin A2.
2. S Phase: In sharp contrast to G1, S phase is processed additively, region by region (Fig. 2a). In this case, we have rather artificially created three "states" (SE, SM, and SL) for illustration in Fig. 3 and explanatory text. In practice, we do not define states by time, but rather by demarcated expression and in S the boundaries are completely arbitrary and not demarcated within the context of DNA content or cyclin A2. Therefore, in practice, S phase would have one state, S1.
   Step: The seven "S" list files from the PROFILES storage area are sequentially processed to enumerate cells and compute the median or mean value for cyclin A2 per region, using the variable list file as above, and the Phase definition line: S;[S1]; {[Rs1,Rs2,Rs3,Rs4,Rs5,Rs6,Rs7]};[2]:{2}:{0.0.0.0.0.0.0}.
3. G2 Phase: Cyclin A2 rises in G2. This can be determined by the broader distribution of cyclin A2 in G2 relative to mitotic cells or a small region in S phase. Further, we have shown that when cells are pulse labeled with BrdU, the label appears first at the bottom of the G2 cyclin A2 distribution, and enters the top of the distribution at a later time (42). We could segment this distribution into tertiles or quartiles, but a more exact approach is to fit the distribution to a series of Gaussian distributions to provide a number of significant levels. This is done by processing a

single region file, "Rg2.txt" with the "Fitted" directive. This will be given its own Subheading 3.4.3. The phase definition file entry looks like this: G2;[G2];{[Rg2]};[1]:{}:{}, where the directive "1" indicates Fitted processing for the phase and the empty state and region directives negate any lower level processing.
Step: see Subheading 3.4.3.

4. M Phase: Mitotic cells generally comprise 1–4% of cycling tissue culture cells. For this example, mitotic cells can be subdivided by setting regions on plots of cyclin B1 vs. cyclin A2 (Fig. 2c). The reason for the complex pattern that cells enter mitosis with high levels of both cyclins and cyclins A2 and B1 are sequentially degraded by the anaphase promoting complex during mitosis. For completeness and clarity, we show the complete segmentation and processing for M resolving both cyclin A2 and B1. However, for cyclin A2 since it reaches a low and uniform level in regions, R_{lm1a}, R_{lm1b}, and R_{lm2}, we could have set one region, $R_{m\text{-}lm}$, and the expression profile would be the same. Here, we have one phase, M, 4 states, and 11 regions. Here, the states begin to mean something. For processing:
Step: The "M" list files from the PROFILES storage area are sequentially processed to enumerate cells and compute the median or mean value for cyclin A2 per region with the phase definition line: M;[M1,M2,M3,M4];{[Rpr].[Rpm1,Rpm2, Rpm3,Rpm4,Rpm5,Rpm6].[Rm].[Rlm1a,Rlm1b,Rlm2]}; [2]:{2.2.2.2}:{0.0.0.0.0.0.0.0.0.0.0}.

3.4.3. Gaussian Fitting

We now explore Gaussian curve-fitting. In our segmentation of multidimensional data, we have essentially ignored statistical variation of the data. For cyclin A2 in G1, variation is unimportant because G1 cells do not express cyclin A2 above our ability to detect it, and therefore, whatever relative value we assign to G1, it is constant and at the postsegmentation stage, a single number. Similarly for S, the covariable DNA content allows us to move along the two-dimensional data center and obtain significant estimates of expression and frequency. However, at the juncture between G1 and S, the failure to address variation results in a mismatch between the expression profile in G1 and expression in early S. This could be resolved by two-dimensional Gaussian fitting, which we have not implemented. We have implemented one–dimensional fitting, which we can apply to any single parameter (variable). That is especially useful to resolve cyclin A2 expression in G2, which in its simplest display, presents as a single "broad" Gaussian distribution. This can be fit to a series of Gaussian distributions, provided we have an estimate of the mean and variation of a more narrow distribution that is biologically significant.

The distribution of cyclin A2 in G2 cells varies as a function of programmed expression and the fact that asynchronous cells exist at points along the kinetic expression profile. That is what we wish to resolve here. The distribution can also be broadened by cells executing programs with different expression profiles. We have not dealt with that here, and populations where this is significant may not be good candidates for this analysis. Additionally, there is variation in the measurements due to electronic noise, photon counting statistics, stray light, auto-fluorescence, nonspecific binding of antibody, specific antibody binding, and instrumental flow characteristics. While we could get some estimates of each of these, in practice, this is not done. It is fair to say that the signal in G2 is heavily dominated by specific antibody binding and therefore the largest source of error is the variation of specific antibody binding. This is likely to be due to the local environment in the area of each instance of an epitope. The total epitope number is distributed between completely masked to "not sterically hindered." At present, the best we can do is measure a population with the least variance from all sources; fit that to a Gaussian distribution, then use those statistics to fit the entire G2 distribution. The cells of the Pr state (Figs. 2 and 3) are such cells.

Our analysis is performed using log values due to the log-normal nature of the data. From this point, we will discuss the problem in terms of a normal distribution, keeping in mind that it is a normal distribution of the log values. We first obtain the standard deviation of common log prophase data (σ_{pr}) (Fig. 5). We then frame a constrained nonlinear optimization problem ("optimization" used in an engineering context) to minimize the mean squared error between the data histogram of cyclin A2 (Fig. 6) and the sum of multiple successive weighted Gaussian fits that a user imposes. During fitting, each model component Gaussian standard deviation is constrained to be "approximately" equal to the standard deviation of the prophase cells (i.e., $\sigma \approx \sigma_{pr}$) (40).

We explored three different approaches in handling the optimization problem – two automated and one manual. The automated

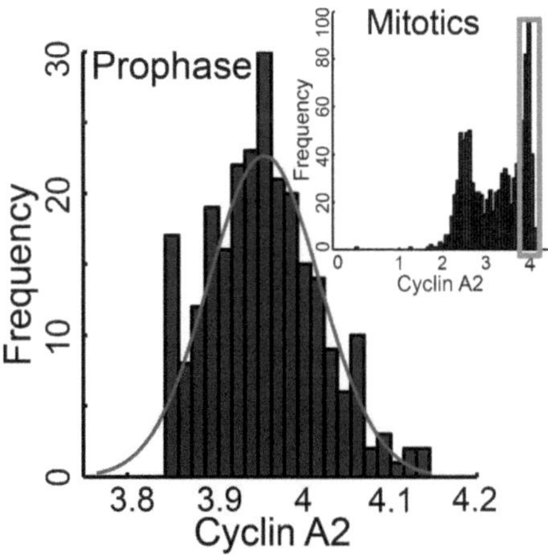

Fig. 5. Prophase cells histogram with mitotics as *inset*. The prophase cells are marked in the mitotics histogram with a *rectangular box*.

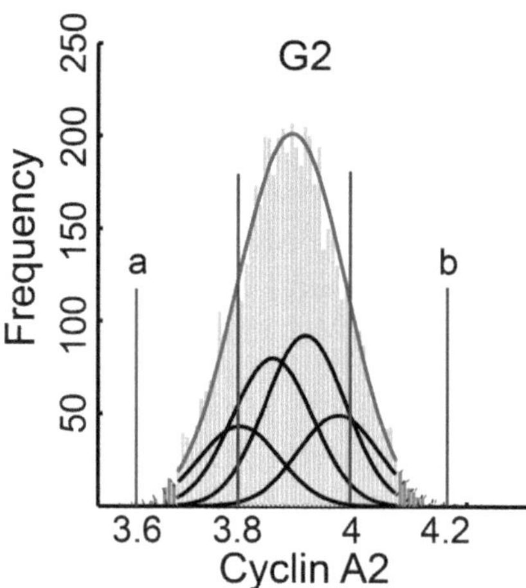

Fig. 6. Example of Gaussian fit to cyclin A2 (log scale) histogram for cells in G2. The *middle vertical lines* mark the entry into G2 and the exit from G2. Lines "**a**" and "**b**" mark the beginning and end of our data distribution.

methods need further improvement before they can be reliably used. The first automated method uses a constrained optimization routine for finding the solution, and this may work in theory but may not provide a meaningful fit. The second automated method is a kernel density estimator, a nonparametric way of fitting the data histogram. Optimization settings are severely curtailed in this option currently.

The manual option, which was used for the results presented here, allows the user to pick the optimal fit by adjusting the number of Gaussians and the distance between them manually and then deciding if the fit is good enough by several checks as discussed below.

The manual-fitting process is started by fitting a single Gaussian "best fit" to the data and computing the error. This is centered at or near the G2 entry value (last value for cyclin A2 at the end of S phase, computed from a special, narrow region) and we optimize the fit to the left side of the G2 histogram (Fig. 6). Next, a number of Gaussian components and the distance between them are defined so that the last component is centered on the G2 exit value. This is defined by a special region in M phase that is defined narrowly as the first elevation in PHH3. A best fit is then found by manipulating the mean, center-to-center distance, weights (area under the curve), and variances assigned to each Gaussian component so as to minimize the mean square error between the data histogram and the sum of the multiple Gaussian fits. In practice, since this is a manual process, we have done this first fitting two components, then adding additional components consecutively until we obtain a fit that is good enough without exploring all possibilities. Special care is taken for the mean of the first component to be close to the G2 entry value, and the last

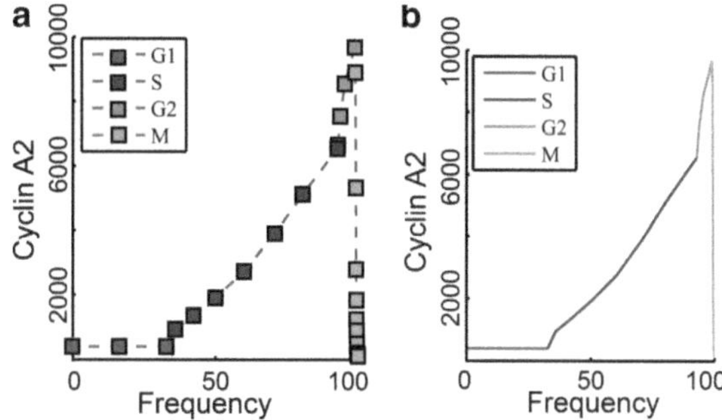

Fig. 7. Some results from CytoSys. The extraction of the cyclin A2 expression profile is presented here. (**a**) The initial time profile extracted is shown. (**b**) A linearly interpolated version of the time profile is presented.

Gaussian to be close to the G2 exit value (Fig. 6). The cyclin A data histogram of Pr cells is shown in Fig. 5. The inset shows the histogram of all the mitotic cells. The prophase cells are marked by the rectangle in the inset, and a resulting fit is shown in Fig. 7. Below, we define the process in mathematical terms.

Problem formulation: monotonic weight constraints: We start with some basic definitions. Let

n_G = number of Gaussians used in the fit

n_{bins} = number of bins used in plotting the histogram of G2 data

a = minimum significant value of data on "x" axis

b = maximum significant value of data on "x" axis

μ_i = mean of the ith Gaussian fit to the data, where $i = 1, 2, \ldots, n_G$

A_{ab} = area under the data histogram between points a and b

$P_i(x)$ = probability density function (PDF) of the ith Gaussian fit to the data

$$P_i(X) = \frac{w_i}{\sigma\sqrt{2\pi}} e^{-\frac{(x-\mu_i)^2}{2\sigma^2}}, \quad i = 1, 2, \ldots, n_G, \tag{1}$$

where w_i is the weight of the ith Gaussian fit.

We also denote by ω_i the cumulatively summed peaks of the Gaussians.

$$\omega_i = \sum_{j=1}^{i} \frac{w_i}{\sigma\sqrt{2\pi}}, \quad i = 1, 2, \ldots, n_G. \tag{2}$$

The optimization of Gaussian fits can be formulated as shown below:

$$\min_{\varepsilon, d, w_i} \left(A_{ab} - \sum_{i}^{n_G} P_i(x) \right), \quad i = 1, 2, \ldots, n_G \tag{3a}$$

and

$$\min_{\varepsilon, d, w_i} (h_j - g_j), \quad i = 1, 2, \ldots, n_G; \forall j = 1, 2, \ldots, n_{bins}, \tag{3b}$$

where g_j is the jth element of the overall Gaussian time curve (which has been linearly interpolated to have n_{bins} values) and h_j is the height (frequency) of the jth bin, subject to the constraints:

$$\begin{aligned} 0 \le w_j \le w_{j_{\max}}, \quad & j = 1,2,\ldots,n_{\mathrm{G}}, \\ w_j \le w_{j+1} \quad \text{for} \quad & j \le n_{\mathrm{H}} - 1, \\ w_j \le w_{j+1} \quad \text{for} \quad & j \le n_{\mathrm{H}}, \end{aligned} \tag{4}$$

where n_{H} is the index of the Gaussian that is assigned the highest weight, and

$$n_{\mathrm{H}} = \begin{cases} \dfrac{n_{\mathrm{G}}}{2} + 1 & \text{for} \quad n_{\mathrm{G}} \text{ even,} \\ \dfrac{n_{\mathrm{G}} + 1}{2} & \text{for} \quad n_{\mathrm{G}} \text{ odd,} \end{cases} \tag{5}$$

i.e., when n_{G} is even, the later of the two central Gaussians is assigned the highest weight, and when n_{G} is odd, the central Gaussian is assigned the highest weight. The weights are represented in vector notation as:

$$\underline{w} = \begin{bmatrix} w_1 \\ w_2 \\ \cdots \\ w_{n_{\mathrm{G}}} \end{bmatrix}_{n_{\mathrm{G}} \times 1}.$$

The inequality constraints (4) can be written in matrix notation as: $A\underline{\gamma} \le \underline{0}$.

Here, γ is the vector of optimization variables: $\underline{\gamma} = \begin{bmatrix} \underline{w} \\ \varepsilon \\ d \end{bmatrix}_{(n_{\mathrm{G}}+2)\times 1}$.

For convenience of notation, we show the matrix equation for the weights portion only.

Let A_{w} denote the matrix of coefficients of the weights from the inequality constraint, i.e., $A_{\mathrm{w}}\underline{w} \le \underline{0}$. Here, A_{w} has the following structure:

$$A_{\mathrm{w}} = \begin{bmatrix} \begin{pmatrix} 1 & -1 & & \\ & \ddots & \ddots & \\ & & 1 & -1 \end{pmatrix} & \underline{0} \\ \underline{0} & \begin{pmatrix} -1 & 1 & & \\ & \ddots & \ddots & \\ & & -1 & 1 \end{pmatrix} \end{bmatrix}_{(n_{\mathrm{G}}-1)\times n_{\mathrm{G}}}.$$

The constraints on ε are $-\sigma_{\mathrm{pr}} / 3 \le \varepsilon \le \sigma_{\mathrm{pr}} / 3$.

The remaining constraints are:

The mean of the first Gaussian (μ_1) is a noninteger multiple (κ) of standard deviations of prophase (σ_{pro}) distant from the minimum significant data value (a).

$$\mu_1 = a + \kappa\sigma_{pr}.$$

The distance between consecutive Gaussians is "d," which is bounded as follows:

$$\sigma_{pr} <= d <= 2\sigma_{pr}.$$

The standard deviation of all the Gaussians is given by:

$$\sigma = \sigma_{pr} + \varepsilon.$$

Solving this optimization problem generates Gaussian means as y values and cumulatively summed Gaussian peaks as x values for the G2 segment of the time curve. This segment must then be incorporated into the overall time curve.

3.4.4. Expression Profile Concatenation, Interpolation, and Scaling

Final expression profile generation requires us to address a few points. These are discussed below:

1. Constant value persisting in a phase/subphase: We impose additional time points for phases that have one (constant) expression value. This is done to preserve the constant shape and to model the phase interfaces to something that better fits the data than interpolation between distant time points (center of the constant region to center of the next region). For this purpose, the mean or median value of the region is repeated at or near the beginning of the state or phase and again at or near the end of the phase.
2. Cell count increase in mitosis: If there is a measured value at the end of the expression profile that is considered biologically different from zero, we divide that value into half to account for reduction at cell division. In other words, taking the example of DNA, right before cytokinesis, if there are x molecules of DNA in the parent cell, then right after cytokinesis – there will be $x/2$ molecules of DNA in each of the daughter cells. This should be true for most of the expression profiles, and hence we can impose the restriction that $y(t_0) = y(t_f)/2$, where t_0 is the point where G1 begins, and t_f is the point at the end of cytokinesis, and $y(t)$ is the computed expression value.
3. Profile completion, interpolation, and scaling: We next focus on refining the constructed expression profile. CytoSys creates four vectors with a single computed expression value for each region correlated with the number of events within each region representing the four cell-cycle phases. The goal is to

concatenate these and replace the event number with the centers of the cumulative frequency represented by the addition of each element to the concatenated vector. Thus, for each expression value, *y* (cyclin A2), the positioning of y is centered within the x time segment, represented by the event number relative to all tallied events. After this has been accomplished, there is a problem at the points of intersection of phases. The expression profile between, for example, the last element in S and the first element in G2 has shape that is often not well represented by standard linear interpolation. To avoid such issues, we have included options, where necessary, to insert additional points at the beginning and ending segments in a phase and assign an arbitrary *y*-value (often, a linear interpolation of the two adjacent values).

The complete expression profile for cyclin A2 is shown in Fig. 7a. CytoSys includes a linear interpolator that allows us to reproduce the extracted expression profile with an arbitrary but defined number of time points. In addition, we can also define an equation in CytoSys to scale the expression profile to any other expression profile (48).

3.5. Conclusion

Our intent here was to present a methodology and a program (Fig. 8) to extract the programmed expression of any biochemical activity that can be measured by cytometry. We have tested this approach for reproducibility (same operator, same file, repeated analyses, and same operator, different files on the same biological sample) and find it to be generally reliable (the full statistical results will be reported elsewhere). Figure 7 is a good representation of the cyclin A2 expression for exponentially growing K562 cells. We have undertaken this work to support modeling efforts as discussed in Subheading 1. We present below our progress in calibrating a published cell-cycle model (15) in Fig. 9.

3.5.1. Cell-Cycle Biology and Modeling

Eukaryotic cells share an ancient cell-cycle control system (18), characterized by a tightly regulated sequence of events that control and coordinate mass doubling, genome duplication, chromosome segregation, etc. This timing is governed by a core set of reactions involving cyclin-dependent kinases and proteolytic complexes that degrade cyclins among many other substrates. The process is monitored by "checkpoint" controls that can inhibit the forward-driving mechanisms until specific conditions are met. All this is manifested through a series of feedback/feedforward mechanisms. Checkpoints govern critical/irreversible events such as the transition between the uncommitted and committed G1 states. Similarly, further checkpoints must be crossed as a cell proceeds from the G2 phase into and through mitosis, ensuring exact chromosome segregation to daughter cells.

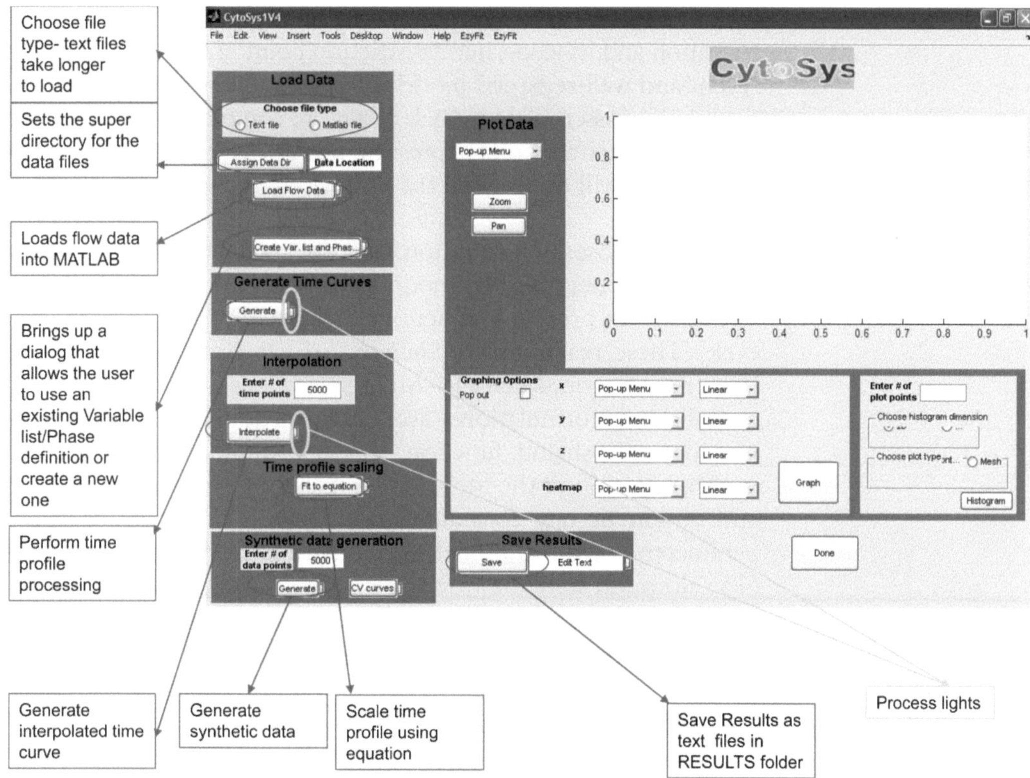

Fig. 8. CytoSys interface: many variable version.

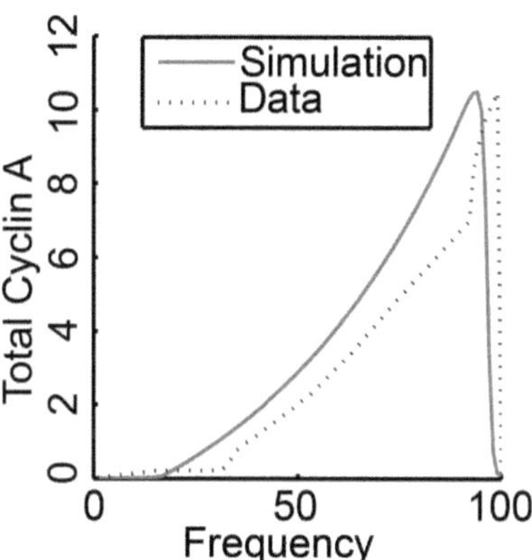

Fig. 9. Preliminary calibration attempts on for a mammalian cell cycle model (18) using a K562 expression profile for cyclin A.

The cell-cycle control system is an attractive problem in systems biology. A review of the literature reveals more than 40 mathematical models published over the past 20 years. Most models have focused on understanding small parts cell cycle, such as a minimal model of Cdc20 activation (43). Other groups have chosen to focus on specific cell-cycle phases and transitions such as the G1/S transition and its associated "restriction point" (17, 45). The most recent and well-regarded models (15, 18) have just begun the difficult task of assembling models of a more complete cell system. We believe that extraction of expression profiles as we have outlined in this chapter will be a significant aid for calibrating such models.

3.5.2. Calibration Using Results from CytoSys

For the purposes of illustration, here we consider a recent and the most complete published model (18). The model encodes many of the known protein interactions that govern the eukaryotic cell cycle. These reactions are encoded with a combination of rate laws (mass action, Michaelis–Menten, etc.) although several more complicated formulations are also introduced, such as the Goldbeter–Koshland function. On the whole, the mammalian implementation of the model comprises 13 differential equations and 66 kinetic rate constants (parameters). The majority of the parameters from this model were inherited from earlier attempts (15) although some new ones were introduced, which were then necessarily adjusted to bring the model into qualitative agreement with limited biological data available. A main challenge addressed in this paper is the acquisition of new and better data, which might be used to improve this model calibration.

The calibration challenge is understood through Fig. 9. While the Csikasz–Nagy model (18) predicts the timing and expected shapes of expression surprisingly well, any relationships to measured expression profiles have not been published. In view of what we know from cytometry, cyclin A is modeled as accumulating in G1, which we would postulate is too soon. Additionally, the model does not capture the sharp increase of cyclin A activity that we see during G2. These issues notwithstanding, our preliminary and modest success in calibrating the Csikasz–Nagy model is an encouraging step in the right direction.

CytoSys is part of our solution to the paucity of quantitative expression over time data that would support computational biology models. While no single systems biology group can claim to have a hold of the entire solution to this problem, we suggest that this methodology for extraction expression profiles of relative high fidelity is a contribution to a field that requires more contributors and more high-quality data, if the promise of systems biology (accurate mechanistic models of biological processes) is to be achieved.

4. Notes

1. Biological sample: The K562 cell line is the first human immortalized myelogenous leukemia cell line to be established (derived from a CML patient in blast crisis)(44, 49). In the

example experiment, K562 cells were cultured in RPMI-1640 with 10% fetal bovine serum. Exponentially growing cells (47) were fixed with MeOH, then labeled with fluorescent antibodies specifically reactive with cyclins A2 and B1, pHH3, and counterstained with DAPI to label DNA. Flow cytometry was on a BD BioSciences LSR II instrument set up with standard filters.

2. Additional CytoSys features: Within CytoSys, expression profiles can be scaled relative to a standard expression profile. For example, independent samples can be labeled with cyclin B1 or cyclin A2 and then reacted with the same fluorochrome-conjugated secondary antibody. With some caveats, the expression profiles derived from these two samples are directly comparable and biologically true, relative to each other. However, in any particular case, the reduced number of parameters may render the complete extraction process difficult or impossible (because of ambiguity). In that case, fragments of the expression profiles that are relevant and nonambiguous can be used to determine the biological correlation and that information can be used to transform the multiparametric data using multiple fluorochromes, which are not directly comparable but are completely resolved. We are preparing a research paper describing this process.

3. Cautions and final comments

   (a) Region setting is currently difficult in the list mode processing software that we use. What is needed are regions that snap to each other and are assured of nonoverlap. Some programs may do this, and therefore, if the type of work described here is something that will be done often, it would be worth investigating which do and which do not.

   (b) The use of Gaussian fitting currently is applied to single-parameter expression data for which regions cannot be set in two dimensions. The reason for this is that in two dimensions, one can see the entry and exit points and, using the geometric pattern of the data, determine the width of the region that will center the entry and exit points, and account for statistical variation at the ends of the phase or statewise expression distribution. This is not possible in one dimension. The next generation of this work should include two-dimensional Gaussian fitting, using the entry and exit points to assist with the third dimension.

   (c) The frequency of cells in the cell cycle for exponentially growing cells are overrepresented by a factor of 2 at the beginning of the cell cycle and that offset decays exponentially to a factor of 1 by the end of the cell cycle. We have not made that adjusted calculation in the output of Figs. 8 and 9.

References

1. Mesarovic, M.D., Sreenath, S.N. and Keene, J.D. (2004)Search for organising principles: understanding in systems biology. *IEE Syst. Biol. (Stevenage)* **1**, 19–27.
2. Chen, W.W., Schoeberl, B., Jasper, P.J., et al (2009) Input–output behavior of ErbB signaling pathways as revealed by a mass action model trained against dynamic data. *Mol. Syst. Biol.* **5**, 1–19.
3. Janes, K.A., Albeck, J.G., Gaudet, S., Sorger, P.K., Lauffenburger, D.A. and Yaffe, M.B. (2005) A systems model of signaling identifies a molecular basis set for cytokine-induced apoptosis. *Science* **310**, 1646–1653.
4. Voit, E.O. and Ferreira, A.E. (2000) Computational Analysis of Biochemical Systems: A Practical Guide for Biochemists and Molecular Biologists, first edition. Cambridge University Press, Cambridge.
5. Burnette, W.N. (1981) "Western blotting": electrophoretic transfer of proteins from sodium dodecyl sulfate – polyacrylamide gels to unmodified nitrocellulose and radiographic detection with antibody and radioiodinated protein A. *Anal. Biochem.* **112**, 195–203.
6. Leng, S., McElhaney, J., Walston, J., Xie, D., Fedarko, N. and Kuchel, G. (2008) Elisa and multiplex technologies for cytokine measurement in inflammation and aging research. *J. Gerontol. A Biol. Sci. Med. Sci.* **63**, 879–884.
7. Hawley, T.S. and Hawley, R.G. (2004) Methods in Molecular Biology: Flow Cytometry Protocols, second edition. Humana Press, Totowa, vol. 263, pp. 1–424.
8. Albieri, R., Barberis, M., Chiaradonna, F., Gaglio, D., Milanesi, L., Vanoni, M., Klipp, E. and Alberghina, L. (2009) Towards a systems biology approach to mammalian cell cycle: modeling the entrance into S phase of quiescent fibroblasts after serum stimulation. *BMC Bioinformatics* **10**, S16.
9. Pomerening, J.R., Kim, S.Y. and Ferrell Jr., J.E. (2005) Systems-level dissection of the cell-cycle oscillator: bypassing positive feedback produces damped oscillations. *Cell* **122**, 565–578.
10. Gonze, D. and Goldbeter, A. (2001) A model for a network of phosphorylation-dephosphorylation cycles displaying the dynamics of dominoes and clocks. *J. Theor. Biol.* **210**, 167–186.
11. Tyson, J.J. and Novak, B. (2001) Regulation of the eukaryotic cell cycle: molecular antagonism, hysteresis and irreversible transitions. *J.Theor. Biol.* **210**, 249–263.
12. Obeyesekere, M.N., Tecarro, E. and Lozano, G. (2004) Model predictions of MDM2 mediated cell regulation. *Cell Cycle* **3**, 655–661.
13. Bai, S., Goodrich, D., Thron, C.D., Tecarro, E. and Obeyesekere, M. (2003) Theoretical and experimental evidence for hysteresis in cell proliferation. *Cell Cycle* **2**, 46–52.
14. Qu, Z., Weiss, J.N. and MacLellan, W.R. (2003) Regulation of the mammalian cell: a model of the G1-to-S transition. *Amer. J. Physiol. Cell Physiol.* **284**, C349–C367.
15. Novak, B. and Tyson, J.J. (2004) A model for restriction point control of the mammalian cell cycle. *J. Theor. Biol.* **230**, 563–579.
16. Swat, M., Kel, A. and Herzel, H. (2004) Bifurcation analysis of the regulation modules of the G(1)/S transition. *Bioinformatics* **20**, 1506–1511.
17. Haberichter, T., Madge, B., Christopher, R.A., Yoshioka, N., Dhiman, A., Miller, R., Gendelman, R., Aksenov, S.V., Khalil, I.G. and Dowdy, S.F. (2007) A systems biology dynamical model of the mammalian G1 cell cycle progression. *Mol. Syst. Biol.* **3**, 84.
18. Csikasz-Nagy, A., Battogtokh, D., Chen, K.C., Novak, B. and Tyson, J.J. (2006) Analysis of a generic model of eukaryotic cell-cycle regulation. *Biophys. J.* **90**, 4361–4379.
19. Chen, K.C., Csikasz-Nagy, A., Gyorffy, B., Val, J., Novak, B. and Tyson, J.J. (2000) Kinetic analysis of a molecular model of the budding yeast cell cycle. *Mol. Biol. Cell* **11**, 369–391.
20. Sveiczer, A., Csikasz-Nagy, A., Gyorffy, B., Tyson, J.J. and Novak, B. (2000) Modeling the fission yeast cell cycle: quantized cycle times in Wee1/Cdc25delta mutant cells. *Proc. Natl. Acad. Sci. U.S.A.* **97**, 7865–7870.
21. Novak, B., Pataki, Z., Ciliberto, A. and Tyson, J.J. (2001) Mathematical model of the cell division cycle of fission yeast. *Chaos* **11**, 277–286.
22. Ciliberto, A., Novak, B. and Tyson, J.J. (2003) Mathematical model of the morphogenesis checkpoint in budding yeast. *J. Cell Biol.* **163**, 1243–1254.
23. Chen, K.C., Calzone, L., Csikasz-Nagy, A., Cross, F.R., Novak, B. and Tyson, J.J. (2004) Integrative analysis of cell cycle control in budding yeast. *Mol. Biol. Cell* **15**, 3841–3862.
24. Sveiczer, A., Tyson, J.J. and Novak, B. (2004) Modelling the fission yeast cell cycle. *Brief. Funct. Genomic Proteomic* **2**, 298–307.

25. Qu, Z., MacLellan, W.R. and Weiss, J.N. (2003) Dynamics of the cell cycle: checkpoints, sizers, and timers. *Biophys. J.* **85**, 3600–3611.
26. Qu, Z., Weiss, J.N. and MacLellan, W.R. (2004) Coordination of cell growth and cell division: a mathematical modeling study. *J. Cell Sci.* **117**, 4199–4207.
27. Steuer, R. (2004). Effects of stochasticity in model of the cell cycle: from quantized cell cycle times to noise-induced oscillations. *J. Theor. Biol.* **228**, 293–301.
28. Srividhya, J. and Gopinathan, M.S. (2006) A simple time delay model for eukaryotic cell cycle. *J. Theor. Biol.* **241**, 617–627.
29. Yang, L., Han, Z., MacLellan, W.R., Weiss, J.N. and Qu, Z. (2006) Linking cell division to cell growth in a spatiotemporal model of the cell cycle. *J. Theor. Biol.* **241**, 120–133.
30. Faure, A., Naldi, A., Chaouiya, C. and Thieffry, D. (2006) Dynamical analysis of a generic boolean model of the control of the mammalian cell cycle. *Bioinformatics* **22**, e124–e131.
31. Melamed, M.R., Lindmo, T. and Mendelsohn, M. L. (1990) Flow Cytometry and Sorting, second edition. Wiley-Liss, New York, pp. 1–824.
32. Shapiro, H.M. (2003) Practical Flow Cytometry, fourth edition. Wiley-Liss, New York, pp. 1–681 (online at http://www.coulterflow.com).
33. Watson, J.V. (1991) Introduction to Flow Cytometry. Cambridge Press, Cambridge, pp. 1–443.
34. Gray, J.W. and Darzynkiewicz, Z. (1987) Techniques in Cell Cycle Analysis. Humana Press, Totowa, pp. 1–407.
35. Darzynkiewicz, Z., Robinson, J.P. and Crissman, H.A. (2001) Methods in Cell Biology: Cytometry, third edition. Academic Press, New York, vols. 63 & 64.
36. Jacobberger, J.W. (1991) Intracellular antigen staining: quantitative immunofluorescence. *Methods* **2**, 207–218.
37. Jacobberger, J.W. (2000) Flow cytometric analysis of intracellular protein epitopes. In Immunophenotyping, Edited by Stewart, C., Nicholson, K. Wiley-Liss, New York, pp. 361–405.
38. Jacobberger, J.W. (2001) Stoichiometry of Immunocytochemical Staining Reactions. Methods in Cell Biology: Cytometry, third edition. Academic Press, New York, vol. 63, pp. 271–298.
39. Jacobberger, J.W., Sramkoski, R.M. and T. Stefan (2011) Multiparameter cell cycle analysis. In Methods in Molecular Biology: Flow Cytometry Protocols, third edition, Edited by Hawley, T.S. and Hawley, R.G., Humana Press, Totowa., vol. 699, pp. 229–249.
40. Jacobberger, J.W., Sramkoski, R.M., Wormsley, S.B., and Bolton, W.E. (1999) Estimation of kinetic cell cycle-related gene expression in G1 and G2 phases from immunofluorescence flow cytometry data. *Cytometry* **35**, 284–289.
41. Frisa, P.S. and Jacobberger, J.W. (2009) Cell cycle-related cyclin B1 quantification. *PLoS One* **4 (9)**, e7064.
42. Jacobberger, J.W., Frisa, P.S., Sramkoski, R.M., Stefan, T., Shults, K.E. and Soni, D.V. (2008) A new biomarker for mitotic cells. *Cytometry A* **73**, 5–15.
43. Goldbeter, A. (1991) A minimal cascade model for the mitotic oscillator involving cyclin and Cdc2 kinase. *Proc. Natl. Acad. Sci. U.S.A.* **88**, 9107–9111.
44. Lozzio, B.B. and Lozzio, C.B. (1979) Properties and usefulness of the original K-562 human myelogenous leukemia cell line. *Leuk. Res.* **3**, 363–370.
45. Frisa, P.S. and Jacobberger, J.W. (2010) Cytometry of chromatin bound Mcm6 and PCNA identifies two states in G1 that are separated functionally by the G1 restriction point. *BMC Cell Biol.* **11**, 26.
46. Flow Cytometry Principles: Biology, University of California, Berkeley (Last accessed February 27, 2010, at http://biology.berkeley.edu/crl/flow_cytometry_basic.html).
47. Sigal, A., Milo, R., Cohen, A., Geva-Zatorsky, N., Klein, Y., Liron, Y., et al (2006) Variability and memory of protein levels in human cell. *Nature* **444**, 643–646.
48. Ezyfit Toolbox – A Free Curve-Fitting Toolbox for Matlab (Last accessed March 16, 2010, at http://www.mathworks.com/matlabcentral/linkexchange/links/1234-ezyfit-toolbox-a-free-curve-fitting-toolbox-for-matlab).
49. Lozzio, B.B., Lozzio, C.B., Bamberger, E.G. and Feliu, A.S. (1981) A multipotential leukemia cell line (K-562) of human orgin. *Proc. Soc. Exper. Biol. Med.* **166**, 546–550.

Part IV

Neuroscience, Cancer, and Stem Cell Research

Chapter 11

Signaling Events Initiated by Kappa Opioid Receptor Activation: Quantification and Immunocolocalization Using Phospho-Selective KOR, p38 MAPK, and K_{IR} 3.1 Antibodies

Julia C. Lemos, Clarisse A. Roth, and Charles Chavkin

Abstract

Psychiatric disorders including anxiety, depression, and addiction are both precipitated and exacerbated by severe or chronic stress exposure. While acutely, stress responses are adaptive, repeated exposure to stress can dysregulate the brain in such a way as to predispose the organism to both physiological and mental illness. Understanding the neuronal chemicals, cell types, and circuits involved in both normal and pathological stress responses are essential in developing new therapeutics for psychiatric diseases. Varying degrees of stressor exposure cause the release of a constellation of chemicals, including neuropeptides such as dynorphin. Neuropeptidergic release can be very difficult to directly measure with adequate spatial and temporal resolution. Moreover, the downstream consequences following release and receptor binding are numerous and also difficult to measure with cellular resolution. Following repeated stressor exposure, dynorphin is released, binds to the kappa opioid receptor (KOR), and causes activation of KOR. Agonist-activated KOR becomes a substrate for G protein receptor kinase (GRK), which phosphorylates the Ser369 residue at the C-terminal tail of the receptor in the first step in the β-Arrestin-dependent desensitization cascade. Through the use of phospho-selective antibodies developed and validated in the laboratory, we have the tools, to assess with fine cellular resolution, the strength of behavioral stimulus required for release, time course of the release, and regional location of release. We have gone on to show that following KOR activation, both ERK 1/2 and p38 MAP kinase phosphorylation are increased through use of commercially available phospho-selective antibodies. Finally, we have identified that one effector of KOR/p38MAP kinase is K_{IR} 3.1 and have developed a phospho-selective antibody against the Y12 motif of this channel. Much like KOR and p38 MAP kinase, phosphorylation of this potassium channel increases following repeated stress. The following chapter discusses immunohistochemical and quantification methods used for phospho-selective antibodies used in various brain regions following behavioral manipulations.

Key words: Kappa opioid receptor, p38 MAP kinase, Phosphorylation, Stress, Fluorescent immunohistochemistry, Brain regional localization

Alexander E. Kalyuzhny (ed.), *Signal Transduction Immunohistochemistry: Methods and Protocols*,
Methods in Molecular Biology, vol. 717, DOI 10.1007/978-1-61779-024-9_11,

1. Introduction

With the advent and implementation of fluorescently tagged viral constructs that can be used to selectively inhibit or activate specific brain regions (i.e., halorhodopsin and channelrhodopsin, respectively) or knock-down, rescue, or overexpress specific receptors, the ability to dissect the role of receptors in different neuronal circuits in affective and motivated behavior has advanced significantly. The use of phospho-selective antibodies in immunohistochemistry (IHC) assays can guide behavioral and physiological experiments and allows for a new level of selectivity beyond immediate early gene immunohistochemistry (e.g., c-fos IHC). In more recent years, the role of both ERK 1/2 and p38 MAPK in controlling cellular excitability, synaptic plasticity, de novo gene transcription, and behavior has been explored (3). In a study examining KOR-mediated conditioned place aversion (CPA) where the animal passively avoids a context paired with prior KOR agonist injection, it was shown that this CPA requires p38 MAPK activation. CPA (for review, see (4)) has been posited as a means to assess dysphoria-like behaviors. It has been demonstrated that in the same cell type, ERK 1/2 and p38 MAPK can be activated in parallel signaling pathways (5); one example is the case of stress-induced dynorphin release. Following KOR activation either with intraperitoneal injection of the exogenous agonist U50,488 or stress-induced dynorphin release, both phospho-ERK 1/2 immunoreactivity (IR) and phospho-p38 MAPK-IR are increased in the same cell types compared to saline-injected control animals (1, 2). While both ERK and p38 MAPK activities increase following stressor exposure in a KOR-dependent fashion, we found that p38 MAPK activation requires G-protein receptor kinase 3 (GRK-3) and β-Arrestin 2 recruitment, whereas ERK 1/2 activation occurs through a GRK-3-independent pathway (6). CPA caused by U50,488 requires phosphorylation of the Ser369 residue of KOR while KOR-dependent stress-induced analgesia does not (7). In contrast, ERK 1/2 activation still occurs in cells where this residue has been mutated to an alanine. Therefore, while both kinases are activated following repeated stress in similar cell types in the same regions, only p38 MAPK appears to play a role in KOR-mediated dysphoric-like behavior (1). This has provided critical insight into uncovering diverging roles for ERK 1/2 and p38 MAPK in mediating affective behavior and could not have been elucidated without the use of phospho-selective antibodies.

Both home-grown and commercially available phospho-selective antibodies are notoriously finicky and require extensive validation at a variety of stages of analysis. Most commonly, problems that occur include high background staining, nonselective immunoreactivity, and high degree of variability between subjects and even between adjacent slices. Methods can be developed to address each of these potential issues, and particular care must be taken to control the quality and consistency in the perfusion conditions, type of blocking buffer, number of washes, incubation time and temperature, and antibody concentration. Critical to producing valid and consistent results is care during imaging and quantification techniques. The technique has inherent subjectivity and therefore a certain degree of rigor and objectivity is required in the analysis.

2. Materials

2.1. Antibody Generation and Affinity Purification

Access to a UV spectrophotometer and high-speed centrifuge is required for several steps throughout the affinity purification process.

2.1.1. Preparation of Antibody Affinity Column

1. CNBr-activated Sepharose 4B.
2. 1 mM HCl.
3. 0.1 M $NaHCO_3$.
4. 0.1 M NaCl.
5. 1 M and 0.05 M Tris.
6. 0.1 M glycine.
7. 5 M $MgCl_2$.
8. 0.02% Sodium azide.

2.1.2. Affinity Purification

1. TBS/azide: 50 mM Tris 0.15 M NaCl, 0.02% sodium azide, pH to 7.4.
2. Serum containing antibody of interest.
3. 5 M $MgCl_2$.
4. 0.1 M glycine.
5. Centriprep-30.

2.1.3. Enzyme-Linked Immunosorbent Assay (ELISA) to Determine Antibody Production

1. PBS: 50 mM NaH_2PO_4 (monobasic), 6.9 g in 1 L 0.15 M NaCl, 8.766 g in 1 L and adjust pH to 7.4.
2. PBS-T (PBS/0.05% Tween-20): Add 250 μl Tween-20 to 500 ml of PBS.
3. 1% nonfat dried milk (NFdM) in PBS: Add 1 g of NFdM powder to 100 ml of PBS. Store at 4°C.
4. 1% BSA (bovine serum albumin) in PBS: Add 1 g BSA to 100 ml of PBS. Store at 4°C.
5. P-NP mix (makes 50 ml): In 40 ml of double distilled water: 10% diethanolamine (5 ml), 1 mg/ml *p*-nitrophenyl phosphate

(50 mg), 100 μg/ml $MgCl_2$ (5 mg), 0.02% sodium azide (10 mg). pH solution to 9.8 and bring to volume. Store in a 50-ml conical tube wrapped tightly in foil and store at 4°C.

2.2. Perfusion Fixation

2.2.1. Solutions

1. 0.4 M Phosphate buffer (0.4 PB): In 1 L distilled water, add 10.5 g Sodium phosphate monobasic, 86.6 g Sodium phosphate dibasic. Filter and store at room temperature.
2. 0.1 M Phosphate buffer (0.1 PB): In 1 L add 250 ml 0.4 PB, 750 ml distilled water.
3. 4 % Paraformaldehyde: In a vented fume hood, add to a glass beaker 250 ml of 0.4 M PB, approximately 600 ml of distilled water, and 40 g of paraformaldehyde (PFA). Stir continuously on a heated magnetic stir plate bringing the temperature to 65°C. Once temperature is reached, reduce heat to 60°C and add 10 M NaOH *dropwise* until all the PFA has gone into solution. pH solution to 7.4. You may have to add a drop or two of HCl, but try to avoid this. Often, the sodium hydroxide takes a while to take effect. After a couple of drops, wait for about a minute to let the PFA go into solution before adding more. Following pH adjustment, filter solution using a funnel that is lined with standard filter paper (Grade 413). For immediate use (<1 month), store at 4°C. For more prolonged use, aliquot into 50 ml conical tubes and store at –80°C (see Note 1).
4. PBS (different from PBS above): In 1 L of 0.1 PB add 8.76 g NaCl.
5. 30% Sucrose: 30 g of sucrose into 100 ml of 0.1 PB.

2.2.2. Instruments

1. Peristaltic pump or 30–60 ml syringe. Peristaltic pumps allow for an initial flush of the blood vessels in the brain with PBS before perfusion with PFA and also are easier to control flow rate, but good fixations can be accomplished with a simple hand-held syringe.
2. Butterfly needle.
3. Sharp medium scissors.
4. Sharp fine scissors.
5. Hemostat.
6. Rongeurs.
7. Fine (#7) forceps.
8. Small spatula.
9. 15 ml conical tubes.

2.3. Immunohistochemistry and Tissue Sectioning

1. For tissue sectioning, prepare 0.1 PB with 0.1% sodium azide (make a 10% stock to be diluted 100-fold in the wells).
2. PBS.

3. Blocking buffer: In PBS, add 5% normal goat serum (NGS) (Vector), 0.3% Triton X-100 (Sigma) (see Note 2).
4. Primary antibody solution: In blocking buffer, add either one primary antibody or cocktail of primary antibodies. For primary phospho-antibodies, typical concentrations range from 1:25 to 1:300. For home-grown antibodies, the concentration is highly dependent on yield. For example, for phospho-KOR we use a range of concentrations from 1:25 (0.0132 μg/ml) to 1:100 (0.0089 μg/ml) for antibodies with yields of 0.33 to 0.89 μg protein/ml affinity purified stock concentration.
5. Secondary antibody solution: In blocking buffer, add either one secondary or cocktail of secondaries raised in the different host species. For example, AlexFluor 488, 555, 633, 647 secondary antibodies from Invitrogen are recommended to be used at a concentration of 1:1000. However, FITC, TRITC, Texas Red, Cy2,Cy3, and Cy5 are available from Invitrogen or Jackson Immunoresearch work well, too. These antibodies may require higher concentrations, for example FITC works best in our hands at 1:250.

2.4. Fluorescence and Confocal Microscopy Imaging

1. Superfrost Plus slides (for mouse brains use, 25 × 75 × 1 mm).
2. Coverslip glass (for slide size above, coverslip size should not exceed 24 × 50 mm).
3. Mounting media: There are several different types of mounting media on the market with different advantages. Vectashield (with and without 4',6-diamidino-2-phenylindole, DAPI) (Vector) is rather inexpensive by volume. However, clear nail polish is often required and there is no antifade technology. ProLong Anti-fade mounting kit (Invitrogen) is more expensive by volume, but reduces fluorescence fading and becomes increasingly more congealed, preventing cover slips from sliding around as easily.
4. Clear nail polish.

2.5. Quantification

While Metamorph software (Molecular Devices, Sunnyvale, CA) allows for more sophisticated quantification of image intensities or cell counts, basic quantification can be achieved with Adobe Photoshop.

3. Methods

The initial validation of phospho-selective antibodies, KORp, and p-Y12-K_{IR} 3.1 developed in this laboratory included a combination of ELISA and immunocytochemistry techniques (8, 9). For ELISAs, different concentrations of antibody (0.005–10 μg/ml) were incubated overnight with separate wells coated by peptides (0.5 μg/ml) corresponding to approximately amino acids having the key residue phosphorylated or nonphosphorylated (8–10). Further validation can be obtained by transfecting DNA constructs in appropriate tissue culture cell lines (e.g., HEK 293, AtT 20, COS-7) to allow expression of the wild-type and mutant form of the protein of interest. For example, treatment of HEK 293 cells expressing KOR-GFP by U50,488 produced concentration-dependent agonist-induced internalization and increased KORp-IR, whereas parallel treatment with HEK cells transfected with KOR having Ser369 mutated to an alanine (KSA-GFP) failed to show agonist-induced internalization or increased KORp-IR. Similarly, untransfected HEK cells and pretreatment with KOR antagonists provided important specificity controls.

A similar approach was used to validate a phospho-K_{IR} 3.1 antibody designed to detect the phospho-tyrosine-12 residue in the cytosolic amino terminal domain of the channel (9). It had been shown previously that TrkB activation by BDNF can phosphorylate tyrosine residues on K_{IR} 3.1 to accelerate deactivation of the channel (11). ELISA confirmed that the antibody reacted more strongly with the phospho-peptide than the nonphospho-peptide (9). K_{IR} 3.1-IR was evident when transfected cells were stimulated with BDNF (9). While the exact methodological protocols are beyond the scope of this chapter, before one can move on to IHC, these validation measures must be conducted when generating a new phospho-antibody.

These in vitro validation methods are usually extended to standard Western blot analysis (9, 10). This stage of validation can be done using cell-culture homogenates treated with various pharmacological manipulations or from brain and/or spinal cord tissue homogenates from animals subjected to either pharmacologic or behavioral manipulation. For example, for further validation of the phospho-K_{IR} 3.1 antibody, Western blots were done using spinal cord tissue from animals that were subjected to sham surgery or partial sciatic nerve ligation (pSNL). Spinal cord tissue from either the ipsi- or contralateral side for each animal was acquired and homogenized, the contralateral side providing a within-animal control. Samples were blotted for either pY12-K_{IR} 3.1 or total K_{IR} 3.1. (First, one must ensure that the phospho-antibodies run at the appropriate weight and that there are no nonspecific bands. The intensities of the bands can be quantified using the Odyssey infrared imaging system (LI-COR Biosciences) and plotted as either fold change over experimental control or loading control (e.g., β-Actin) (9, 10)). As expected, there was no difference in total K_{IR} 3.1 between sham and pSNL-treated animals, nor between ipsi- and contralateral sides of the same animal. pY12-K_{IR} 3.1-IR was evident only in tissue from the ipsilateral spinal cord of the nerve-ligated subjects (9). This set of experiments not only validates the antibody, but acts as a good compliment to IHC.

Generally, these in vitro methods are used prior to in vivo analysis. Perfusing, sectioning, staining, and imaging brain slices using phospho-selective antibodies are not drastically different than many other staining protocols that look at the localization and amount of total protein across brain regions. However, these primary antibodies seem particularly sensitive to poor fixation and sectioning as well as variations in incubation time and temperature. The following description of our current protocols may be a very useful starting point for the individual optimization processes required for each new analysis.

3.1. Generation, Validation, and Affinity Purification of Phospho-Selective Antibodies

3.1.1. Antibody Generation

1. Indentify an amino acid sequence within the protein of interest containing a known phosphorylation site. A typical amino acid sequence length would be approximately 12–15 amino acids with an added N-terminal lysine that enables conjugation to the BSA and Sepharose column (12). Using the GenBank™ data base, ensures that the sequence chosen is unique, no other proteins have >50% homology or four identical amino acids in a row.
2. The phospho-peptide is conjugated to KLH and injected into rabbits. Again, contract commercial services are available. 500 μg of antigen peptide was combined with Complete Freund's Adjuvant for initial inoculation, then animals received boost inoculations of 250 μg of antigen peptide on days 14, 21, 49, 56, and monthly thereafter. Bleeds (approximately 6–7 ml) are collected once a month (9).

3.1.2. Preparation of the Affinity Column

1. Weigh 1 g of CNBr-activation Sepharose 4B. Swell for 15 min in 1 mM HCl and transfer to a sintered glass funnel. Wash with 200 ml of 1 mM HCl, then wash with 20 ml of 0.1 M $NaHCO_3$.
2. Suspend 2–5 μmol peptide in 0.1 M $NaHCO_3$, 0.1 M NaCl, pH 8.3. If the peptides are not soluble in this buffer, dissolve in water or acid and then adjust pH to 8.3. Bring the final volume to 5 ml. Measure and record OD_{280} (optical density) to check concentration.
3. Add peptide solution to Sepharose and incubate overnight at 4°C with mild rocking.
4. Add 0.55 ml of 1.0 M Tris, pH 8.0. Rotate for 2 h at room temperature.
5. Pour slurry into a 0.7 × 15-cm column. Collect flow-through and measure OD_{280} to determine peptide-coupling efficacy.
6. Wash 3× with 15 ml of 0.05 M Tris, pH 7.4, 0.15 M NaCl, then 15 ml of 0.1 M glycine, pH 3.0, then 1.5 ml of 5 M

$MgCl_2$, and finally 15 ml of 0.05 M Tris, 0.15 M NaCl, pH 7.4.

7. For storage, wash column with 0.05 M Tris, pH 7.4, 1.5 M NaCl + 0.02% sodium azide. Store at 4°C.

3.1.3. Affinity Purification of Antipeptide Antiserum

1. Wash column before use with 20 ml TBS/Az.
2. Add 3–4 ml of serum to column. Seal the top and bottom with parafilm and gently rotate for 72 h at 4°C.
3. Wash unbound material with 20 ml TBS/Az, taking care not to let the resin dry.
4. Elute antibody with 5 ml 5 M $MgCl_2$, collecting into the outer vessel of a Centriprep-30 containing 5 ml TBS/Az.
5. Concentrate antibody in Centriprep to ~1 ml by spinning 1,5000 × *g* for 20 min (use Beckman J6-B centrifuge, set to 4°C and speed = 2,300 rpm). Discard fluid in central vessel. Add 5 ml TBS/Az to concentrated antibody in outer vessel, mix, and concentrate again to ~1 ml. Repeat twice more, adding cold TBS/Az each time (final concentration $MgCl_2$: 20 mM, after four additions of TBS/Az).
6. Measure OD_{280} of concentrated antibody. Using a standard containing a known concentration of protein (i.e., BSA), get an OD_{280} reading that will be used as the conversion factor for assessing the antibody concentration. For example, 1 mg/ml may yield an OD_{280} reading of 1.4. Thus, if you get a reading of 1.2 for the purified antibody, the yield would be 0.857 mg/ml.
7. Dilute concentrated antibody by about half with glycerol and store at −20°C.
8. Following affinity purification, regenerate column (allow each of these to drip through) with 30 ml TBS/Az, then 30 ml 0.1 M glycine, then 5 ml 5 M $MgCl_2$. Add 35 ml TBS/Az and cap the top and bottom of the column and wrap ends with parafilm. It may not be necessary to regenerate the column after every affinity purification.

3.1.4. ELISA to Determine Antibody Production

Adopted from the method of Karen DeJongh and Aaron Patillo. This protocol assumes tested antibodies were raised in rabbit; if not, be sure to match the IgG-AP Ab used in Step 7 *to the species* (see Note 3).

1. Coat 96-well plates with 50 μl per well of a 10-μg/ml peptide solution, prepared in PBS. Prepare 5 ml per plate to be coated. Incubate overnight at 4°C or for 4 h at room temperature. Remove peptide solution from well.
2. Block unbound sites with 50 μl 1% skim milk powder in PBS. Incubate for 2 h at 37°C or overnight at 4°C.

3. Wash wells 3× 150 μl PBS-T. Swirl and remove solution by aspiration each time.
4. Identify and prepare the antibody dilutions in PBS-T for each bleed as desired. It is useful to create a standard data entry form to record the results for each concentration and bleed.
5. In triplicate, add 50 μl of diluted antibody to wells. Leave the upper three wells of each plate empty to use as a blank (fill with 50 μl PBS-T alone). Incubate for 2 h at room temperature or overnight for 4°C.
6. Wash wells 3× with 150 μl of PBS-T.
7. Add 50 μl of a 1:3,000 dilution of alkaline phosphatase-conjugated anti-rabbit IgG in PBS/1% BSA to each well. Incubate for 2 h at room temperature or overnight at 4°C.
8. Wash wells 6× with 150 μl PBS-T.
9. Add 50 μl of 1 mg/ml P-NP mix to each well. Incubate for 30 min at room temperature, protected from light. (You can place the plate(s) inside a lab bench drawer.) Do NOT empty wells when complete.
10. Add 50 μl 2.5 M NaOH to terminate reaction.
11. Using a plate reader, promptly read absorbance at 410 nM.
12. The phospho-selective antibody should react severalfold more strongly to a phosphopeptide than an equivalent nonphosphorylated sequence.

3.2. Perfusion Fixation

The following instructions are for intracardiac perfusion of mice or rats. Phospho-selective antibodies appear to be particularly sensitive to the quality of perfusion and postfixation protocol (see Note 4).

1. Set up the perfusion area with tools, butterfly needle, 50-ml tube of PBS, and either 50-ml tube of PFA for mice or 250-ml container for rats.
2. 15-ml tubes containing 4% PFA should be labeled accordingly and set aside.
3. If you have access to a peristaltic pump, attach butterfly needle and accompanying tubing to the tubing of the peristaltic pump so that there is tight seal between the two; ensure there is no leaking. Wrap parafilm around the needle, only leaving about 0.5 mm of the needle exposed to ensure that the need does not puncture the underside of the heart. If using syringe, simply attach needle to syringe full of PFA and skip Step 4.
4. Run the pump so that the tubing fills with PBS to the point where PBS is coming out of the needle. Then, transfer the input end of the peristaltic pump's tubing to the 4% PFA. If the pump has a three-way stop-cock mechanism, then simply

switch the dial so that the PFA will now flow through the tubing when switched on again.

5. Anesthetize the animal with Nembutal or isoflurane and proceed only when the animal is unresponsive to painful stimuli (foot pinch).
6. Secure the arms with pins or tape onto whatever surface has been designated for perfusions.
7. Use a pair of large-to-medium sharp scissors, pinch the fur and muscle around the abdomen upward, and make a deep angled cut into the flesh through the abdominal muscle, exposing, but not damaging the liver.
8. Cut up each side of the animal from the original abdominal incision to right below the arm pits exposing the liver, intestines, and chest cavity.
9. Use the small sharp scissors and make a horizontal cut across the diaphragm exposing the heart and lungs.
10. Use a hemostat and clamp the sternum and roll the hemostat toward the head of the animal exposing the heart further.

    Some people prefer to reverse Steps 11 and 12. This is a matter of preference and personal comfort.
11. Use small sharp scissors and cut the right atrium. Dark oxygenated blood should come out.
12. Insert butterfly needle into the left ventricle making sure to insert the needle perpendicular to the heart and into the apex of the ventricle. This is to avoid puncturing the septum and spilling PFA into the right ventricle.
13. Begin perfusing using syringe or peristaltic pump. For the first 5–10 ml (50 ml for rats) of PFA, the perfusion should be running at a rate of approximately 3–5 ml/min. Following this, reduce speed to 2 ml/min and perfuse the animal with the remaining PFA.
14. Decapitate the animal and dissect out the brain and place the brain in 4% PFA in 15 ml conical tube.
15. Following the perfusion, postfixate for anywhere from 1 to 12 h depending on the perfusion quality (more time for bad perfusions). Then, transfer the brains to 30% sucrose and incubate at 4°C until brain has sunk to the bottom of the tube.

3.3. Brain Sectioning

There are several different sectioning devices on the market from floor-standing cryostats to counter-top microtomes. Counter-top microtomes work well for free floating sections that are ≥30 μm. Thinner sections should be made using a cryostat and should be thaw-mounted on superfrost plus slides. The use of the above-mentioned phospho-selective antibodies has only been tested on free floating sections. It is not known if they are effective on thaw-mounted tissue sections. The following instructions are for counter-top microtomes.

1. Fill the cavity surrounding the mounting stage with dry ice that has been smashed into fine powder and wait until frost accumulates around the outside of the stage.
2. Remove brain from 15 ml conical tube. For coronal sectioning, cut off the brainstem and small portion of the cerebellum to create a flat surface. For sagittal sections, cut down the midline. Horizontal slices should not need any tissue blocking.
3. Using a transfer pipette, take the 30% sucrose solution and drip the solution onto the frozen stage until an appropriately sized bed layer forms. Do not let the solution freeze all the way before placing the brain on the stage with forceps. For coronal sections, the olfactory bulbs should be facing up. For sagittal sections, the most lateral temporal lobe should be facing up. For horizontal sections, either the most dorsal or ventral side up should work and should depend on regions of interest.
4. Drip more sucrose solution around the brain, waiting intermittently for each layer to freeze. Tissue-Tek ® OCT compound should also work and logistically is better to use if using a cryostat; however, we have not tested the efficacy of the antibodies following sectioning in OCT. (In the case of cryostat, create the bed layer of OCT on the removable circular block and mount the brain. Once the brain is affixed to the cryostat block, surround the tissue in OCT compound and quickly return to the cryostat to freeze. You may find that freezing the brain directly in dry ice before mounting on the cryostat block is more effective. However, this may affect the phospho-staining and cell morphology adversely).
5. Once the brain is surrounded in frozen sucrose solution, take a 50-ml conical tube that has had the closed end cut off and place over the brain. Then, take the fine dry ice powder and with a spoon pour it over the brain and wait for 45–60 s.
6. Remove tube and dust off dry ice exposing the frozen brain.
7. Using sharp disposable or reusable blades, begin sectioning brain. It is useful to moisten the blade with 0.1 PB to ensure smooth sectioning. If you still notice shredding of the tissue, place your finger on top of the tissue for 1 s. With these methods, the brain should not be too cold or the tissue will not section evenly.
8. Place sections into wells containing 0.1 PB with 0.1 % sodium azide and store at 4°C until you are ready to use slices. If you

are interested in a singular brain region, then it is perfectly fine to section and confine the brain according to regions of interest. However, if you are interested in studying the entire brain, then the brain should be sectioned serially, placing one slice in well #1-X and then starting back at 1 such that each well contains a "representative brain." Consult your mouse or rat atlas. Given a certain region of interest and thickness of section, you can calculate how many wells to use to ensure that each well contains approximately the same number of slices through the rostro-caudal axis of region.

3.4. Immunohistochemistry

For information regarding *immunocytochemistry* with these antibodies, please refer to:

1. In a set of 12 well plates, fill each well with PBS and select at least three representative sections along the rostro-caudal axis for each brain region of interest for each treatment group. There are plastic baskets with mesh at the bottom that are designed to fit in 12-well plates. However, these can be easily crafted using cell-culture baskets. It is highly recommended that you use these baskets rather than placing the sections directly into the wells and using a fine paint brush to transfer the sections from well to well. This avoids damaging the tissue (see Note 5).

   A common mistake is to overcrowd the sections. For robust antibodies that are used as cell-type markers such as NeuN or GAD67, this is not as large a problem. However, for phospho-selective antibodies, it is important that each section gets similar penetration. Therefore, select 3–6 sections per well depending on the size of the brain section. Each well may contain a sample of one brain region or if you wish to include a more complete representation of a single brain region, you may need to use 2–3 wells. Given the number of treatment groups, you may have to run your experiment in batches. This is fine as long as one conducts parallel staining of matching brain regions for each treatment group (e.g., Batch #1 Hippocampus from Treatment groups X–Z). Managing 4—6 12-well plates is approximately the upper limit. Depending on your question, it may not be possible to process sections from more than one subject per treatment group. This is not a problem as long as each treatment group is represented. While one would like to think that the staining should not differ from round to round, the reality is that it most definitely can vary. For each round of staining, the imaging will also be collected in parallel (see Subheading 3.5). The quantification averages either the intensity or cell counts from at least three sections from one animal to create a single N (see Subheading 3.6). Therefore, each batch will ultimately create a single N for each treatment group (if one animal per treatment group was used in a round of staining) and the representative images should be selected based on the mean intensity or cell count for each group.

2. Once the sections from each subject are selected, place sections in the first well (each column should be reserved for one set of sections from one subject) into the plastic basket. Make sure that at least 2 ml of PBS are in each well and sufficiently covering the sections.
3. Sections should be washed in PBS 3× 10 min with mild agitation on a shaker, transferring the basket containing the sections to the second and then the third well every 10 min. Make sure to remove excess PBS from previous well by gently touching the basket to the side of the well.
4. During the washes, prepare the blocking buffer.
5. Aspirate off the PBS from the first well.
6. Pipette 1 ml of blocking buffer into each well. Remove basket from third wash and touch a Kim® wipe to the bottom of the basket removing excess PBS and place the basket into the blocking buffer.
7. Once the basket is in the well with the blocking buffer, add an additional 1 ml of blocking buffer to each well.
8. Block for 1 h with mild agitation. During this time, prepare cocktail of primary antibodies making sure that each primary has been raised in a different host animal.
9. Both home-grown and commercial phospho-selective antibodies can be precious and expensive. To reduce the volume necessary for each set of sections, primary antibody staining can be done in 1.7 ml Eppendorf® tubes. For each set of sections, use 500 μl of blocking buffer containing primary antibody. Prepare one stock of primary antibody solution (e.g., for four Eppendorf ® tubes, prepare 2 ml of stock solution) than aliquot 500 μl of the solution into each carefully labeled Eppendorf ® tube.
10. Following the blocking step, place sections into their respective Eppendorf ® tubes containing the primary antibody solution with a fine paint brush. Phospho-selective antibodies often need at least 36 h of incubation at room temperature or 48–72 h of incubation at 4°C with mild agitation. Phospho-p38 MAPK and phospho-ERK 1/2 antibodies work in both conditions. Incubating for longer duration at 4°C appears to reduce background. Every 12–24 h, gently rotate each tube upside down and then place back into holder in upright position. KORp and pY12-K_{IR} 3.1 antibodies are typically incubated for 72 h at 4°C. If you notice uneven staining on a given

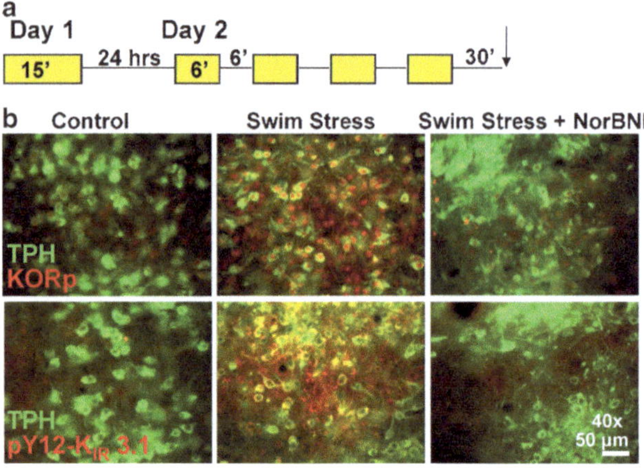

Fig. 1. *Anatomical localization of phospho-KOR-IR and phospho-pY12-K$_{IR}$3.1-IR in the serotonergic dorsal-raphe nucleus (DRN) following repeated swim stress.* (**a**) Schematic of the repeated swim stress paradigm used for all studies. (**b**) Fluorescent images showing repeated swim stress increases both KORp-IR (*red, top*) and pY12-K$_{IR}$ 3.1-IR (*red, bottom*) on TPH+ (*green*) and TPH- DRN neurons above basal activation found in "no stress" control animals and is sensitive to NorBNI (10 mg/kg) pretreatment.

section or across sections from the same set, you can also use a 24-well plate (no baskets) using a similar volume for each well that is used for the Eppendorf® tubes (see Note 6).

11. The use of cell markers (e.g., TPH for serotonin, TH for dopamine/norepinephrine, GAD 67 for GABA, vGLUT1,2,3, for glutamate, GFAP for astrocytes, NeuN for neurons, etc.) is extremely useful for indentifying the colocalization of activated proteins to specific cell types (see Figs. 1 and 2). Most of these markers are very robust, often are mouse monoclonal antibodies (although some are raised in both rabbit and mouse such as GFAP and GAD) and work well in several different incubation conditions including those described above.
12. Most affinity-purified antibodies are raised in rabbit, while commercially available antibodies such as phospho-p38 MAPK are available in both mouse and rabbit hosts. Thus, it is possible to colocalize phospho-p38 MAPK with a homegrown antibody such as KORp as shown in Fig. 2. However, to assess the pattern of staining in a brain region of a subject for two rabbit polyclonal antibodies, one must take alternating serial sections from a brain region for staining of the two antibodies, respectively (Fig. 1).
13. For cell count or intensity quantification against total or nonphosphorylated protein, the antibodies for phosphorylated and nonphosphorylated protein must be raised in different host species in order to colabel the same tissue. If this is not

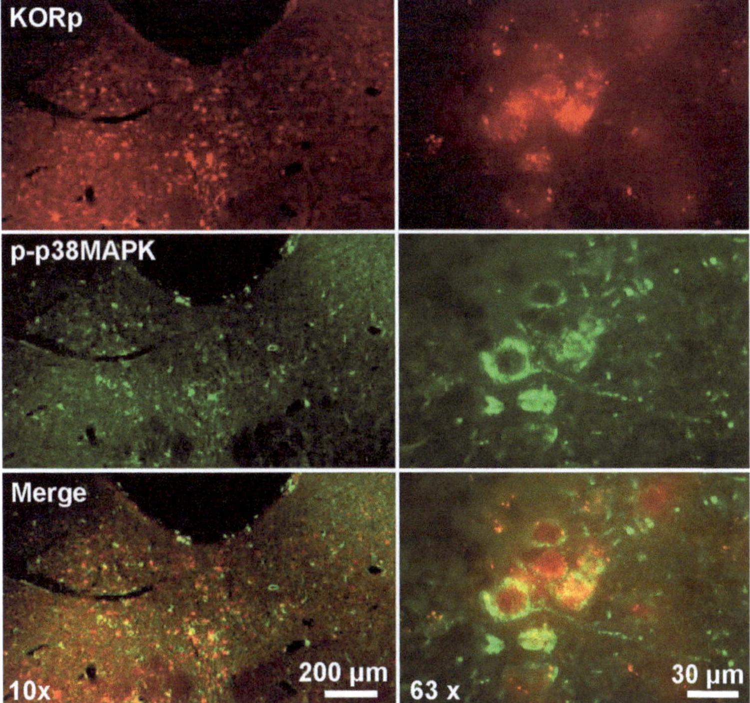

Fig. 2. *Stress-induced activation of KOR and p38 MAPK in same cells of the serotonergic dorsal-raphe nucleus*. Fluorescent images of the dorsal-raphe nucleus following co-immunofluorescent staining with rabbit KORp and mouse phospho-p38 MAPK at 10× (*right*) and 63× (*left*) magnification. The image demonstrates that these proteins are activated following repeated stress and colocalized to the same cells.

possible, a ratio of intensities or cell counts can be constructed between activated and nonactivated regions present on the same section.

14. Following primary antibody incubation, fill 12-well plates (it is okay to use the same plates from Step 1 that have been rinsed and dried) with PBS.
15. Remove sections from their respective tubes and place into baskets in the first well containing PBS. Wash at least 3×10 min with mild agitation just as in Step 3. If you are noticing high levels of background staining, it may be necessary to increase this washing step to 4–6×10 min.
16. During this wash, prepare secondary antibody solution. If you are using natural goat serum, ensure that your secondary antibodies are goat antibodies (i.e., they should read goat anti-mouse or goat anti-rabbit) (see Notes 7 and 8).
17. Similar to Steps 5 and 6, aspirate off PBS from the first well and add 1 ml of secondary antibody-blocking buffer solution. Transfer baskets containing sections to wells containing

secondary antibody solution and then add an additional 1 ml to each well (see Note 9).

18. Cover 12 plates with tin foil to avoid fluorescent photobleaching.
19. Incubate sections in secondary antibody solution for 2 h at room temperature with mild agitation. If necessary, you may incubate at 37°C for 1 h; however, generally this does not work as well for phospho-selective antibodies.
20. Wash sections 3 × 10 min in PBS with mild agitation.
21. Wash sections 2 × 10 min in 0.1 PB with mild agitation. Again, if you notice high background staining, you may want to increase your number of washes.

    Steps 20 and 21 should also be done with the plates covered in tin foil.
22. Mount sections on superfrost plus® slides in 0.1 PB and place slides on paper towel in a dark place (e.g., in a drawer or cabinet). Allow excess moisture to evaporate off the slide and sections. Usually, this occurs within 30 min to 1 h. The goal is to prevent diluting out the mounting media without allowing the sections to dry out. If the sections dry out, salt crystals can form on the tissue causing autofluorescent artifact.

3.5. Confocal and Epifluorescent Imaging

Epifluorescent microscopes are usually better for images taken at lower magnification (4–20×) that allow you to see gross neuroanatomical structure, while confocal microscopes tend to better for higher magnifications (40–100×) and often work better with oil lenses (see Notes 10 and 11). Confocal microscopy is better when one is interested in subcellular localization of the protein of interest. Regardless of which microscope is being used, the general principles of image acquisition are the same.

3.5.1. Epifluorescent Microscopy

1. Most epifluorscent microscopes are equipped with CCD camera and accompanying software. Part of the software package contains a preview and look-up table component that can be used when acquiring the image. The look-up table contains measurements including exposure time, gain, and offset.
2. Take your set of slides and scan each slide by eye to quickly assess what section of what slide is the brightest/highest intensity of nonartifactual staining (if the perfusion was bad, some sections may have heavy blood vessel staining that will autofluoresce).
3. The look-up table will contain a histogram of the intensities for every pixel in the image. The exposure time, gain, and offset should be set such that the pixel intensity histogram is approximately Gaussian or slightly skewed to the left with few pixels oversaturated (see Note 12). It is important that the

contrast is not adjusted so much so that the tissue texture is not discernable.

4. Once an exposure time, gain, and offset has been decided on *for a given brain region*, keep those settings consistent for all the subsequent image acquisition for all sections of all treatment groups *for a given magnification*. If you change to a different brain region or magnification, you must adjust the exposure time, gain, and offset appropriately. Sections with severe artifacts (see Note 13) should be excluded from quantification if using intensity measures (see Fig. 1).

3.5.2. Confocal Microscopy

1. Rather than a look-up table with exposure time, gain, and offset, confocal microscopes simply have dials to adjust gain and offset. A useful tool with the confocal software is the QLUT button that false colors the image based on the pixel intensity such that "black" pixels are labeled green, saturated pixels appear blue, and oversaturated pixels appear white. This is very helpful for making sure the image is balanced. A balanced image should have a moderately dense number of green pixels (about half the pixels) with some blue pixels and very few if any white pixels.
2. Once again an offset has been decided upon save that setting and use it for all the subsequent sections from all treatment groups (see Note 14).

3.6. Quantification

Analysis of phospho-selective immunoreactivity can be performed in two ways: intensity measurements and cell counts. Each method has specific drawbacks and therefore to obtain the clearest data, it is best to perform both these measurements for comparison (see Note 15).

3.6.1. Intensity Measurements

1. These instructions are based on the software MetaMorph Version 6.2r6 (Molecular Devices, Sunnyvale, CA), and assume use of the monoclonal antiphospho-p38 MAPK that detects dual phosphorylation of Thr180 and Tyr182, KORp, or pY12-K_{IR} 3.1. A minimum of three anatomically parallel tissue slices are necessary for a single data point.
2. Create and save a region that encompasses the area where you would expect to see a change in the phospho-antibody-IR. Create and save a second region of the same area where you would not expect to see a change in phospho-antibody-IR. Use these same regions on all analysis of tissue from the same experimental group.
3. Log the average pixel intensity in each of the regions. Repeat for all slices.
4. Calculate the ratio between the two regions, and average all anatomically parallel slices of a single individual (Fig. 3a) (see Note 16).

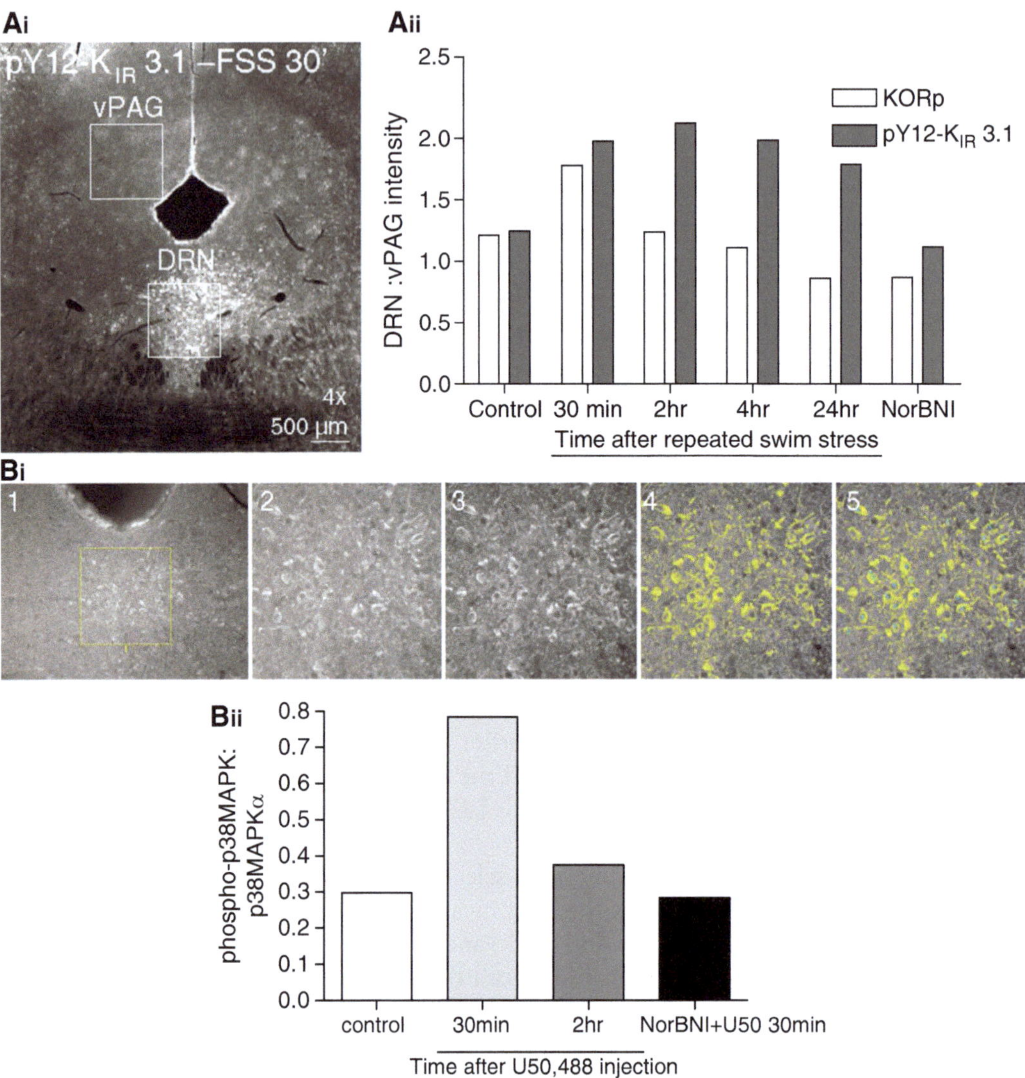

Fig. 3. *Two different methods for quantifying immunohistochemistry data*. (**Ai**) An example image in which the intensities of two identically sized areas in the same section are compared from a stressor exposed animal. The serotonergic dorsal raphe demonstrates increases in pY12-K$_{IR}$3.1-IR while the ventral periaqueductal gray is unchanged. (**Aii**). A ratio is constructed the KORp or pY12-K$_{IR}$ 3.1 DRN:vPAG intensity over a for three slices from a subject an averaged to generate a single N. The graph demonstrates a time course of KORp and pY12-K$_{IR}$ 3.1 activation following repeated swim stress. Activation of both proteins is not present in subjects treated with the selective KOR antagonist norBNI. (**Bi**) Outlines the steps necessary when using the cell-counting technique to characterize the amount of phospho-immunoreactivity compared to the nonphospho-immunoreactivity. *1*. Select region, *2*. Zoom in on region, *3*. "auto-enhance," *4*. set intensity threshold to one standard deviation above mean, 5. Toggle between threshold on/off and count cells that meet BOTH morphological AND intensity criterion. (**Bii**) An example experiment looking at the time course of p38 MAPK activation following intraperitoneal injection of the selective KOR agonist U50,488.

3.6.2. Cell Count

1. These instructions are based on the software MetaMorph Version 6.2r6 and can be used for any antibody that marks the activated form of a protein, costained with the antibody for the nonactivated form of the protein, both of which produce neuronal cell body staining. A minimum of three

anatomically parallel tissue slices are necessary for a single data point.

2. Create and save a region that contains the area of interest. Use this same region on all analysis of tissue from the same experimental group.
3. Zoom in on the region and use "auto-enhance" to accentuate the contrast level of the image if need be. Set the inclusive threshold so that only pixels that have an intensity one standard deviation above the mean pixel intensity will be highlighted.
4. Count cells as active if they meet two criteria:
   (a) Perimeter of the cell must be highlighted by the threshold (what percentage of the perimeter must be highlighted depends on the quality of staining, but must remain constant across an experimental group).
   (b) Morphology of the cell must be appropriate and in focus (i.e., a visible nucleus).
5. Log the cell counts of each region and create a ratio of the number of cells with the activated form of the protein to the number of cells with the nonactivated form of the protein (Fig. 3b) (see Note 17).

4. Notes

1. 4% PFA is standard, however, particularly for thinner sectioning (<30 μm), lowering the concentration of PFA to 3% may be desirable. Similarly, you may find that initially placing brains in 10% sucrose and then transferring them to 30% sucrose after 24 h provides better staining. Both these are highly dependent on the antibody.
2. Some antibodies work better with different blocking conditions. For example, the total guinea-pig dynorphin protein works with donkey serum, but not goat serum. If NGS does not work, try donkey serum or fish skin gelatin.
3. Make sure to have ample buffer made up in advance. Moreover, when adding the contents to the wells, do not pipette directly into the well, rather drizzle the fluid down the side into the bottom.
4. Depending on the protein of interest, one might consider rapidly freezing fresh frozen tissue rather than perfusion fixation. This method is especially useful when staining for cytoskeleton-anchored proteins. For protocols using this methodology, refer to (13).

5. While it may seem obvious, make sure to label (using marker or lab tape), both the top and base of each of the 12 plates uniquely. It is very easy to mix up tops and bottoms when using more than one 12-well plate and this can cause major problems when trying to compare different behavioral manipulations.
6. Commonly, one uses a maximum of three primary antibodies, often a mouse monoclonal antibody, rabbit polyclonal, and guinea pig polyclonal. If one channel contains robust green or red fluorescent protein, only two primary antibodies may be required. There are other host species on the market including chicken, rat, goat, etc. However, these are less common. If your microscope has four channel capabilities, it is possible to use a cocktail of four primary antibodies. It is very difficult in this situation to prevent overlapping emission spectra, so be aware of this problem when selecting primary and secondary antibodies. For example, for two fluorophores in proximity of each other along the emission spectra, it may be wise to choose two accompanying primaries with different patterns of staining.
7. Despite cryoprotection in sucrose, if the morphology of the cells does not look appropriate, this could be due to a poor perfusion; however, it is also attributable to long-term exposure to triton. One troubleshooting trick may be to either reduce or exclude triton from the secondary antibody solution.
8. If the staining appears very weak, you may consider amplifying the signal with an additional streptavidin step. In this case, use a goat anti-X biotin antibody and incubate for 2 h with cocktail of secondary antibodies that accompany the other primary antibodies (e.g., goat anti-Y and goat anti-Z). Run an extra 3×10 min in PBS washing step, then incubate sections for 60–90 min in streptavidin conjugated to the appropriate fluorophore. If your staining is still very weak, you may consider conjugating your primary antibody to quantum dots (there are kits to do this available through Invitrogen).
9. You may notice bright sedimentation on your tissue when you image. This is due to the sediment or crystallization of the secondary antibody, which can occur over time and also at the bottom of the secondary antibody container. You can vortex the tubes containing the concentrated secondary antibody to avoid this, but generally you should not use secondaries that are over 6–8 months old. Another source of artifact is bleed through from another channel. This is only really a problem with epifluorescent microscopes. If a counterstain is particularly robust, it may bleed through to your experimental antibody. This is particularly a problem when

the counterstain used a 555/Rhodamine/TRITC fluorophore for its secondary antibody. You may consider switching fluorophores, reducing the concentration of the primary antibody for the counterstain, or using a confocal microscope.

10. Epifluorescent microscopes do not have pinhole-like confocal microscopes and thus there is scattered light above and below the plane of focus that becomes more obvious at a higher magnification. At lower magnitude, allowing in more light allows the tissue to look more like tissue. However, an epifluorescent microscope would be inappropriate to use if one is interested in imaging small subcellular components such as dendritic spines or nodes of Ranvier or is interested in examining receptor internalization. If you do not have access to a confocal microscope, it is possible to attach a Z-motor to a regular epifluorescent microscope to allow generation of the Z-stack as well as deconvolution software. Rather than through use of a pinhole, deconvolution software uses an algorithm to essentially trace back the point source of scattered light, thereby removing it. This software is also useful with confocal images to even further the level of resolution.

11. While you should consult with your confocal representative or core facility manager about the exact confocal parameters appropriate for your sample, generally you should set your formatting and line or section averaging based on Nyquist sampling theory. Essentially, the pixel resolution (this information is often in the control window of the confocal software) should be appropriate for the size of the smallest subcellular structure one wishes to image. Most confocal formats include 512 × 512 or 1,024 × 1,024, with line averaging from 1 to 16. A greater level of pixel resolution increases the time per image and also risks bleaching of the sample. However, for imaging small puncta (e.g., internalized receptor) or small structures (e.g., dendritic spines), it is most likely necessary to use a higher number of pixels.

12. Depending on the subcellular region of interest, some oversaturation may be desirable. For example, if one is interested in examining the localization of a specific protein at spines, then oversaturation of the cell soma may be required. As long as this is kept consistent across treatment groups, this does not pose a problem.

13. Most often, artifacts include autofluorescence from blood vessels, accumulation of fluorescent sediment, and autofluorescence from cell organelles. It may be necessary when first working with a new antibody to run a "no primary" control to assess the degree of autofluorescence inherent in your system with your brain regions.

14. The use of Photoshop for adjusting the contrast is somewhat controversial and potentially unethical; check with the policy of the journal to ensure compliance. For images that will ultimately end up in the publication, it is reasonable to adjust the contrast using the Levels tool if the following requirements are met:
    (a) The images being compared across treatment groups were stained and imaged in parallel.
    (b) The contrast for all the images is adjusted in the same way.
15. Generally, it is better to do the analysis (and if possible the imaging) blind to treatment. Once you mount and coverslip the sections, have another member of your lab code the slides. For example 1, 2, 3 = vehicle, agonist, and agonist + antagonist, respectively. You will have to set your exposure time, gain, and offset based on the brightest slice rather than treatment group. Maintain this coding system during the analysis. One caveat to this is that it may become very obvious to you which set of slices is the experimental group if the finding is robust. If possible, have someone else in the lab who is unfamiliar with the hypothesis being tested to analyze the data. This will ensure the most rigorous and unbiased results.
16. One drawback to intensity measurements is nonspecific binding, or autophosphorylation due to damaged tissue. Therefore, clean and consistent perfusions are necessary to obtain solid data.
17. One drawback to the cell-count methodology is that it excludes noncell body fluorescence (i.e., immunoreactivity on dendrites or fibers).

References

1. Bruchas, M. R., Land, B. B., Aita, M., Xu, M., Barot, S. K., Li, S., and Chavkin, C. (2007) Stress-induced p38 mitogen-activated protein kinase activation mediates kappa-opioid-dependent dysphoria. *J. Neurosci.* **27**, 11614–11623.
2. Bruchas, M. R., Xu, M., and Chavkin, C. (2008) Repeated swim stress induces kappa opioid-mediated activation of extracellular signal-regulated kinase 1/2. *Neuroreport* **19**, 1417–1422.
3. Thomas, G. M., and Huganir, R. L. (2004) MAPK cascade signalling and synaptic plasticity. *Nat. Rev. Neurosci.* **5**, 173–83.
4. Bardo, M. T., and Bevins, R. A. (2000) Conditioned place preference: what does it add to our preclinical understanding of drug reward? *Psychopharmacology (Berl)* **153**, 31–43.
5. Bolshakov, V. Y., Carboni, L., Cobb, M. H., Siegelbaum, S. A., and Belardetti, F. (2000) Dual MAP kinase pathways mediate opposing forms of long-term plasticity at CA3-CA1 synapses. *Nat. Neurosci.* **3**, 1107–1112.
6. Bruchas, M. R., Macey, T. A., Lowe, J. D., and Chavkin, C. (2006) Kappa opioid receptor activation of p38 MAPK is GRK3- and arrestin-dependent in neurons and astrocytes. *J. Biol. Chem.* **281**, 18081–18089.
7. Land, B. B., Bruchas, M. R., Schattauer, S., Giardino, W. J., Aita, M., Messinger, D., et al. (2009) Activation of the kappa opioid receptor in the dorsal raphe nucleus mediates the aversive effects of stress and reinstates drug seeking. *Proc. Natl. Acad. Sci. U S A* **106**, 19168–19173.
8. McLaughlin, J. P., Xu, M., Mackie, K., and Chavkin, C. (2003) Phosphorylation of a

carboxyl-terminal serine within the kappa-opioid receptor produces desensitization and internalization. *J. Biol. Chem.* **278**, 34631–34640.

9. Ippolito, D. L., Xu, M., Bruchas, M. R., Wickman, K., and Chavkin, C. (2005) Tyrosine phosphorylation of K(ir)3.1 in spinal cord is induced by acute inflammation, chronic neuropathic pain, and behavioral stress. *J. Biol. Chem.* **280**, 41683–41693.
10. Clayton, C. C., Bruchas, M. R., Lee, M. L., and Chavkin, C. (2010) Phosphorylation of the mu-opioid receptor at tyrosine 166 (Tyr3.51) in the DRY motif reduces agonist efficacy. *Mol. Pharmacol.* 77, 339–347.
11. Ippolito, D. L., Temkin, P. A., Rogalski, S. L., and Chavkin, C. (2002) N-terminal tyrosine residues within the potassium channel Kir3 modulate GTPase activity of Galphai. *J. Biol. Chem.* **277**, 32692–32696.
12. Bausch, S. B., Patterson, T. A., Appleyard, S. M., and Chavkin, C. (1995) Immunocytochemical localization of delta opioid receptors in mouse brain. *J. Chem. Neuroanat.* **8**, 175–189.
13. Pan, Z., Kao, T., Horvath, Z., Lemos, J., Sul, J. Y., Cranstoun, S. D., et al. (2006) A common ankyrin-G-based mechanism retains KCNQ and NaV channels at electrically active domains of the axon. *J. Neurosci.* **26**, 2599–25613.

Chapter 12

Immunohistochemical Assessment of Signal Transduction and Cell-Cycle Networks in Neural Tumors

Daniel Ciznadija, Afsar Barlas, and Katia Manova

Abstract

The ability to detect transient changes in molecular networks lies at the heart of cancer biology research. This is especially apparent during tumorigenesis, where initiating mutations typically affect mitogens and cell-cycle molecules such as PDGF or retinoblastoma protein (Rb). One of the primary consequences of such processes is the inappropriate stimulation of downstream targets, normally through posttranslational modification. Immunohistochemistry (IHC) provides an important tool for assessing such changes in situ, permitting different aspects of tumor biology to be examined as a tissue undergoes transformation. Nevertheless, this can be difficult to achieve, particularly in complex environments like the brain. Here, we provide the automated methodology we have employed for the successful detection of phosphorylation of S6 ribosomal protein (S6-RP) and the retinoblastoma protein (Rb) in response to PDGF stimulation in a mouse model of glial brain tumor development.

Key words: Glioma, Cell-cycle, Cell growth, Phospho-Rb, Phospho-S6 ribosomal protein, Immunohistochemistry, Brain

1. Introduction

Cancer is a disease process that can be initiated by mutations affecting mitogenic pathways, leading to oncogenic transformation and aberrant cellular growth. This is particularly evident in brain tumors such as oligodendroglioma and glioblastoma multiforme, where the primary transformative event frequently affects the EFG receptor or the PDGF/PDGF receptor loop (1). These mutations lead to abnormal signaling in glial stem and progenitor cells, with downstream molecules such as p70S6 kinase transducing information to cell-cycle and growth regulators

Alexander E. Kalyuzhny (ed.), *Signal Transduction Immunohistochemistry: Methods and Protocols*, Methods in Molecular Biology, vol. 717, DOI 10.1007/978-1-61779-024-9_12,

like cyclin D1, Rb, and S6 ribosomal protein, causing them to become inappropriately activated.

Immunohistochemistry (IHC) is a convenient and nondestructive method of detecting endogenous antigens within architecturally intact tissue. This has the advantage of permitting differential changes in molecular networks to be assessed in situ during processes such as tumorigenesis or drug interventions. Nevertheless, detection of labile, low abundance molecules by IHC poses a particular challenge, especially in those instances where a molecular function is controlled by posttranslational modification such as phosphorylation. In addition, undertaking this within the structurally intricate and highly lipidated environment of the brain adds a further layer of complexity. We describe in the following chapter methods we have successfully employed to assess the phosphorylation of the growth and the cell-cycle regulators S6 ribosomal protein and Rb in a mouse model of PDGF-induced glial brain tumorigenesis (2). While the technique described herein is based on automated detection procedures, it may also be applied manually, although we would strongly urge investigators to use machine-based protocols to minimize variability and achieve robust signals (Fig. 1).

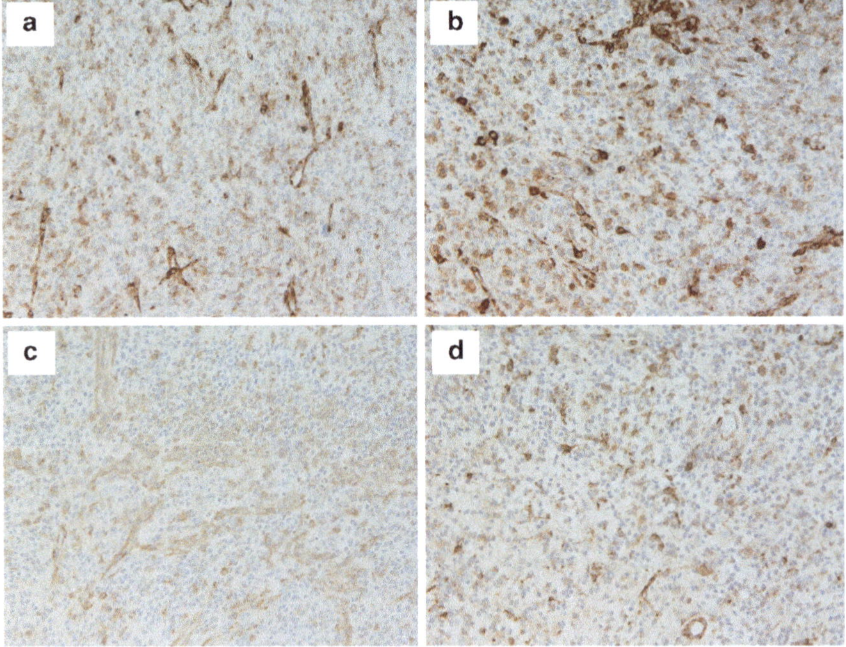

Fig. 1. Automated detection techniques provide superior detection of phosphorylated S6 ribosomal protein in murine glial tumors induced by chronic PDGF signaling. Murine gliomas induced by chronic PDGF signaling were stained for cytoplasmic phosphorylated S6 ribosomal protein using either the automated protocol described in this chapter (**a** and **b**) or manually (**c** and **d**) (all 200×). It is clear that the staining achieved using the automated system is more intense and reproducible than that obtained using manual IHC techniques (although staining is still clearly evident).

2. Materials

2.1. Fixation of Mouse Brain Tissue

1. Straight forceps and scissors (surgical grade).
2. 10% Neutral buffered formalin (see Note 1).
3. Round bottom tubes (10–15 ml volume).

2.2. Embedding Mouse Brain Tissue in Paraffin and Slide Preparation

1. Straight forceps and scalpel (surgical grade).
2. Styrofoam container (for dissecting the brain).
3. Kimwipes.
4. Plastic tissue embedding molds.
5. Ethanol (50, 70, 95% and absolute).
6. Paraffinization machine.
7. Paraffin for embedding.
8. Metal embedding blocks.
9. Cryoplatform.
10. Paraffin microtome.
11. Ultrastick glass slides (25 × 75 mm).
12. Curved forceps.
13. Waterbath.
14. Slide drying rack.

2.3. Automated Immunohistochemical Staining of Mouse Brain

1. Immunohistochemical staining is carried out using the Discovery XT automated system from Ventana Medical Systems. All the necessary solutions (listed below; a–f), except the primary, secondary, and signal development reagents are provided by the supplier:
   (a) EZ Prep/EZ buffer.
   (b) Cell conditioner #1.
   (c) Reaction buffer.
   (d) Avidin and biotin block.
   (e) Blocker D.
   (f) Streptavidin-HRP.
2. BSA/PBS: 2% BSA (w/v) dissolved in PBS, pH 7.4.
3. Blocking solution: 10% goat serum diluted in 2%BSA/PBS.
4. Phospho-Rb^{Ser780} rabbit polyclonal antibody (Cell Signaling Technology, #9307).
5. Phospho-S6 ribosomal protein$^{Ser235/236}$ rabbit polyclonal antibody (Cell Signaling Technology, #2211).
6. Biotinylated goat anti-rabbit IgG from Vectastain ABC kit (Vector Labs).

7. Hemotoxylin.
8. Permount mounting medium.
9. Coverslips (22 × 40 mm).
10. Nail varnish.

3. Methods

3.1. Fixation of Mouse Brain Tissue

All procedures for this section may be carried out at room temperature.

1. Fill enough round-bottomed tubes for your experiment with 10 ml 10% neutral buffered formalin (formalin is toxic and should be used in a well-ventilated area) (see Note 2).
2. Kill any sick or symptomatic mouse as required by your institutional animal use protocols.
3. Using surgical scissors, remove the head from the body.
4. Holding the head at the snout with the forceps, insert the point of the scissors underneath the skin flap at the base of the skull, adjoining the neck, and cut the skin lengthwise across the top of the skull to expose the underlying cranium. It is sometimes easier to cut away the skull-cap by placing the tips of the forceps through the orbital (eye) sockets and gripping the skull this way prior to cutting.
5. At the base of the skull, adjoining the neck you should see the exposed and severed spinal column (white bone). Insert the point of the scissors into the spinal column and cut upwards and across the cranial cap to expose the brain (see Note 3).
6. Use the forceps to separate the two halves of the cranial cap.
7. Remove the brain from the cranial space by inserting the closed forceps *between* the brain and the bottom of the cranial space *at the front end of the head* (towards the eyes) and levering the brain upwards and out.
8. Using the forceps to gently grasp the brain, place it into a tube of fixative.
9. Repeat Steps 2–8 for as many animals as you have.
10. Once all the brains have been collected, secure the tubes into a tube holder and rotate or rock gently for 24–48 h (see Note 4).

3.2. Embedding Mouse Brain Tissue in Paraffin and Slide Preparation

At this stage, the tissue needs to be embedded in paraffin to enable sectioning and subsequent antibody staining.

1. Using a pencil, label as many plastic tissue molds as you require.
2. Decant the fixative into an appropriate waste container and place the brain on the styrofoam container.

3. Use a Kim-wipe to remove any excess fixative.
4. Hold the brain in place by pressing down GENTLY on top using closed forceps.
5. Remove and discard the cerebellum (located at the back end of the tissue, near the region where the spinal column would join the brain) using the scalpel (unless the cerebellum is of interest).
6. Again using the forceps to hold the brain in place, section the brain tissue into four pieces through the coronal plane (see Note 5).
7. Place the pieces of brain into a plastic tissue mold and close the snap-lock lid to ensure that the tissue does not fall out of the mold.
8. Place the tissue molds into 70% ethanol until ready to continue with tissue paraffinization (see Note 6).
9. Tissue molds containing the brain slices should be placed into a receptacle in the paraffinization machine.
10. After the cycles are complete, the brain slices are ready to be embedded in paraffin to generate the tissue blocks from which slides will be made.
11. Take the tissue molds containing the paraffinized tissue to the embedding machine.
12. Remove and discard the lid from the tissue mold.
13. Place a small amount of hot liquid paraffin into a metal-embedding mold.
14. Take the brain tissue slices and arrange face-down in the liquid paraffin as desired. *Do not discard the base of the plastic tissue mold – this will be used again shortly.*
15. Place the metal tissue mold with the liquid paraffin and brain slices on top of a cryoplatform to freeze the tissue within the paraffin.
16. Fill the metal tissue mold to the top with more hot liquid paraffin and place the plastic tissue mold base on top, base-down (see Note 7).
17. Place the metal mold base-down on a cryoplatform to freeze the paraffin and create the tissue block (see Note 8).
18. Repeat Steps 12–17 for the rest of your samples.
19. Leave the molds on the cryoplatform for 30 min until the paraffin has frozen solid.
20. Carefully remove the paraffin tissue blocks from the metal molds.
21. Using a scalpel, trim away the excess paraffin from around the block, being careful not to cut yourself.

22. Place the tissue block into a microtome and cut away 5 μm sections of paraffin until the tissue is exposed.
23. Place a drop of water on the cryoplatform and put the tissue block tissue-side down on top (see Note 9).
24. Leave the tissue block on the cryoplatform for 20–30 min (see Note 10). During this time, switch on the waterbath and set it to between 40 and 45°C. Also, label the slides on which you wish to place the tissue sections. Use a pencil for marking so that the label does not come off during the staining procedure.
25. Place the tissue block back into the microtome and cut away 5 μm sections of paraffin-harboring tissue. The sections should not be wrinkled and you should see opaque tissue within them (see Note 11).
26. Once you have paraffin sections that are wrinkle-free, use curved forceps to carefully remove them from the microtome and place them into the waterbath. The sections should flatten out and become straight at this point, with as few wrinkles in the tissue regions as possible (see Notes 12 and 13).
27. Take a slide, with the side on which you wish to place the section facing up, and maneuver it underneath a tissue section.
28. Carefully raise the slide up out of the water, placing the section on the slide and ensuring it sits as near to the middle of the slide as you can get it (see Note 14).
29. Drain the excess water by tapping the slide against paper towel and place the slide in a drying rack (see Note 15).
30. Repeat Steps 20–29 for as many tissue blocks and slides as you require.

3.3. Automated Immunohistochemical Staining of Mouse Brain for Phosphorylation of Rb and S6 Ribosomal Protein

Once you have generated all your slides, you can now stain them for your antigen of interest. Again, while it is possible to perform this manually, we highly recommend use of an automated system to reduce variability and increase signal strength. At this stage, it is important that you have enough slides from each tissue block for antigen staining, as well as any slides that will be for used controls (see Note 16). Staining for phospho-Rb^{Ser780} (Fig. 2) and phospho-S6 $RP^{Ser235/236}$ was performed using a Discovery XT machine (Ventana). The staining protocol used for both antigens is presented below. All steps are automated, except the application of the primary antibody, which must be done manually when required. All other solutions are made prior to starting the staining procedure.

1. Deparaffinization.
2. Conditioner #1 (using standard conditions–for antigen retrieval (32 min)).

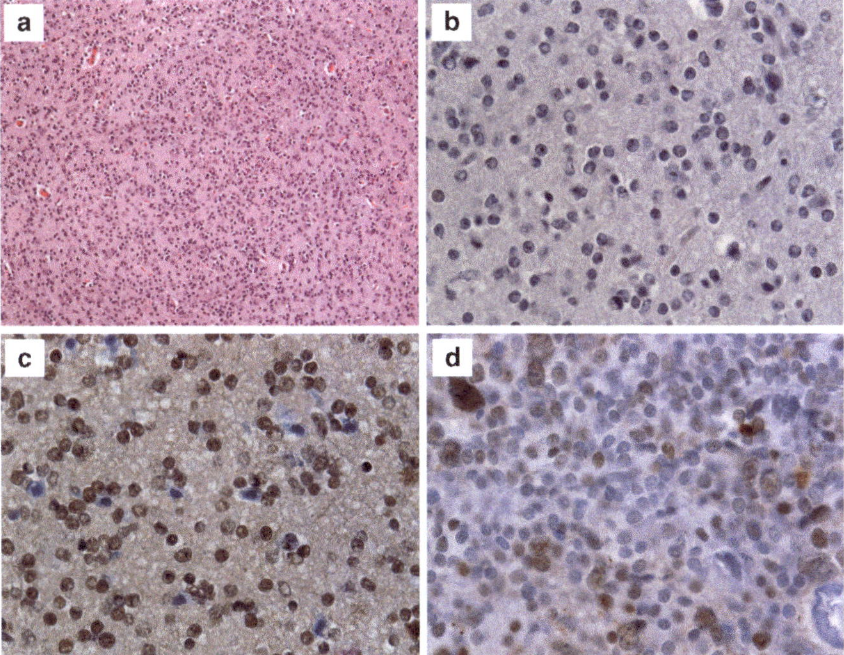

Fig. 2. Retinoblastoma protein (Rb) is phosphorylated specifically in tumorigenic glial cells in gliomas induced by chronic PDGF signaling. (**a**) Hemotoxylin–eosin staining of a mouse brain section harboring a glial tumor induced by chronic PDGF signaling. Tumorigenic glial cells are identified by their small, hyperchromatic nuclei (50×). (**b**) IgG-negative control for (**a**) (400×). (**c**) phospho-Rb^{Ser780} staining of a murine glial tumor generated by chronic PDGF signaling. The glial tumor was stained using the protocol outlined in this chapter. Notice that the brown staining is restricted to the proliferative glial cells that surround the quiescent and unstained neurons in a classic secondary structure defining oligodendroglioma, termed perineuronal satellitosis (400×) (2). (**d**) Rb$^{-/-}$ pituitary tissue used as a negative control for the phospho-Rb^{Ser780} antibody. This tissue was stained at the same time as the glial tumor tissue. Note that most of the tissue is negative, although some background remains. Compare this to the IgG control serum used in (**b**). This illustrates the necessity of appropriate negative controls to ensure specificity of staining.

3. Block sections for 32 min with blocking solution (see Notes 17 and 18).
4. Avidin/biotin block (12 min each).
5. Dilute the primary antibody (1/300 for phospho-S6-RP$^{Ser235/236}$ and 1/100 for phospho-Rb^{Ser780}) in 2%BSA/PBS and apply manually (see Note 19).
6. Incubate for 3 h for phospho-S6-RP and 4.5 h for phospho-Rb.
7. Wash in reaction buffer.
8. Secondary antibody is added to the sections (diluted to 7.5 μg/ml in 2% BSA/PBS) and incubated for 12 min for phospho-S6-RP and 60 min for phospho-Rb.
9. Wash in reaction buffer.
10. Detection reagent is added (Streptavidin-HRP/DAB-D/DAB-H_2O_2/Copper-D for phospho-S6-RP; the same for phospho-Rb with Blocker-D added to the mixture).

11. Detection reagent is inactivated and hemotoxylin added (4 min).
12. Slides are postcounterstained with bluing reagent (4 min).
13. Slides are removed from the machine.
14. A drop of Permount mounting medium is added, followed by cover-slipping (see Note 20).
15. Slides can be sealed by applying a thin layer of nail varnish around the slide edges and drying for 1 h at room temperature.

4. Notes

1. For the antigens we have described in this protocol, we use 10% buffered formalin. This may not be suitable for all antigens. The type of fixative required (formalin, paraformaldehyde, frozen in OCT, etc.) will need to be determined empirically for each antigen.
2. The nature of the model we work with requires us to process many animals at a time. As a result, we do not use perfusion fixation techniques and instead rely on postfixation of brain tissue after killing of the animal. For the antigens we have highlighted, this works well. However, if this is unsuitable for the antigen you are examining, it may be worth considering perfusion fixation. If postfixation is used, it is important to minimize the amount of time between removal of the brain and fixing.
3. Cutting across the skull-cap without damaging the brain tissue can be difficult. We find it helpful to cut across in short snips rather than a single unbroken cut.
4. The brain will require at least overnight incubation in fixative to ensure adequate tissue penetration. We have (inadvertently) fixed brains for up to 96 h without any difference in results with these antigens, but this may not hold true for all antigens of interest. Fixation conditions are important and will need to be determined empirically. Further, conical bottom tubes (such as 15-ml falcon tubes) should be avoided as tissue may become wedged in the bottom, leading to uneven fixation. Finally, we have found no difference in antigen staining between fixing a sample by rocking side-to-side or rotating end-over-end.
5. Dissecting a brain in the coronal plane requires cutting tissue sections like a loaf of bread, beginning from the frontal cortex at the front of the brain (where the eyes would be) and sectioning towards the cerebellum at the back (where the spinal column would join the brain).

6. Tissue molds with brain slices may be left in 70% ethanol for up to 1 week prior to processing.
7. The base plate from the plastic tissue mold is used to form the foundation on which the paraffin tissue block will be formed. Paraffin should be added such that it levels out ABOVE the base plate you place on top. If paraffin spills over the sides of the metal tissue mold after you place the plastic base on top, add more paraffin. This is to ensure that the plastic base plate does not come off as you remove the block from the mold.
8. Avoid leaving any air bubbles in the paraffin before placing it on the cryoplatform as this will lead to fracturing of the block when you try to remove it from the mold after the paraffin sets.
9. It is important to leave the blocks on the cryoplatform for a sufficient time to enable them to become cold enough to section properly. Should they not be cold enough, the sections generated will be folded and creased and unusable for IHC.
10. Leaving the blocks on the cryoplatform for an extended period of time is inadvisable as the tissue will begin to absorb the water and swell. This will cause the tissue to protrude from the block, making it difficult to cut properly and wasting tissue.
11. Using sections that harbor folds or creases will lead to trapping of reagents and aberrant background signals.
12. To accelerate the process of making tissue slides, it is possible to remove a string of attached paraffin sections using the forceps and carefully placing the entire set in the waterbath (avoid allowing the sections to fall on top of each other). To separate the sections, close the forceps and then place them gently (curve-down) on the break-point between two sections. Carefully allow the forceps to flick open, forcing the two sections apart. This takes some practice, so use a paraffin block with no tissue at first until you become accustomed to the procedure.
13. Try to keep your forceps dry and remove any attached paraffin; otherwise, the sections will stick to the forceps and tear as you remove them from the microtome and try to place them in the waterbath. It will also make it difficult to use the technique outlined in Note 12 to speed up slide-making.
14. You want to ensure that tissue sections are as centered on the slide as possible. Areas of tissue that sit near the edges of the slide are more susceptible to erroneous signals due to "edge effects," whereby the tissues are not exposed as effectively to all treatments.
15. We prefer to use fresh slides for each staining run, but it is possible for multiple slides from each block to be made and stored for up to 1 week at 4°C without any diminution in signal.

16. Having proper control slides will assist in the interpretation of results, as well as guide you should any part of the staining procedure not work optimally. We suggest you use the following control slides:
    (a) Positive control – tissue in which expression of antigen has been shown and detected previously. If this is not available, cells from culture, where antigen expression is known from Western blotting can be a good substitute (generate these slides by growing cells on chambered slides or cytospinning culture cells directly onto slides).
    (b) Negative controls – these controls come in various forms. It is common to omit the primary antibody and use the secondary antibody alone. While this is not a *bona fide* negative control, this may help you understand the source of any background noise. A better negative control (in conjunction with the secondary alone control) is nonimmune serum from the same species in which the primary antibody was made (for a polyclonal antibody) or an antibody to an irrelevant antigen of the same class and subclass as the primary (for a monoclonal). The best negative control (which can be difficult to obtain) is the tissue you are examining from an animal in which the antigen of interest has been knocked out.
17. The blocking solution is used to reduce nonspecific binding of the antibodies during the staining procedure. It is normally composed of inactivated serum from the same species as the secondary antibody. Should your background staining remain high, you can alter the concentration of the blocking serum (from anywhere between 2 and 20%) or incubate for a longer period of time. It is also possible to change the type of serum. Donkey serum tends to work well as a general blocking reagent.
18. Staining mouse tissue with a primary antibody that was generated in mice can be particularly difficult because of high backgrounds. If this is the case, it is worth blocking your sections with mouse IgG blocking reagent (MOM Basic Kit, Vector Laboratories) in 0.1 M glycine for 30 min.
19. In order to find the optimal antibody concentration and incubation times that maximize your signal-to-noise ratio, it is highly recommended that you begin by first using three primary antibody concentrations, one high (1/25 or 10 μg/ml if you know the stock antibody concentration), one medium (1/100 or 5 μg/ml), and one low (1/300 or 1 μg/ml). You can then optimize further based on the results of this antibody trial. In addition, you can manipulate the length of time you incubate the antibody for, as well as the temperature. Some antibodies work better when incubated at 4°C overnight, others at 37°C

for 30 min. You will have to determine what works best for your antibody empirically.

20. When applying cover-slips, try to avoid trapping air-bubbles underneath. Any trapped air will make imaging of the tissue in that area difficult. To ensure no air-bubbles become trapped, press down firmly on the center of the applied cover-slip with your finger until all the air-bubbles escape from the sides. If you need to remove the cover-slip, do not try to force it off; otherwise, the tissue may tear. Cover-slips can be removed by placing the slide in distilled water, which should allow the cover-slip to float off freely.

Acknowledgments

We acknowledge the hard work of everyone in the Molecular Cytology Core Facility at Memorial Sloan-Kettering Cancer Center for helping make our experiments just that little bit easier. We also thank Massimo Squattiro (MSKCC) for allowing us to use some of his results in this protocol chapter.

References

1. Holland, E.C. (2001) Gliomagenesis: genetic alterations and mouse models. *Nature Genetics* **2**, 120–129.
2. Dai, C., Celestino, J.C., Okada, Y., Louis, D.N., Fuller, G.N., and Holland, E.C. (2001) PDGF autocrine stimulation dedifferentiates cultured astrocytes and induces oligodendrogliomas and oligoastrocytomas from neural progenitors and astrocytes in vivo. *Genes and Development* **15**, 1913–1925.

Chapter 13

Novel Multicolor Immunofluorescence Technique Using Primary Antibodies Raised in the Same Host Species

Jillian Frisch, J.P. Houchins, Michael Grahek, Jordan Schoephoerster, Jodi Hagen, Joseph Sweet, Leopoldo Mendoza, David Schwartz, and Alexander E. Kalyuzhny

Abstract

Simultaneous detection of multiple tissue antigens is one of the most frequently used immunohistochemical (IHC) techniques. In order to avoid cross-reactivity of each secondary antibody with multiple primary antibodies when doing either dual- or triple-labeling immunofluorescence, it is necessary to use primary antibodies raised in different host species such as mouse, rabbit, and goat. However, in many cases, suitable primary antibodies raised in different species are unavailable. We have developed a novel technique for triple-labeling immunofluorescence that can be used with primary antibodies derived from a single host source. This technique includes modification of one primary antibody with biotin (ChromaLink™ Biotin) and a second primary antibody with DIG (ChromaLink™ Digoxigenin). For IHC staining, cells or tissue sections are incubated first with unconjugated primary antibody against the first target protein followed by detection with antiprimary secondary antibody conjugated to NorthernLights™ NL-637 tag (fluorescence in the far-red spectral region). Subsequently, the same tissue sections are incubated with a mixture of same species biotin-labeled primary antibody (against the second target protein) and DIG-labeled primary antibody (against the third target protein) followed by detection using a mixture of Streptavidin NorthernLights™ NL-493 tag (green fluorescence) and anti-DIG secondary antibody conjugated to a Rhodamine Red X™ tag (red fluorescence). This technique provides good spectral separation of colors depicting different antigens of interest while avoiding cross-reactivity between irrelevant primary and secondary antibodies. In addition, this multiplexed IHC technique provides significant convenience to researchers who have only primary antibodies raised in the same host species at their disposal.

Key words: Immunofluorescence, Immunohistochemistry, Phospho-specific antibodies, Triple-labeling, Antibody conjugation, Biotinylation, Digoxigenin

Alexander E. Kalyuzhny (ed.), *Signal Transduction Immunohistochemistry: Methods and Protocols*, Methods in Molecular Biology, vol. 717, DOI 10.1007/978-1-61779-024-9_13,

1. Introduction

Multicolor immunofluorescence histochemistry is one of the most frequently used techniques in neuroscience research (1). It allows researchers to study temporal and spatial colocalization of different neuronal targets (2). When targeting two or three antigens simultaneously, researchers must employ primary antibodies raised in different host species to avoid cross-reactivity of each secondary antibody with multiple primary antibodies (3). However, in many cases the most specific primary antibodies available are raised in the same host species precluding investigators from doing traditional multicolor immunofluorescence and requiring them to employ complicated protocols (4, 5). In order to use primary antibodies raised in the same host species and avoid the risk of their cross-reaction with secondary antibodies, it would be necessary to modify the primary antibodies so that they interact only with the secondary antibody of choice.

The purpose of this study was to develop a simple IHC technique allowing unmodified primary antibody to be combined with primary antibodies raised in the same host species but modified via either biotinylation or labeling with digoxigenin (DIG) for triple-labeling multicolor fluorescence (Fig. 1).

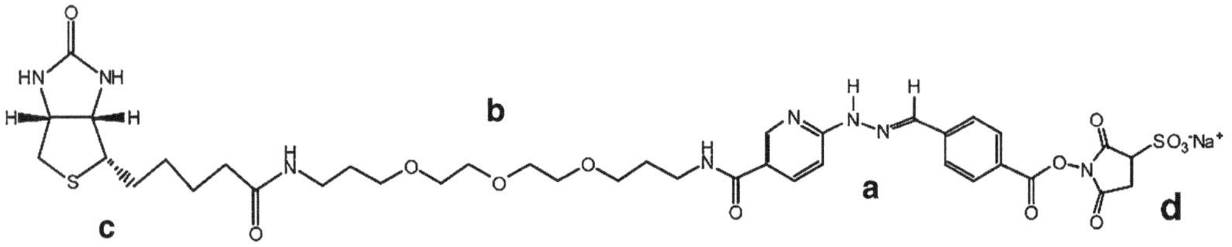

ChromaLink™ Biotin (1)

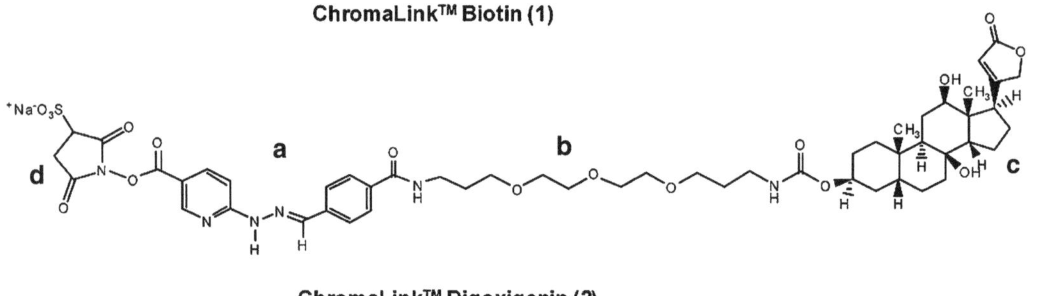

ChromaLink™ Digoxigenin (2)

Fig. 1. The structures of ChromaLink Biotin (*1*) and ChromaLink Digoxigenin (*2*). The conjugation kits include all the components needed to label rapidly 100 μg of antibody (~90 min) with >80% yield and quantify the number of haptens incorporated into the antibody. Molecules' structure contains a bisaryl hydrazone chromophore (**a**) linked by a PEG3 linker arm (**b**) to the hapten (**c**) and a water-solubilizing sulfo-NHS ester (**d**). This reagent allows for direct spectroscopic quantification of incorporated biotin using as little as 2 μL in a NanoDrop™ spectrophotometer.

2. Materials

2.1. Animals

Rats (Sprague–Dawley, 200 g) were deeply anesthetized with sodium phenobarbital and transcardially perfused with PBS followed by Lana's fixative and 10% sucrose. Brain was cut into nominal 10 μm thick cryostat sections (see Note 1).

2.2. Primary Antibodies

1. Affinity-purified rabbit anti-human, mouse, rat phospho-RSK1(S221)/RSK2(S227) (R&D Systems Cat # AF892). See ref. (6).
2. Affinity-purified rabbit anti-human, mouse, rat phospho-ERK1 (T202/Y204)/ERK2 (T185/Y187) (R&D Systems Cat # AF1018). See ref. (7).
3. Affinity-purified rabbit anti-human phospho-Ret (Y905) (R&D Systems Cat # AF3269). See ref. (8).

2.3. Solulink Conjugation Kits

1. ChromaLink Biotin One-Shot Antibody Labeling Kit (Cat # B-9007-009K).
   - Sulfo-ChromaLink™ Biotin 8.0 μg.
   - 1× Modification buffer (pH 7.4)1.5 mL.
   - Collection tubes1.5 mL.
   - Zeba™ Spin Columns 0.5 mL.
   - Biotinylated IgG Control 100 μg.
   - 1 M Tris–HCl (pH 8.9) 100 μL.
2. ChromaLink Digoxigenin One-Shot Antibody Labeling Kit (Cat # B-9014-009K).
   - ChromaLink™ Digoxigenin 11 μg.
   - 1× Modification buffer (pH 7.4) 1.5 mL.
   - 1× PBS (pH 7.2) 1.5 mL.
   - DMF (anhydrous) 25 μL.
   - Collection tubes 1.5 mL.
   - Zeba™ Spin Columns 0.5 mL.
   - 1 M Tris–HCl (pH 8.9) 100 μL.

2.4. Fluorescent Secondary Antibody and Detection Reagents

1. NorthernLights™ Anti-rabbit NL-637 (R&D Systems; Cat # NL005).
2. NorthernLights™ Streptavidin NL-493 (R&D Systems; Cat # NL997).
3. Anti-DIG Rhodamine Red X™ (Jackson ImmunoResearch; Cat # 200-292-156).
4. DAPI.

2.5. Immunohistochemistry

1. Dako PAP Pen: to draw a hydrophobic circular line around tissue sections to prevent a leakage of primary and secondary antibodies applied to tissue sections.
2. Phosphate-buffered saline (PBS): Fill a 1-L beaker with 900 mL of distilled water and dissolve 0.23 g of NaH_2PO_4 (anhydrous), 1.15 g Na_2HPO_4 (anhydrous), and 9 g NaCl. Adjust pH to 7.4 using 1 M NaOH and/or 1 M HCl. Adjust volume to 1 L with distilled water.
3. Fixative: 4% formaldehyde in Sorenson's Phosphate Buffer. Wear mask and gloves and use chemical fume hood when preparing paraformaldehyde fixative. Start by making the Sorenson's Phosphate Buffer by dissolving 8.06 g potassium phosphate and 19.99 g dibasic sodium phosphate in 900 mL deionized water. pH to 7.2 and fill with deionized water to 1 L. Then make 8% formaldehyde solution by dissolving 10 g of paraformaldehyde powder (Sigma, St. Louis, MO) in 95 mL of deionized water using heating stir plate. Heat this solution during stirring. Turn the heat off after temperature reaches 56–58°C and add 1–2 drops of 1 M NaOH to clear solution. Continue stirring for another 20–30 min and then filter this solution using regular filter paper (for example Whatman #1). Add the paraformaldehyde solution to 125 mL Sorenson's Phosphate Buffer, and fill with deionized water to 250 mL.
4. Antibody diluent: PBS containing 1% bovine serum albumin, 1% normal donkey serum, 0.3% Triton X-100 (v/v), and 0.01% sodium azide. Store at –20°C.
5. Antifade mounting media: NorthernLights™ Guard (Catalog # NL996; R&D Systems, Inc.).
6. Coverslips for histological slides: 24 × 50 mm, thickness #1.
7. Nuclear counterstain: 300 nM DAPI in PBS.
8. Microscopy: fluorescence microscope Provis equipped with cooled DP71 color digital camera (Olympus, Melville, NY) and fluorescence filter set to visualize Rhodamine Red™ X (570 nm excitation and 590 nm emission), NL-637 (637 nm excitation and 658 nm emission), NL-493 (493 nm excitation and 514 nm emission), and DAPI (345–360 nm excitation and 456–460 nm emission).

3. Methods

3.1. Biotinylation of Anti-human Phospho-Ret (Y905)

1. Initial sample – lyophilized vial of 100 μg of antibody; Resuspend the sample in 100 μL 1× Modification Buffer (pH 7.4) to yield a 1.0 mg/mL solution.

3.1.1. Sample Preparation

2. Initial sample – >1 mg/mL of antibody in liquid form; Adjust the antibody concentration by diluting an aliquot to 1 mg/mL and 100 μL with 1× Modification Buffer (pH 7.4). Transfer this volume (100 μL) to a new 1.5-mL microfuge tube and refrigerate the unused portion of the concentrated sample.

3.1.2. First Buffer Exchange

1. Prepare spin columns by twisting off bottom closures and loosening caps.
2. Put each spin column into a collection tube and place in microcentrifuge in a balanced position. Spin for 1 min in the microcentrifuge.
3. Remove columns from the centrifuge and discard the collection tube solution. Mark on the side of the column where resin is slanted upward (see Note 5).
4. Mark one column with the letter A and the other column with the letter B.
5. Add 300 μL of 1× Modification Buffer to column A and 300 μL of 1× PBS to column B (see Note 2). Loosely recap lids and put back in centrifuge.
6. Centrifuge columns in their collection tubes for 1 min.
7. Repeat Steps 5 and 6 two additional times for a total of three washes.
8. Transfer spin column A into a new collection tube.
9. Add 100 μL of antibody sample to column A and 100 μL of 1× PBS to column B, recapping the lids loosely.
10. Centrifuge for 2 min and collect eluate at the bottom of the collection tubes.
11. Collect antibody in tube A and set aside. Discard eluate from column B.
12. Rehydrate spin columns with 300 μL of 1× PBS in column B and molecular grade water in column A (see Note 3).
13. Using a NanoDrop™ spectrophotometer, scan the antibody sample from collection tube A to determine the amount of recovered antibody, which should be at a concentration of 0.8–1.2 mg/mL in a volume of 95–105 μL.

3.1.3. Biotinylation Procedure

1. Transfer buffer-exchanged antibody solution directly to the vial of Sulfo-ChromoLink Biotin labeling reagent.
2. Mix solution thoroughly by pipetting solution up and down and vortexing for a few seconds. Centrifuge for 5 s to collect reaction mixture at the bottom of the tube.
3. Allow reaction to proceed for 60 min at room temperature.
4. When reaction is complete, quench reaction by adding 10 μL 1 M Tris to vial, mix well.

5. Place previously rehydrated spin columns in centrifuge and spin for 1 min to remove buffer.
6. Transfer spin column B to a new collection tube.

3.1.4. Second Buffer Exchange

1. Add entire contents of quenched biotin labeling reaction to column B.
2. Add 100 μL of molecular grade water to column A.
3. Place columns in centrifuge and spin for 2 min.
4. Transfer biotin-labeled antibody to a new 1.5-mL microfuge tube. Refrigerate labeled sample.

3.1.5. Determining Biotin Incorporation

1. Turn on NanoDrop™ spectrophotometer and launch its software.
2. Select A280 option on the menu.
3. Place a 2-μL drop of molecular grade water on clean pedestal, click "OK."
4. Click off the 340 nm normalization option using the mouse.
5. In the window labeled Sample Type, select "IgG" option from the pull-down menu.
6. Blank the NanoDrop™ spectrophotometer by placing a 2-μL drop of 1× PBS on the pedestal and click "Blank" by using the mouse (see Note 4).
7. Clean off the pedestal and transfer 2 μL of labeled antibody solution to the pedestal and click "Measure."
8. Record the absorbance at 280 nm from the wavelength absorbance window.
9. Record the absorbance at 354 nm by typing "354" into the wavelength window.
10. Use Molar Substitution Ratio (MSR) formula in product insert to calculate the number of biotin molecules attached to one molecule of the antibody. Typical MSRs range from 3 to 8 biotins per antibody.

3.2. Digoxigenin Conjugation of Anti-human, Mouse, Rat Phospho-RSK1(S221)/RSK2(S227)

3.2.1. Sample Preparation

1. Initial sample: vial of 100 μg of lyophilized antibody; Resuspend the sample in 100 μL 1× Modification Buffer (pH 7.4) to yield a 0 mg/mL solution.
2. Initial sample: if >1 mg/mL of antibody in liquid form adjust the antibody concentration by diluting an aliquot to 1 mg/mL and 100 μL with 1× Modification Buffer (pH 7.4). Transfer this volume (100 μL) to a new 1.5-mL microfuge tube and refrigerate the unused portion of the concentrated sample.

3.2.2. First Buffer Exchange

1. Prepare spin columns by twisting off bottom closures and loosening caps.
2. Place each spin column into collection tube and put in microcentrifuge opposite to each other. Spin for 1 min in microcentrufuge (see Note 5).
3. Remove columns from centrifuge and discard collection tube solution. Mark on the side of the column where resin is slanted upward.
4. Mark one column with the letter A and the other column with the letter B.
5. Add 300 μL of 1× Modification Buffer to column A and 300 μL of 1× PBS to column B. Loosely recap lids and put back in centrifuge.
6. Centrifuge columns in their collection tubes for 1 min.
7. Repeat Steps 5 and 6 twice for a total of three washes.
8. Transfer spin column A into a new collection tube.
9. Add 100 μL of antibody sample to column A and 100 μL of 1× PBS to column B, loosely recapping the lids.
10. Centrifuge for 2 min to collect eluate at the bottom of the collection tubes.
11. Collect antibody in tube A and set aside. Discard eluate from column B.
12. Rehydrate spin columns with 300 μL of 1× PBS in B and molecular grade water in A.
13. Using NanoDrop™ spectrophotometer, scan the antibody sample from collection tube A to find the amount of recovered antibody. The antibody should be at a concentration of 0.8–1.2 mg/mL in a volume of 95–105 μL.

3.2.3. Digoxigenin Labeling Procedure

1. Add 5 μL of DMF directly to the bottom of ChromaLink Digoxigenin labeling reagent vial.
2. With a P-10 pipette, pipette the DMF solution up and down, rinsing the side wall of the tube several times to dissolve the Digoxigenin. The solution should appear clear and slightly yellow.
3. Add the entire volume of antibody solution to the DMF/ChromaLink Digoxigenin vial; mix by pipetting up and down several times.
4. Incubate for 60 min at room temperature.
5. After the reaction is complete, quench the reaction by adding 10 μL 1 M Tris.

6. Centrifuge quenched reaction for 30 s.
7. Place the previously rehydrated spin columns in the centrifuge and spin for 1 min. Discard flow-through buffer.
8. Transfer column B into a new collection tube.

3.2.4. Second Buffer Exchange

1. Add the entire contents of quenched digoxigenin labeling reaction to column B.
2. Add 100 μL of molecular grade water to column A.
3. Place columns in centrifuge and spin for 2 min.
4. Transfer DIG-labeled antibody to a new 1.5-mL microfuge tube.

3.2.5. Determining Digoxigenin Incorporation

1. Turn on NanoDrop™ spectrophotometer and start its software.
2. Select A280 option on the menu.
3. Place a 2-μL drop of molecular grade water on clean pedestal, click "OK."
4. Click off the 340 nm normalization option using the mouse.
5. In the window labeled Sample Type, select "IgG" option from the pull-down menu.
6. Blank NanoDrop™ spectrophotometer by placing a 2-μL drop of 1× PBS on the pedestal and click "Blank" by using the mouse.
7. Clean off pedestal and transfer a 2-μL volume of labeled antibody solution to pedestal and click "Measure."
8. Record the absorbance at 280 nm from the wavelength absorbance window.
9. Record the absorbance at 354 nm by typing "354" into the wavelength window. Refrigerate labeled sample.
10. Use MSR formula in product insert to calculate the number of digoxigenin molecules attached to one molecule of the antibody. Typical MSRs range from 2 to 8 digoxigenin molecules per antibody.

3.3. Immunofluorescence

Day 1

1. Remove slides from storage (see Note 6) and draw a PAP pen ring around tissue sections.
2. Load ~150 μL of 1× PBS onto each slide and incubate for 10 min to remove excess OCT compound (see Note 7) and to rehydrate tissue sections.
3. Drain off as much liquid as possible by tapping the slide on its edge and touch up PAP pen as needed.

4. Load ~150 μL of unconjugated primary antibody (rabbit anti-human, mouse, rat phospho-ERK1 (T202/Y204)/ ERK2 (T185/Y187) onto tissue section and incubate overnight at 4°C (see Note 8).

Day 2

5. Rinse sections three times, 1 min each time, in 1× PBS + Tween-20.
6. Drain off as much liquid as possible by tapping the slide on its edge and touch up PAP pen as needed.
7. Load ~150 μL of secondary antibody corresponding to species of primary antibodies (anti-rabbit NL-637); incubate 1 h at room temperature.
8. Rinse sections three times, 1 min each time, in 1× PBS + Tween-20.
9. Drain off as much liquid as possible by tapping the slide on its edge and touch up PAP pen as needed.
10. Load ~150 μL of a mixture of biotinylated primary antibody and DIG-conjugated primary antibody (Btn rabbit anti-human phospho-Ret (Y905) and DIG rabbit anti-human/ mouse/rat phospho-RSK1(S221)/RSK2(S227)); Incubate overnight at 4°C (see Note 9).

Day 3

11. Rinse sections three times, 1 min each time, in 1× PBS + Tween-20.
12. Tap off as much liquid as possible and touch up PAP pen as needed.
13. Load ~150 μL of mixture of Streptavidin NL493 and Anti-DIG Rhodamine Red X™; Incubate 1 h at room temperature.
14. Rinse sections three times, 1 min each time, in 1× PBS + Tween-20.
15. Counterstain tissues using DAPI (or other nuclei-labeling dye) and coverslip slides under antifade mounting media NorthernLights™ Guard (Catalog # NL996; R&D Systems, Inc.) (see Fig. 2 and Note 10).

4. Notes

1. For immunofluorescence in CNS, it is recommended to use younger rats that have not accumulated large amounts of the "aging" pigment lipofuscin in neuronal cells (9). Lipofuscin has a broad emission spectra that overlaps with a large number

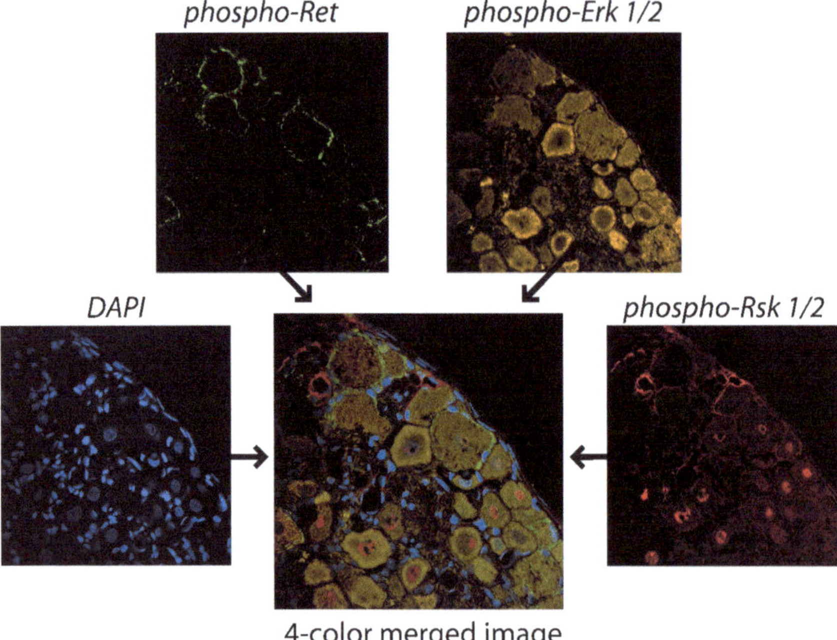

Fig. 2. Confocal image of immunofluorescence detection of phosphorylated proteins Ret, ERK 1/2, and RSK 1/2 on 10 μm thick cryostat sections of the rat dorsal root ganglion (DRG) tissue. Four-color image in the center of the panel was generated by merging four individual single-color digital images.

of fluorophores and obscures their visualization (10). If there is no choice but to use older animals, then lipofuscin's autofluorescence can be quenched by incubating tissue sections with 1–10 mM $CuSO_4$ in 50 mM ammonium acetate buffer (pH 5.0) or with Sudan Black B (SB) in 70% ethanol (11).

2. Do not disrupt the resin bed of the spin column with the pipette tip. This will ensure that the entire column matrix is available for consistency in centrifugation.
3. Keep spin columns hydrated at all times until end of conjugation to maximize antibody recovery.
4. Blanking with a solution other than PBS on NanoDrop™ spectrophotometer will give inaccurate results because the antibody is originally diluted with PBS.
5. When placing spin columns in microcenterfuge, arrange columns so that slanted resin beds point away from the rotor to ensure consistency of centrifugation.
6. Cryosections are often kept in slide boxes at -20°C for long-term storage. During freeze–thaw cycles, moisture can accumulate in boxes and damage tissue sections. Sealing box seams with 1 in. wide stretchable tape (Lab Safety Supply Cat # 5474Y) can eliminate this problem if boxes are allowed to

warm to room temperature before the tape is removed. Tape should be replaced before boxes are returned to the freezer.

7. To prepare tissue blocks for cutting on the cryostat, it is recommended to use Andwin Scientific Tissue-Tek* CRYO-OCT Compound (Fisher Scientific Cat # 14-373-65). This embedding media is easy to handle and it produces good and consistent results that minimize the appearance of tissue-freezing artifacts. Freezing artifacts can be seen as "swiss cheese" holes resulting from the loss of cell nuclei during freezing. Besides affecting cellular morphology, freezing artifacts also affect the profile of IHC labeling: edges of the holes tend to absorb more primary and secondary antibodies causing the appearance of numerous bright halos that obscure specific staining.
8. All incubations should take place in a moist environment. Slide Show 20 incubation chambers (Newcomer Supply Cat # 6844-20cl) were used in this experiment and are recommended. The amount of antibody needed per slide depends on the size of the PAP pen rings. Too little antibody can dry up during incubation and too much antibody can leakage over the PAP pen ring off the slide. Generally, something between 50 and 200 μL is appropriate. Avoid tissue sections drying during the incubation because this will result in high nonspecific background staining. If, after incubation, some tissue sections become dry, rehydrating them does not help, and such tissue sections should be discarded. Watch out for partially dry margins of the tissue sections because staining in these areas will appear stronger than in the rest of the tissue section.
9. Secondary antibodies bound to unconjugated primary antibodies may, under some conditions, dissociate from them and bind with biotin- and DIG-labeled primary antibodies added during later subsequent incubation steps. This can be detected as unexpected tissue labeling with either biotin- and/or DIG-labeled antibodies. To block the unwanted reaction of dissociated secondary antibodies, before applying biotin- and DIG-labeled primary antibodies, incubate tissue sections with 1–5% normal rabbit serum. After 30 min, drain it from the slides, rinse tissue sections with PBS and add the mixture of biotin- and DIG-labeled primary antibodies.

   To ensure that tissue labeling does not result from nonspecific binding of secondary antibodies, it is recommended to employ a secondary antibody control. This is done by incubating tissue sections with antibody diluent that does not have primary antibodies, and then incubate with secondary antibodies: a lack of labeling will indicate that there is no nonspecific binding of secondary antibodies to tissue. However, if nonspecific labeling is observed, additional steps are required to abolish it. For example, nonspecific labeling may be caused via crosslinking interaction of nonreduced aldehyde groups of the formaldehyde-based fixatives with IgGs of secondary antibodies. To block nonreduced aldehyde groups, tissues, before adding primary antibodies, can be incubated with 0.5 mg/mL of sodium borohydrate ($NaBH_4$) for 10–20 min at room temperature. Alternatively, before adding primary antibodies, nonspecific tissue-binding sites may be blocked by incubating tissues with 10% normal serum (horse, swine, or donkey) for 5–30 min at room temperature. It is of critical importance to use blocking normal serum from species other than the host of primary antibodies; otherwise, secondary antibodies will cross-react with blocking serum retained by tissues causing nonspecific labeling: for example, blocking tissue with goat serum and adding antigoat secondary antibodies will result in strong background staining.

10. NorthernLights™ Guard is a glycerol-based mounting media that protects fluorescent dyes from fading. It can be used alone or mixed 1:1 with counterstaining reagents like DAPI.

References

1. Kalyuzhny, A.E. (2009) The dark side of the immunohistochemical moon: industry. *J. Histochem. Cytochem.* **57**, 1099–1101.
2. Wessendorf, M.W., Appel, N.M., Molitor, T.W., and Elde, R.P. (1990) A method for immunofluorescent demonstration of three coexisting neurotransmitters in rat brain and spinal cord, using the fluorophores fluorescein, lissamine rhodamine, and 7-amino-4-methylcoumarin-3-acetic acid. *J. Histochem. Cytochem.* **38**, 1859–1877.
3. Staines, W.A., Meister, B., Melander, T., Nagy, J.I., and Hökfelt, T. (1988) Three-color immunofluorescence histochemistry allowing triple labeling within a single section. *J. Histochem. Cytochem.* **36**,145–151.
4. Wang, B.L., and Larsson, L.I. (1985) Simultaneous demonstration of multiple antigens by indirect immunofluorescence or immunogold staining. Novel light and electron microscopical double and triple staining method employing primary antibodies from the same species. *Histochemistry.* **83**, 47–56.
5. Van der Loos, C.M., Das, P.K., Van den Oord, J.J., and Houthoff, H.J. (1989) Multiple immunoenzyme staining techniques. Use of fluoresceinated, biotinylated and unlabelled monoclonal antibodies. *J. Immunol. Methods.* **117**, 45–52.
6. Cavet, M.E., Lehoux, S., and Berk, B.C. (2003) 14-3-3beta is a p90 ribosomal S6 kinase (RSK) isoform 1-binding protein that negatively regulates RSK kinase activity. *J. Biol. Chem.* **278**, 18376–1883.
7. Casar, B., Sanz-Moreno, V., Yazicioglu, M.N., Rodríguez, J., Berciano, M.T., Lafarga, M., et al., (2007) Mxi2 promotes stimulus-independent ERK nuclear translocation. *EMBO J.* **26**, 635–646.
8. Takahashi, M., Buma, Y., Iwamoto, T., Inaguma, Y., Ikeda, H., and Hiai, H. (1988) Cloning and expression of the ret proto-oncogene encoding a tyrosine kinase with two potential transmembrane domains. *Oncogene.* **3**, 571–578.
9. Brizzee, K.R., Ordy, J.M., and Kaack. B. (1974) Early appearance and regional differences in intraneuronal and extraneuronal lipofuscin accumulation with age in the brain of a nonhuman primate (Macaca mulatta). *J. Gerontol.* **29**, 366–381.
10. Dowson, J.H., Armstrong, D., Koppang, N., Lake, B.D., and Jolly, R.D. (1982) Autofluorescence emission spectra of neuronal lipopigment in animal and human ceroidoses (ceroid-lipofuscinoses). *Acta Neuropathol.* **58**, 152–156.
11. Schnell, S.A., Staines, W.A., and Wessendorf, M.W. (1999) Reduction of lipofuscin-like autofluorescence in fluorescently labeled tissue. *J. Histochem. Cytochem.* **47**, 719–730.

Chapter 14

Activation and Differentiation of Mesenchymal Stem Cells

Pravin J. Mishra and Debabrata Banerjee

Abstract

Mesenchymal stem cells (MSCs) are multipotent cells and exhibit two main characteristics that define stem cells: self-renewal and differentiation. MSCs can migrate to sites of injury, inflammation, and tumor. Moreover, MSCs undergo myofibroblast-like differentiation, including increased production of alpha smooth muscle actin (α-SMA) in response to transforming growth factor-β (TGF-β), a growth factor commonly secreted by tumor cells to evade immune surveillance. Based on our previous finding, hMSCs become activated and resemble carcinoma-associated myofibroblasts upon prolonged exposure to conditioned medium from MDAMB231 human breast cancer cells. Here, we show that keratinocyte-conditioned medium (KCM) induces differentiation of MSCs to resemble dermal myofibroblast-like cells using immunofluorescence techniques demonstrating punctate vinculin staining, and F-actin filaments.

Key words: Myofibroblast, Dermal myofibroblast, Stem cells, Breast cancer, Colorectal cancer, Mesenchymal stem cells, Immunofluorescence, Cell differentiation, Migration, SDF-1

1. Introduction

The adult human bone marrow harbors two populations of progenitor cells, the hematopoietic stem/progenitor cells and non-hematopoietic or mesenchymal/bone marrow stromal cells (MSCs). MSCs are multipotent stromal cells that differentiate into cell lineages of bone, cartilage, fat, and fibrous connective tissues under appropriate inductive conditions (1–6).

MSCs, although initially characterized from bone marrow, are also distributed in various other tissues such as muscle connective tissue, perichondrium, adipose tissue, periosteum, and fetal tissues (7–10). MSCs have also been found in amniotic fluid and placenta (11, 12). In addition, MSCs can be isolated from umbilical cord blood (UCB), but the success rate of this isolation

Alexander E. Kalyuzhny (ed.), *Signal Transduction Immunohistochemistry: Methods and Protocols*, Methods in Molecular Biology, vol. 717, DOI 10.1007/978-1-61779-024-9_14,

is limited due to the low frequency of MSCs in UCB (13, 14). Given these attributes, MSCs are strikingly of great interest for cancer therapeutics and tissue regeneration (15).

MSCs have been characterized by flow cytometry based on expression of several characteristic surface markers. hMSCs express CD105 (SH2 or endoglin), CD73 (SH3, SH4, or ecto-5'-nucleotidase), CD90 (THY1), CD44, CD71 (transferrin receptor), and CD271 (low-affinity nerve growth factor receptor) (16–18). In addition, bone marrow derived MSCs can also be characterized by lack of expression of hematopoietic CD45, CD34, CD14, CD80, CD86, and CD40 markers (17).

Currently, for in vitro culture, MSCs are expanded in alpha MEM in the presence of 10% fetal calf serum (FCS) or fetal bovine serum (FBS). The use of MSCs cultured in FCS has raised concerns about the biosafety of these cells for clinical applications due to the possible transmission of prions and stimulation of immunogenic responses within the host (19). Consequently, the use of serum-free media in the expansion of MSCs is currently being tested. Besides serum-free medium, the use of autologous human serum for the in vitro expansion of MSCs have been tested and shown to yield MSCs with greater proliferative capacity as compared to that of allogeneic human serum or FCS (20). For routine cell culture for non-regenerative/nontherapeutic use, MSCs can be cultured in FCS.

We present experimental methods for studying activation and/or differentiation of MSCs using immunofluorescence techniques. These are based on expression of markers associated with the activated or the differentiated state of MSCs. Examples we have chosen include activation by tumor-conditioned medium, Stromal-derived factor-1 (SDF-1), which activate MSCs to migrate to tumor sites and promote tumor growth. The example of differentiation used here is keratinocyte-conditioned medium (KCM) induced differentiation of MSCs to resemble dermal myofibroblasts. This is thought to be involved in the process of wound healing. In the process of wound healing, myofibroblasts are responsible for generation of mechanical forces that allow proper granulation, tissue contraction, and wound healing. Matrix contraction depends on both alpha-smooth muscle actin (α-SMA) expression within cellular stress fibers, and assembly of large focal adhesions linking myofibroblasts to the matrix.

2. Materials

2.1. Cell Culture

1. Dulbecco's Modified Eagle's Medium (DMEM) (Gibco Grand Island, NY) Supplemented with 10% FBS (Gibco) and one penicillin-streptomycin (Gibco).
2. Solution of trypsin (0.25%) and ethylenediamine tetraacetic acid (EDTA) (Sigma; St. Louis, MO) (see Note 1).

3. α-MEM Medium (Gibco Grand Island, NY) supplemented with 10% FBS (Gibco) and 1% penicillin-streptomycin (Gibco).
4. RPMI medium (Invitrogen Corporation, Carlsbad, CA, http://www.invitrogen.com) supplemented with 10% FBS and 1% penicillin-streptomycin (Gibco).
5. Phosphate buffered saline (PBS) (Gibco, NY).
6. Human bone marrow derived mesenchymal stem cell (hMSC) pooled donor cell line (Lonza, MD).
7. Unprocessed bone marrow (36×10^6 cells/ml) (Lonza, MD).
8. Colorectal cancer cell C85 & breast cancer cell MDAMB231 (American Type Culture Collection; Manassas, VA, http://www.atcc.org).
9. Normal human epithelial primary keratinocyte cell line (NHEK; C-12001) derived from foreskin (~500,000 cells) (Promocell GmbH; Heidelberg, Germany) and cultured in Keratinocyte Growth medium (KGM; C-20011) (see Note 2).
10. Millipore sterile 50 ml filtration system with 0.45 μm PVDF membrane.

2.2. Immunofluorescence

1. Coverslips for histological slides.
2. Histological slides ($25 \times 75 \times 1$ mm).
3. Phosphate buffered saline (PBS).
4. Paraformaldehyde (Sigma): 4% (w/v) solution in PBS fresh for each experiment (see Note 3). The solution may need to be carefully heated (use a stirring hot-plate in a fume hood) to dissolve, and then cool to room temperature for use.
5. Permeabilization solution: 0.1% (v/v) Triton X-100 in PBS.
6. Blocking buffer: 3% (w/v) BSA in PBS (see Note 4).
7. Monoclonal Anti-Vinculin antibody (1:200, P1951; Sigma-Aldrich); α-Smooth Muscle Actin (1:250; mouse monoclonal clone 1A4, A2547); Fibroblast Surface Protein (1:250; mouse monoclonal clone 1B10, F4771); Vimentin (1:200, clone VIM-13.2, V5255; Sigma-Aldrich).
8. Secondary antibody: Alexa Fluor488P (Ab') 2, IgG (H+L) (1:400; Molecular Probes) and Alexa Fluor 555 goat anti-mouse IgM (1:400; Invitrogen).
9. Phalloidin–Tetramethylrhodamine B isothiocyanate (50 ng/ml), (Sigma-Aldrich).
10. α-Tubulin (Sigma, St. Louis, MO) diluted 1:2,000.
11. Counterstaining: nuclear dye TOPRO-3 iodide (1:1,000; Invitrogen, Molecular Probes) in PBS (Life Technologies).
12. Mounting Medium with 4',6-diamidino-2-phenylindole (DAPI) (VectaShield, Vector Laboratories).

3. Methods

3.1. Cell Culture

1. hMSC pooled donor cells were cultured in α-MEM and maintained below passage 15. In addition, MSCs are isolated from unprocessed bone marrow (36×10^6 cells/ml).
2. A Ficoll gradient is used to eliminate non-MSC bone marrow cells and the cells are placed in a 5% CO_2 incubator at 37°C.
3. Media is supplemented daily and aliquots from passages two to five are frozen in the liquid nitrogen tank for use in future experiments.
4. Flow cytometry is utilized to determine cell surface markers expressed on hMSCs and these cells are negative for CD45, HLA-DR, and CD11b (see Note 5).

3.2. Immuno-fluorescence Staining of F Actin

1. hMSCs are placed on glass slides and allowed to adhere overnight. Breast cancer cell MDA-MB231 conditioned medium is added onto the cells the following day and the incubation continued for an additional 24 h.
2. Removed medium, washed with PBS 2 times at RT, cells are fixed with 4% Paraformaldehyde for 10 min, washed with PBS (5 times; no incubation needed), and processed for immunofluorescence staining of F-actin according to previously published procedures, briefly permeablized cells with 0.1% Triton for 5 min, washed twice with PBS (5 min), and blocked with blocking buffer for 15 min at RT, washed with PBS 5 times.
3. Phalloidin-TRITC is used at final concentration of 50 ng/ml; TOPRO-3 is used for nuclear staining at 1:250 dilutions.
4. Actin stress fiber formation is observed in hMSCs stimulated with tumor cell- conditioned medium.
5. This organization is completely disrupted by Jak2 inhibitor, MEK inhibitor, or a combination of the two inhibitors (see Note 6).
6. Inhibitor treatment not only disrupted the F-actin reorganization induced by tumor cell conditioned medium but also destabilized the cytoskeleton network. Cells were unable to migrate in response to SDF-1 gradients, consistent with our previous results (21).

3.3. Immuno-fluorescence Staining for MTOC

1. MSCs are placed on glass slides immediately adjacent to C85 tumor cells for 24 h and allowed to adhere. Fresh growth medium is added gently to cover the MSCs and tumor cell spots.
2. Following incubation for 24 h, cells are processed for immunofluorescence staining for F actin and microtubule organization

center (MTOC) according to standard techniques using fluorescent-labeled reagents obtained from Molecular Probes (Invitrogen, Carlsbad, CA).

3. Cells are fixed with 3.7–4% paraformaldehyde for 10 min.
4. Washed three times in 1× PBS and permeabilized with 0.1% Triton for 5 min, washed twice with PBS (5 min).
5. Blocked with blocking buffer for 15 min at RT.
6. Primary antibody is then added for 1 h in dark at RT (room temp), washed twice in PBS (5 min).
7. Secondary antibody added for 1 h in dark at RT.
8. Samples are then washed twice in PBS and nuclear stain added for 10 min in complete media, washed in PBS and cover slips are mounted, dried overnight, and ends sealed with nail polish.
9. Phalloidin-TRITC is used at final concentration of 50 ng/ml; α-tubulin is diluted 1:2,000 dilution while the nuclear stain is diluted 1:250.
10. Exposure to CM from C85 cells (human colon carcinoma cells metastatic to liver) induced characteristic reorganization of actin filaments and MTOC in rat MSCs as detected by immunofluorescence staining (see Note 7). The MTOC also becomes polarized in the direction of migration. This indicated that changes in morphology correlated with enhanced migration (22).

3.4. Exposure of hMSCs to Keratinocyte-Conditioned Medium (KCM)

1. Normal human epithelial primary keratinocyte cell line cultured in KGM. Conditioned medium (CM) from these human keratinocytes is harvested following overnight culture (see Note 8).
2. Centrifuged (Eppendrof: 5,810) at 1,811 ×*g* for 5 min and supernatant passed through sterile 50 ml filtration system.
3. KCM-induced expression of cytoskeletal markers vinculin and F-actin filaments in differentiated hMSCs further indicated dermal myofibroblast-like differentiation in KCMSCs. KCMSCs also show punctate vinculin staining, characteristic of focal adhesions (see Note 9).
4. Phalloidin-positive visible stress fibers were also positive for alpha smooth muscle actin. KGMSCs expressed less alpha smooth muscle actin, while KCMSC expressed increased amounts of vinculin and alpha smooth muscle actin (Fig. 1a, b).
5. Hence, we suggest that prolonged exposure to KCM induces differentiation of BMD-hMSCs with expression of dermal myofibroblast markers and increased expression of cytokines.

a Keratinocytes-conditioned media exposed MSCs (KCMSCs)

Vinculin (green) Phalloidin (red) DAPI (Blue)

3-color merged image

b Keratinocytes-conditioned media exposed MSCs (KCMSCs) as a control

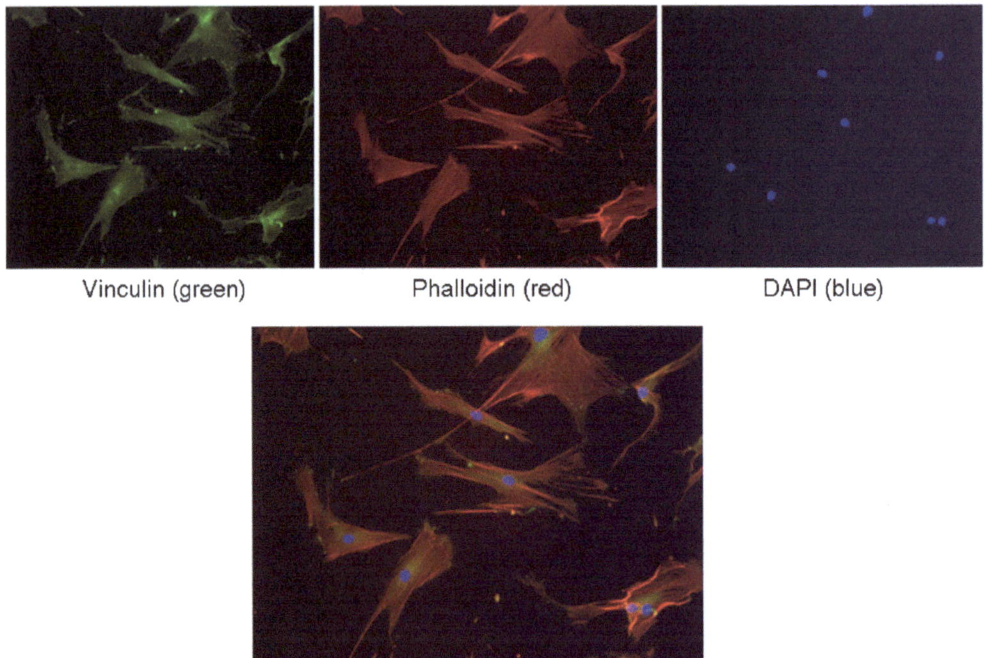

3-color merged image

Fig. 1. Dermal myofibroblast-like differentiation of hMSCs after KCM exposure. (**a**) Merge immunofluorescence image of KCMSCs stained for vinculin (*green*) and phalloidin (*red*). The focal adhesions (*green*) appear to hold down actin stress fibers (*red*). (**b**) KGMSCs showing diffused vinculin staining when compared with KCMSCs. Naïve hMSCs stained for vinculin and phalloidin as a control, additionally differentiated KCMSCs and KGMSCs stained for a-SMA, FSP, and Vimentin (data not shown).

3.5. Immunofluorescence Analysis for Markers of Myofibroblast Differentiation

1. Immunostaining is performed on cells grown on sterilized coverslips (with 70% alcohol followed by flaming, then place them in 12-well plate, which again can be incubated in the tissue culture hood with UV light on for 1 h) in 12-well plates.
2. The cells are fixed in 4% paraformaldehyde (at room temperature, 10 min), washed with 1× PBS (see Note 10) followed by permeabilization with 0.1% Triton X-100 for 10 min.
3. Cells are again washed, exposed to blocking medium (α-MEM) with 10% FBS, and then incubated with primary antibodies (Vinculin, α-SMA, FSP, and vimentin) for 1 h at room temperature.
4. After five subsequent washes with PBS for 5 min each, cells are immunostained with secondary antibodies (IgG (Vinculin, α-SMA, FSP) and IgM (vimentin)) at a dilution of 1:400 in a blocking medium.
5. When cells are concomitantly stained for actin stress fibers, they are incubated with Phalloidin–Tetramethylrhodamine B isothiocyanate mixed with the secondary antibody.
6. Following further washes, the cells are counterstained with the nuclear dye TOPRO-3 iodide in PBS at room temperature in the dark, followed by subsequent washing.
7. Cells are embedded in VectaShield mounting medium with DAPI and examined by fluorescence and confocal microscopy.
8. The naïve and differentiated hMSCs are quantitated for expression of myofibroblast specific markers.
9. Total cell number is obtained by counting the total number of DAPI-stained nuclei under the microscope. Percentage of marker expressing cells to the total number of the cells is then calculated.
10. Myofibroblast markers such as fibroblast surface protein and vimentin expression was observed in both KCMSCs and KGMSCs (see Note 11).

4. Notes

1. Dilute the Trypsin with the PBS in 1:1 ratio for MSCs or else one can buy less concentrated Trypsin.
2. Plate the frozen cells as per the manufacturer's protocol, briefly preheat the serum free media, and thaw the vial containing the cell, make sure cell is not completely thaw and at that stage fill the media in the T25 size flask and then gently plate the cells and then incubate.

3. One can store the 20% of Paraformaldehyde at –20°C for future use.
4. Complete growth medium can be used instead of blocking medium. It is better to match blocking serum with your antibody source.
5. Cell surface markers for flow cytometry are Stro-1, CD105, CD90, HLA-ABC, and CD44 using FITC-labeled antibodies (BD Biosciences).
6. Our initial results indicate that Jak2 and MEK are involved in cytoskeletal reorganization.
7. MSCs exposed to CM have F actin filaments organized along the length of the cell in keeping with the general appearance of the cell as an elongated rod shape (see (22)).
8. hMSCs are exposed to fresh keratinocyte-conditioned media (KCM) repeatedly for 30 days with freshly harvested KCM being added every third day. RT-PCR analysis was also performed to verify increased production of SDF-1 and CXCL5 mRNA in KCMSCs vs. KGMSCs and also by QRT-PCR for SDF-1 (not shown).
9. The focal adhesions appear to hold down actin stress fibers, as evidenced by colocalization of punctate vinculin on phalloidin-stained actin fibers.
10. At this stage, one can store the plate by wrapping it in plastic saran wrap at 4°C. Fixation is the most important step. The best condition and fixation time should be determined for each antibody.
11. The induction of α-SMA, F-actin, and punctate vinculin staining are consistent with induction of differentiation of hMSCs into dermal myofibroblast-like cells by KCM exposure.

References

1. Folkman, J. (1971) Tumor angiogenesis: therapeutic implications. *N. Engl. J. Med.* **285**, 1182–1186.
2. Pugh, C.W., and Ratcliffe, P.J. (2003) Regulation of angiogenesis by hypoxia: role of the HIF system. *Nat. Med.* **9**, 677–684.
3. Mishra, P.J., Mishra, P.J., Humeniuk, R., Medina, D.J., Alexe, G., Mesirov, J.P., et al. (2008) Carcinoma-associated fibroblast-like differentiation of human mesenchymal stem cells. *Cancer Res.* **68**, 4331–4339.
4. Hung, S.C., Deng, W.P., Yang, W.K., Liu, R.S., Lee, C.C., Su, T.C., et al. (2005) Mesenchymal stem cell targeting of microscopic tumors and tumor stroma development monitored by noninvasive in vivo positron emission tomography imaging. *Clin. Cancer Res.* **11**, 7749–7756.
5. Caplan, A.I. (1994) The mesengenic process. *Clin. Plast. Surg.* **21**, 429–435.
6. Knospe, W.H., Gregory, S.A., Husseini, S.G., Fried, W., and Trobaugh, F.E. Jr. (1972) Origin and recovery of colony-forming units in locally curetted bone marrow of mice. *Blood.* **39**, 331–340.
7. Friedenstein, A.J., Petrakova, K.V., Kurolesova, A.I., Frolova, G.P. (1968) Heterotopic of bone marrow. Analysis of precursor cells for osteogenic and hematopoietic tissues. *Transplantation.* **6**, 230–247.
8. Arai, F., Ohneda, O., Miyamoto, T., Zhang, X.Q., and Suda, T. (2002) Mesenchymal stem

cells in perichondrium express activated leukocyte cell adhesion molecule and participate in bone marrow formation. *J. Exp. Med.* **195**, 1549–1563.

9. Campagnoli, C., Roberts, I.A., Kumar, S., Bennett, P.R., Bellantuono, I., and Fisk, N.M. (2001) Identification of mesenchymal stem/progenitor cells in human first-trimester fetal blood, liver, and bone marrow. *Blood.* **98**, 2396–2402.
10. Young, H.E., Steele, T.A., Bray, R.A., Hudson, J., Floyd, J.A., Hawkins, K., et al. (2001) Human reserve pluripotent mesenchymal stem cells are present in the connective tissues of skeletal muscle and dermis derived from fetal, adult, and geriatric donors. *Anat. Rec.* **264**, 51–62.
11. In't Anker, P.S., Scherjon, S.A., Kleijburg-van der Keur, C., de Groot-Swings, G.M., Claas, F.H., Fibbe, W.E., et al. (2004) Isolation of mesenchymal stem cells of fetal or maternal origin from human placenta. *Stem Cells.* **22**, 1338–1345.
12. In't Anker, P.S., Scherjon, S.A., Kleijburg-van der Keur, C., Noort, W.A., Claas, F.H., Willemze, R., et al. (2003) Amniotic fluid as a novel source of mesenchymal stem cells for therapeutic transplantation. *Blood.* **102**, 1548–1549.
13. Erices, A., Conget, P., and Minguell, J.J. (2000) Mesenchymal progenitor cells in human umbilical cord blood. *Br. J. Haematol.* **109**, 235–242.
14. Bieback, K., Kern, S., Kluter, H., and Eichler, H. (2004) Critical parameters for the isolation of mesenchymal stem cells from umbilical cord blood. *Stem Cells.* **22**, 625–634.
15. Kinnaird, T., Stabile, E., Burnett, M.S., Epstein, S.E. (2004) Bone-marrow-derived cells for enhancing collateral development: mechanisms, animal data, and initial clinical experiences. *Circ. Res.* **95**, 354–363.
16. Pittenger, M.F., Mackay, A.M., Beck, S.C., Jaiswal, R.K., Douglas, R., Mosca, J.D., et al. (1999) Multilineage potential of adult human mesenchymal stem cells. *Science.* **284**, 143–147.
17. Uccelli, A., Moretta, L., Pistoia, V. (2008) Mesenchymal stem cells in health and disease. *Nat. Rev.* **8**, 726–736.
18. Bernardo, M.E., Locatelli, F., and Fibbe, W.E. (2009) Mesenchymal stromal cells. *Ann. N. Y. Acad. Sci.* **1176**, 101–117.
19. Horwitz, E.M., Gordon, P.L., Koo, W.K., Marx, J.C., Neel, M.D., McNall, R.Y., et al. (2002) Isolated allogeneic bone marrow-derived mesenchymal cells engraft and stimulate growth in children with osteogenesis imperfecta: implications for cell therapy of bone. *Proc. Natl. Acad. Sci. USA.* **99**, 8932–8937.
20. Shahdadfar, A., Fronsdal, K., Haug, T., Reinholt, F.P., Brinchmann, J.E. (2005) In vitro expansion of human mesenchymal stem cells: choice of serum is a determinant of cell proliferation, differentiation, gene expression, and transcriptome stability. *Stem Cells.* **23**, 1357–1366.
21. Gao, H., Priebe, W., Glod, J., and Banerjee, D. (2009) Activation of signal transducers and activators of transcription 3 and focal adhesion kinase by stromal cell-derived factor 1 is required for migration of human mesenchymal stem cells in response to tumor cell-conditioned medium. *Stem Cells.* **27**, 857–865.
22. Menon, L.G., Picinich, S., Koneru, R., Gao, H., Lin, S.Y., Koneru, M., et al. (2007) Differential gene expression associated with migration of mesenchymal stem cells to conditioned medium from tumor cells or bone marrow cells. *Stem Cells.* **25**, 520–528.

Part V

Novel Assays and Techniques

Chapter 15

Double *In Situ* Detection of Sonic Hedgehog mRNA and pMAPK Protein in Examining the Cell Proliferation Signaling Pathway in Mouse Embryo

Sho Fujisawa, Mesruh Turkekul, Afsar Barlas, Ning Fan, and Katia Manova

Abstract

Double *in situ* detection of RNA molecules and proteins in tissue sections is not trivial. A successful experiment heavily depends on the preparation of the tissue as well as the quality of the probes and antibodies. Detection of two or more molecular markers also requires reagents and experimental conditions that will preserve authenticity (accuracy) of the single staining patterns. Here, we describe in detail the protocols used to detect sonic hedgehog (*Shh*) mRNA by *in situ* hybridization and immunofluorescence staining for phosphorylated mitogen-activated protein kinase (pMAPK) in the same mouse embryonic tissue sections. In addition to protocols for manual immuno-staining, we provide data from automated machine-based staining protocols and highly recommend it to achieve strong signal and reproducible results.

Key words: *Shh* mRNA *in situ* hybridization, Colorimetric and fluorescence mRNA detection, Immunofluoresence (IF), pMAPK, Mouse embryo, Cell signaling

1. Introduction

Detection of RNA molecules in addition to protein is often unsuccessful. The sample must be prepared specifically to preserve both the target RNA and the protein. The *in situ* hybridization signal must survive the procedure required for protein detection or, conversely, the protein signal must remain unchanged after the completion of the RNA detection. The quality of the probes and the antibodies as well as the type of tissue can influence the outcome of the staining. However, studying the interaction between RNA molecules and proteins is often crucial in elucidating

Alexander E. Kalyuzhny (ed.), *Signal Transduction Immunohistochemistry: Methods and Protocols*, Methods in Molecular Biology, vol. 717, DOI 10.1007/978-1-61779-024-9_15,

the signaling pathways. Here, we describe detailed protocols for detecting Sonic Hedgehog mRNA (*Shh*) by *in situ* hybridization and phosphorylated mitogen-activated protein kinase (pMAPK) protein by immunofluorescence. As a core-facility laboratory that caters to a wide variety of researchers and projects, we have an extensive experience trouble-shooting staining issues. Our hope is to make available the techniques we use to tackle difficult experiments.

Shh plays a major role in embryonic development as well as cancer progression. Together with Gli transcriptional factors, *Shh* regulates a wide variety of cellular processes, including but not limited to, organogenesis, cell-cycle regulation, and cell differentiation (1–3). *Shh* is secreted and can initiate signal transduction in autocrine and paracrine fashion (4). *Shh*-induced cell proliferation in healthy as well as cancerous tissues is mediated in part by increased activation of MAPK (5, 6). While many laboratories study *Shh*-activated signal transduction pathways in cultured cells, we sought to understand their expression level and localization in embryonic tissues. The cells producing *Shh* are labeled by detection of *Shh* mRNA and cells influenced by *Shh* are elucidated by staining for the activated form of MAPK, pMAPK by immunofluorescence.

While we provide protocols for manual detection of *Shh* mRNA and pMAPK protein, we will also describe detection using automated machine-based protocols with Ventana Discovery Systems (Ventana Medical Systems). At our core-facility laboratory, we have successfully conducted experiments with strong, specific, and reproducible results. Automation has minimized variability and human error during molecular detection in tissue sections. We encourage readers to review the results from the automated staining experiments shown in Fig. 1.

The double detection experiment is divided into four major steps: (1) preparation of the sample, (2) manual and automated hybridization of DIG-conjugated probe onto *Shh* mRNA, (3) manual and automated detection of DIG with fluorescence or enzyme histochemistry (alkaline phosphatase), and (4) manual and automated immuno-fluorescence detection of pMAPK. The stained samples are imaged using microscopes or entirely digitized using a scanner. Advancement in microscopy technology has allowed us to capture the entire stained sections at very high resolution. Images in Fig. 1 are examples taken with Mirax Scanner (Carl Zeiss).

Fig. 1. (continued) scanned images of the entire mouse embryo show that *Shh* mRNA signal and pMAPK immuno-fluorescence are most prominently observed in spinal cord and notochord. Liver tissue and oral cavity also express significant levels of both. High magnification images of the brain (**B**), spinal cord (**C**), and tooth buds (**D**) reveal that the distribution of *Shh* mRNA (*black arrows*) and pMAPK (*green arrows*) are mutually exclusive but adjacent to each other. The images support the gradient effects of secreted *Shh* onto nearby regions and the resulting activation of MAPK-mediated signal transduction. Scale bars: A = 500 μm, B and C = 200 μm, D = 100 μm.

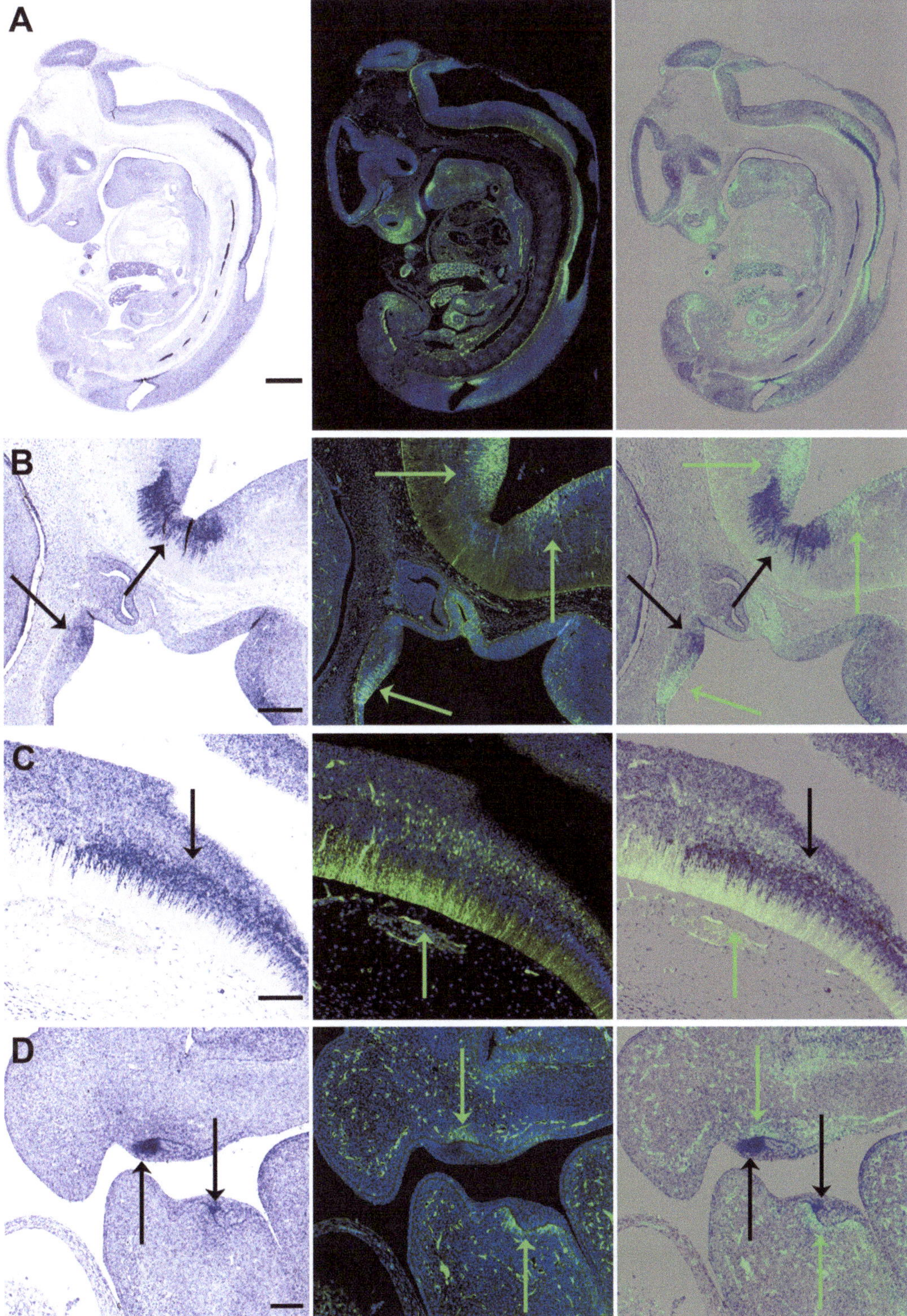

Fig. 1. Mutually exclusive distribution of *Shh* mRNA and pMAPK. E13.5 mouse embryonic sections were probed for *Shh* mRNA (*bluemap*, *left* column), then immuno-stained for pMAPK (Alexa488, *middle* column). Both detection methods were conducted using the automated protocols. *Right* column shows the overlay of the two signals. (**A**) Low magnification

2. Materials

In *an situ* hybridization experiment, it is crucial that target RNA molecules are not degraded before staining. All reagents should be molecular biology grade, and always use RNase-free utensils/glassware to weigh reagents and prepare solutions. All solutions must be made using DEPC-treated H_2O, void of any RNases. It is considered a good practice to allocate an area and equipment/tools for RNA work only!

2.1. Preparation of Mouse Embryo for In Situ Detection

1. 4% Paraformaldehyde (PFA), reagent grade, crystalline prepared in PBS at pH = 7.4.
2. Filter System, 0.22 μm pore, CA membrane.
3. ParaPlast Plus Tissue Embedding Medium.
4. Ethanol: 50%, 70%, 95%, and absolute.
5. Histoclear, a xylene substitute.
6. Tissue-embedding molds.
7. 30% sucrose in PBS.
8. Tissue-Tek OCT compound.
9. 0.1% poly-L-lysine solution in dH_2O.
10. FisherBrand Superfrost/Plus slides 2″ × 3″.
11. Isopentane (2-methylbutane).
12. 10× Difco FA Buffer (PBS).
13. Dewer flask containing liquid nitrogen.
14. Microtome for cutting paraffin-embedded tissue sections (Leica).
15. Cryomicrotome (Leica).

2.2. Manual In Situ Hybridization with DIG-Labeled RNA Probe to Detect Mouse Shh mRNA

1. Deionized water treated with diethylpyrocarbonate (DEPC) (see Note 1).
2. PBS.
3. Proteinase K (Sigma) diluted to 20 μg/ml in PBS, pH 7.4. Reconstitute the Proteinase K powder in DEPC-treated water to 20 mg/ml, aliquot, and store at −20°C. Dilute in PBS as required immediately before use.
4. 56°C slide warmer (or oven).
5. RNase Zap (Ambion).
6. Histoclear, a xylene substitute.
7. 95% ethanol and absolute ethanol.
8. Coplin jars.
9. Individually wrapped sterile tongue depressors for weighing chemicals.

10. Box of RNase-free weighing paper.
11. Acetylation buffer: 0.25% acetic anhydride in 0.1 M triethanolamine (pH = 8.0). Add acetic anhydride immediately before use and always use freshly prepared buffer. Use the basket to mix the solution several times. This step neutralizes the amine groups and decreases the nonspecific binding of the probe.
12. Hybridization Buffer: 50% deionized formamide, 1× Denhardt' s reagent, 10% dextran sulfate, 250 μg/ml yeast tRNA, 0.3 M NaCl , 20 mM Tris–HCl pH = 8.0, 5 mM EDTA pH = 8.0, 10 mM Sodium phosphate buffer, and DEPC-treated water.
13. High stringency wash: 50% formamide in 2× SSC.
14. DIG-conjugated RNA probe: use 100–500 ng RNA probe for each slide, prepared in 100 μl hybridization buffer. Heat the diluted probe for 2 min at 65°C to detangle any secondary structures. Hot water bath is preferable. Chill the diluted probe for 5 min on ice before use to prevent reannealing (see Note 2).
15. RNAse buffer: 0.5 M NaCl, 10 mM Tris–HCl pH = 7.5, 5 mM EDTA pH = 8.0, and dH_2O.
16. 5×, 2×, and 0.1× SSC (Dilute 20× SSC with dH_2O to prepare the solutions).
17. Plastic RNase-free cover slips, Rinzyl, 24 × 50 mm.
18. Black slide box, holds 25 slides.
19. Black electrical tape.
20. RNase A: 20 μg/ml RNase A diluted in RNase buffer.

2.2.1. Manual Fluorescence Detection of DIG in Shh Probe After Hybridization

1. Blocking solution: 10% heat inactivated normal goat serum and 2% Bovine serum albumin in PBS.
2. PBT: 0.1% Tween-20 in PBS.
3. Anti-DIG mouse Fab fragment- HRP (POD) conjugated (Roche).
4. TSA PLUS Cy3 kit (Perkin Elmer) or TSA AlexaFluor kit (Invitrogen).
5. FisherFinest® Premium Cover Glass 24 × 50 mm.
6. DAPI solution: 5 μg/ml DAPI (4′,6-Diamidino-2-phenylindole dihydrochloride) in PBS.
7. Fluorescence mounting media: Mowiol® 4-88 Reagent (Calbiochem, Darmstadt, Germany) prepared in glycerol and Tris–HCl buffer according to the vendor protocol.
8. Peroxidase blocking buffer: 1% H_2O_2 diluted in PBT (PBS + 0.1% Tween-20).

2.2.2. Manual Colorimetric Detection of DIG in the Shh After Hybridization

1. Blocking solution: 10% heat inactivated normal goat serum and 2% Bovine serum albumin in PBS.
2. PBT: 0.1% Tween-20 in PBS.
3. Anti-DIG-alkaline phosphatase (AP) conjugated sheep Fab fragment (Roche).
4. NTMT buffer: 100 mM NaCl, 100 mM Tris–HCl pH = 9.5, 50 mM $MgCl_2$, 0.1% Tween-20, and 0.5 mg/ml tetramisole hydrochloride (levamisole) (see Note 3).
5. BM Purple (NBT/BCIP ready-to-use) (Roche).
6. Plastic screw-top coplin jar.
7. Plastic five-slide mailer.
8. Nuclear fast red: 1% Nuclear fast red in 5% Aluminum sulfate.
9. 70%, 95% ethanol and absolute ethanol.
10. Histoclear, a xylene substitute.
11. Permount.

2.3. Automated In Situ Hybridization with DIG-Labeled Shh mRNA Probe

1. All the following reagents are provided by Ventana Medical Systems:
   (a) EZ Prep/EZ buffer.
   (b) Reaction buffer.
   (c) Antibody diluents/pAB Block.
   (d) RiboCC/Cell conditioning.
   (e) RiboPrep/Ribofix.
   (f) RiboHybe.
   (g) RiboClear.
   (h) RiboWash/CC Wash.
   (i) Protease 3.
   (j) Streptavidin - Alkaline phosphatase.
   (k) Activator (component of Blue Map kit).
   (l) Blue Map NBT (component of Blue Map kit).
   (m) Blue Map BCIP (component of Blue Map kit).
2. Anti-DIG, a mouse monoclonal antibody, conjugated to biotin (Sigma).
3. 70%, 95%, and absolute ethanol in dH_2O.
4. Histoclear, a xylene substitute.
5. Permount.
6. Nuclear fast red: 1% Nuclear fast red in 5% Aluminum sulfate.

2.4. Manual Immuno-Fluorescence Detection of pMAPK

1. 4% PFA, reagent grade, crystalline, prepared in PBS at pH = 7.4.
2. BSA- Bovine Serum Albumin Fraction V. It is important to use the most pure form of bovine serum albumin for immuno-staining to avoid contamination with bovine immunoglobulins.
3. 10× Difco FA Buffer (PBS).
4. Washing buffer: 0.1% BSA in PBS.
5. 2%BSA/PBS (Dissolve BSA in PBS).
6. Peroxidase blocking buffer: 1% H_2O_2 in washing buffer.
7. 10 mM citric acid buffer pH = 6.0 (mix citric acid and sodium citrate to prepare the buffer at pH = 6).
8. Blocking solution: 10% Heat inactivated normal goat serum, 2% BSA, and 0.1 M glycine in PBS.
9. Phospho-p44/42 MAPK (Tyr202/Tyr204), rabbit monoclonal antibody (Cell Signaling Technology).
10. Goat anti-rabbit IgG from Vectastain ABC kit anti-rabbit (Vector Labs).
11. Streptavidin-HRP.
12. TSA AlexaFluor 488 and 568 kits (Invitrogen).
13. DAPI solution: 5 μg/ml DAPI (4′,6-Diamidino-2-phenylindole dihydrochloride) in PBS.
14. Fluorescence mounting media: Mowiol® 4-88 Reagent (Calbiochem, Darmstadt, Germany) prepared in glycerol and Tris–HCl buffer according to the vendor protocol.

2.5. Automated Immunofluorescence Detection of pMAPK

1. All reagents and solutions, except for the primary, secondary antibody, and the reagents for signal detection were purchased from Ventana Medical Systems. These include:
   (a) EZ Prep/EZ buffer.
   (b) Cell Conditioner #1.
   (c) Reaction buffer.
   (d) Avidin and Biotin Block.
   (e) Blocker D.
   (f) Streptavidin-HRP.
2. 2%BSA/PBS (Dissolve BSA in PBS).
3. Blocking solution: 10% Heat inactivated normal goat serum and 2% BSA/PBS.
4. Primary antibody [phospho-p44/42 MAPK (Tyr202/Tyr204)], rabbit monoclonal antibody (Cell Signaling Technology).
5. DAPI solution: 5 μg/ml DAPI in PBS.

6. Biotinylated goat anti-rabbit IgG from Vectastain ABC kit (Vector Labs).
7. TSA AlexaFluor 488 and 568 kits (Invitrogen).
8. Fluorescence mounting media: Mowiol® 4-88 Reagent (Calbiochem, Darmstadt, Germany) prepared in glycerol and Tris–HCl buffer according to the vendor protocol.

2.6. Data Acquisition and Review

1. Zeiss Axioplan 2 Imaging microscope, equipped with QImaging Retiga EX Camera and Zeiss AxioCam MRm Camera.
2. Mirax Scanner (Carl Zeiss, Germany).

3. Methods

3.1. Preparation of Mouse Embryo for In Situ Detection

All steps should be carried on ice or at 4°C, where specified, using reagents prechilled on ice.

1. Sacrifice the pregnant mouse in accordance to the Animal Protocol at your institution.
2. Dissect out E13.5 embryo in PBS and fix by immersion in freshly prepared 4% PFA made in PBS (see Note 4).
3. Gently rock the sample in a flat-bottom tube filled with 4% PFA in PBS overnight at 4°C. The volume of PFA solution should be at least 20 times the volume of the sample (see Notes 5 and 6).
4. Transfer the embryo into 30% sucrose in PBS and rock at 4°C for 1 day or until the tissues sink to the bottom of the vial/tube (see Note 7).
5. Transfer the embryonic tissue into 1:1 mixture of OCT and 30% Sucrose in PBS; rock gently at 4°C for 1–3 h (see Note 8).
6. Transfer the embryonic tissue into chilled OCT and incubate on ice for 30 min (see Note 9).
7. Continue working on ice. Place the samples in embedding molds filled with OCT. Orient the samples as required.
8. Pour isopentane (2-methylbutane) in a plastic beaker and place it into a Dewar flask containing liquid nitrogen. The volume of the isopentane should be sufficient to allow the OCT mold to be completely submerged in it. Precool the isopentane until crystals form on the bottom of the isopentane-containing beaker (temperature of isopentane should be −150°C). Do not let liquid nitrogen get into the isopentane beaker. Using cold long metal forceps, submerge the molds into the beaker with isopentane (see Note 10).

9. OCT blocks will freeze immediately (approximately 10–15 s) in isopentane. Transfer the OCT blocks on dry ice and store in a -80°C freezer.
10. Samples are sectioned at 4–10 μm using a cryostat (see Note 11).
11. If you are continuing with *in situ* hybridization right away, the slides should be dried in a vertical position in a chemical hood for 20 min, and then baked for 1 h on a 56°C hot plate (see Note 12).
12. If you need to keep the slides for later use, they can be stored in a -80°C freezer.

Alternatively, tissue sections can be embedded in paraffin. Follow Steps 1–3 above, then continue with the protocol below.

4. Wash twice in ice cold PBS, 30 min each.
5. Dehydrate the sample through ethanol series (70%, 95%, and twice in absolute ethanol) for 30 min each.
6. Clear the sample in Histoclear three times, 5–30 min each. The time depends on the tissue size. Clear until the tissue becomes transparent.
7. Incubate in 1:1 mixture of Histoclear:Paraffin for 45 min followed by three changes in Paraffin (use a vacuum oven at 59°C). The temperature should be 3–4°C higher than the melting temperature of the paraffin used.
8. Paraffin-embedded samples can be stored in 4°C refrigerator.

3.2. Manual In Situ Hybridization

All solutions in Steps 2–17 of the protocol must be made with DEPC-treated deionized H_2O (see Note 1).

1. Retrieve the slides from -80°C freezer and air-dry them for 20 min at room temperature. Then, bake the slides on a 56°C hot plate for 1 h (see Note 13).

   All the following procedures are performed at room temperature unless otherwise noted.
2. Treat the slides with 20 μg/ml Proteinase K in PBS (pH 7.4) for 15 min at 37°C water bath (see Note 14).
3. Wash in PBS, 2 × 5 min each.
4. Refix the sample in 4% PFA in PBS for 10 min.
5. Wash in PBS, 2 × 5 min each.
6. Dip the slides in acetylation buffer for 10 min; use the basket to mix the solution several times. This step neutralizes the amine groups and decreases the nonspecific binding of the probe.
7. Wash 3 × 5 min each in PBS.

8. Dehydrate the slides in the following order: DEPC-treated deionized water for 3 min, 95% ethanol for 3 min, and then absolute ethanol for 2 × 5 min each. Air-dry the slides completely on paper towel.
9. Apply 100–500 ng/slide of the RNA probe in 100 μl of hybridization buffer. Hybridization buffer can be stored at −20°C for up to 6 months (see Note 15).
10. Slowly lower plastic RNase-free cover slips onto the sections, avoiding bubbles (see Note 16).
11. Place the sections horizontally in a black slide box. Humidify the slide box by placing at the bottom paper towels soaked in the high-stringency wash buffer. It is always a good practice to group slides with the same probe together to avoid cross-contamination.
12. Incubate the slides in horizontal position overnight (16–18 h) in a hybridization oven set at 55°C (see Note 17).
13. In copling jars, incubate the slides in 5x SSC prewarmed at 55°C, in order to float off the cover slips (see Note 18).
14. Transfer the slides into a different coplin jar containing prewarmed high-stringency wash buffer. Incubate at 65°C for 30 min (see Note 19).
15. Equilibrate with RNase buffer, 2 × 5 min each at 37°C.
16. Treat with 20 μg/ml RNase A for 30 min at 37°C (see Note 20).
17. Wash with high-stringency wash buffer, 2 times 20 min each at 65°C.
18. Wash with RNase buffer 10 min.
19. Wash with 2× SSC for 15 min.
20. Wash with 0.1× SSC for 15 min.

3.2.1. Manual Fluorescence Detection

1. In coplin jars, incubate the slides in peroxidase blocking buffer for 15 min (see Note 21).
2. Wash with PBT, 3 × 5 min each.
3. Drain excess fluid and lay the slides flat in a humidified slide box or tray (make sure the slides do not dry out).
4. Prepare blocking solution (see Note 22).
5. Apply 400 μl/slide of the blocking solution and incubate for 30 min.
6. Drain excess liquid from the slides.
7. Apply 100 μl/slide of anti-DIG-HRP (POD) mouse Fab fragments, diluted 1:100 in blocking solution (see Note 23).
8. Incubate for 1 h at room temperature or overnight at 4°C.

9. Wash with PBT, 3 × 5 min each.
10. Prepare TSA PLUS Cy3 by diluting it 1:50 in amplification diluent (Perkin Elmer). Apply 100 µl/slide and use plastic cover slip to cover the sections. Incubate for 10 min (see Note 24).
11. Transfer the slides into coplin jars. Wash with PBT, 3 × 5 min.
12. Apply DAPI staining solution for 5 min.
13. Wash with PBT, 3 × 5 min each.
14. Fix in 4% PFA for 10 min.
15. Wash with PBT, 3 × 5 min each. The slides are now ready to be processed according to the protocols for pMAPK detection.
16. In any double-detection experiment, we recommend a subset of slides to remain singly stained for comparison. It is also important to check that *in situ* hybridization was successful before continuing to immuno-fluorescent staining. Mount the slides with Fluorescence mounting media and store at −20°C (see Note 25).

3.2.2. Manual Colorimetric Detection

Steps 1–6 are the same as described above for the fluorescence detection.

7. Apply 100 µl/slide anti-DIG alkaline phosphatase-coupled antibody, diluted in blocking solution.
8. Incubate overnight at 4°C.
9. Wash with PBT, 3 × 5 min.
10. Wash in freshly prepared NTMT buffer, 2 × 10 min.
11. Place slides horizontally, add enough substrate solution (BM purple) to cover all sections.
12. Incubate in the dark at RT until the signal develops (could take 1 h or more, sometimes 2–3 days).
13. Wash with PBT, 3 × 5 min each.
14. Fix in 4% PFA for 10 min.
15. Wash with PBT, 3 x 5 min. The slides are ready for pMAPK protein detection.
16. In any double-detection experiment, we recommend a subset of slides to remain singly stained for comparison. These single-stained slides should be dried, dehydrated in ethanol series, cleared with histoclear, and then mounted with Permount. It is also important to check that *in situ* hybridization was successful before continuing to immuno-fluorescent staining. Counterstaining with Nuclear fast red allows the blue signal of mRNA to stand out.

3.3. Automated In Situ Hybridization and Colorimetric Detection

1. Deparaffinization.
2. Fixation.
3. Acetylation.
4. Proteolytic retrieval.
5. Washes.
6. Manual application of the RNA probe.
7. Hybridization for 6 h at 66°C.
8. High-stringency washes.
9. Refixation.
10. Washes.
11. Application of the anti-DIG antibody, diluted 1:200 in 2% BSA/PBS.
12. Blocking.
13. Streptavidin-Alkaline phosphatase.
14. Colorimetric reaction with Blue Map kit.
15. In any double-detection experiment, we recommend a subset of slides to remain singly stained for comparison. These single-stained slides should be dehydrated in ethanol series, then mounted with Permount. It is also important to check that *in situ* hybridization was successful before continuing to immuno-fluorescent staining. Counterstaining with Nuclear fast red for 5 min before dehydration and mounting allows the blue signal of mRNA to stand out.

3.4. Manual Immuno-Fluorescence Detection of pMAPK

If the slides were cover-slipped at the end of *in situ* hybridization, float them off by placing the slides in a coplin jar filled with PBS until the coverslips come off naturally. Do not use force to lift the coverslips; this may cause damage to the tissue sections.

1. Wash with washing buffer for 3 × 5 min each.
2. In coplin jars, incubate the slides in peroxidase blocking buffer 15 min (see Note 21).
3. Wash with washing buffer for 3 × 5 min each.
4. Perform antigen retrieval: In coplin jars, incubate the slides in 10 mM citric acid buffer pH = 6.0 for 15 min in a microwave at 98°C. Equilibrate the slides to room temperature, for about 30–60 min (see Note 26).
5. Wash with washing buffer for 3 × 5 min each.
6. Prepare blocking solution (see Note 27).
7. Transfer the slides in horizontal position in a humidified box or tray, and apply the blocking solution for 30 min (see Notes 28 and 29).

8. Prepare the primary antibody solution by diluting the antibody in blocking solution (IgG concentration 0.2 mg/ml) 1:200 to obtain 1 μg/ml final concentration of the antibody. The IgG concentration of the antibody was reported to be 0.2 mg/ml by the vendor for pMAPK (see Note 30).
9. Remove the blocking solution by tilting the slides onto a paper towel, and then apply 100 μl/slide of the primary antibody. Place a plastic cover slip over the slide, being careful not to form bubbles (see Note 31).
10. Incubate in anti-pMAPK antibody overnight in moist chamber at 4°C (see Note 32).
11. Flush the cover slip off by using a squeeze bottle with PBS.
12. Wash with washing buffer for 3 × 5 min each.
13. Prepare the secondary antibody: to 10 ml 2% BSA/PBS, add 50 μl biotinylated goat anti-rabbit IgG from Vector (final concentration 7.5 μg/ml) and 150 μl of heat-inactivated normal goat serum.
14. Apply 200 μl/slide of secondary antibody and incubate for 1 h in moist chamber.
15. Wash with washing buffer for 3 × 5 min each.
16. Incubate for 20 min at RT in 1:100 Streptavidin-HRP (Invitrogen), made in 2% BSA/PBS.
17. Wash with washing buffer for 3 × 5 min each.
18. Incubate for 15 min at RT in the dark with 1:100 TSA AlexaFluor conjugate (Invitrogen), made in amplification buffer containing 0.0015% H_2O_2 or 1:200 Tyramide-FITC, Rhodamin, or Cy5 (Perkin Elmer) made in amplification diluents (see Note 33).
19. Wash with washing buffer for 3 × 5 min each.
20. Stain slides with DAPI solution for 5 min.
21. Wash with washing buffer for 3 × 5 min each.
22. Mount cover slips using Fluorescence mounting media (see Note 25), and keep the slides in the dark at –20°C.

3.5. Automated Immuno-Fluorescence Detection of pMAPK

Automated immuno-fluorescence detection was performed using Discovery XT processor with the following steps:

1. Block the sections for 30 min with blocking solution.
2. Primary antibody is diluted to 5 μg/ml in 2%BSA/PBS and is applied manually.
3. The incubation lasts for 3 h.
4. Washes with reaction buffer.
5. Incubation for 16 min with biotinylated anti-rabbit IgG, diluted to 7.5 μg/ml in 2%BSA/PBS.

6. Washes with reaction buffer.
7. Detection is performed with Blocker D, Streptavidin-HRP, followed by incubation with TSA AlexaFluor 488 or other, for 12 min.
8. Washes with reaction buffer.
9. Stain with DAPI solution for 5 min.
10. Mount cover slips using Fluorescence mounting media and keep the slides in the dark at –20°C (see Note 25).

3.6. Data Acquisition and Review

1. Scan the slides using Mirax Scanner from Carl Zeiss.
2. View the scanned slide with Mirax Viewer and take sample images.

4. Notes

1. In order to preserve the RNA molecules from degradation by RNases, all solutions until after the hybridization step of the *in situ* hybridization procedure should be made with DEPC-treated dH_2O. Add 0.1% DEPC (harmful and toxic) to deionized water, shake vigorously until DEPC forms fine suspension, leave overnight, and sterilize by autoclaving. DEPC decomposes to CO_2 and ethyl alcohol by the heat in the autoclave and DEPC-treated water is not toxic.
2. To generate DIG-RNA probe, use 1 μl of 1 mg/ml linearized template cDNA with 5′ overhangs (or 100–200 ng of extra pure PCR product), 19 μl DEPC dH_2O, 3 μl 10× DIG-labeling mix (Roche), 3 μl 10× transcription buffer, 1 μl RNasin, 3 μl T3, T7 or SP6 RNA polymerase (Roche). Mix the components at RT and incubate for 2 hours at 37°. Add 0.5 μl of DNAse I (Roche) and incubate for 15 min at 37°C. Run the reaction mix through RNA clean-up column (Roche) or perform ethanol precipitation by adding 1/10th of the volume 8M $LiCl_2$ (Sigma). Precipitate probe with 2.5 times the volume absolute ethanol. Vortex and leave at –20°C (probe can be stored at this stage). Next day, spin probe maximum speed (14,000 rpm for 20 min at 4°C). Remove the alcohol, be careful to not disturb pellet (probe). Wash pellet in 50 μl chilled 70% ethanol. Spin again. Remove alcohol. Let pellet dry, resuspend in 30 μl DEPC water, vortex very well, and spin down. Run 1 μl of probe on a formaldehyde RNA gel to check probe. Aliquot the probe in 1–2 μl/tube and store at –20°C (remains stable for years). Never do phenol/chloroform extraction of DIG-labeled product because it will partition into organic phase.

3. To prepare 100 ml NTMT buffer, use 2 ml 5 M NaCl, 10 ml 1 M Tris–HCl, pH 9.5, 5 ml 1 M $MgCl_2$, 0.1 ml Tween-20, 82.9 ml H_2O. Add 0.5 mg/ml levamisole before use.
4. 4% PFA fixative must be used fresh, maximum 48 h after being made. It should be kept at 4°C. It is not recommended to freeze aliquots of PFA. Most commercially available preparations are not suitable for *in situ* hybridization, especially if they contain methanol to prolong shelf-life. Buffered formalin is not optimal fixative for *in situ* hybridization. Never use warm PFA; always chill it on ice before use. To prepare PFA, use a chemical hood and follow the steps:
   (a) Weigh PFA powder (4 g PFA for 100 ml fixative).
   (b) Preheat deionized water to 68°C in Erlenmeyer flask (~60 ml).
   (c) Take flask off the hot plate, add PFA powder, and mix quickly by agitation.
   (d) Clear milky PFA solution with 0.1 N NaOH (4–5 drops will be sufficient, keep continued agitation of the flask).
   (e) Add 25 ml water to precool and then add 10 ml of 10× PBS.
   (f) Use pH paper to make sure the pH = 7.4. (Fixative can damage pH meter electrodes).
   (g) Bring the volume to 100 ml with deionized water.
   (h) Filter through 0.22 μm filter system, cool on ice before use.
5. Most adult tissues will require overnight incubation in PFA. For mouse embryos between the ages of E5.5 and E10.5, shorter incubation of 30 min to 5 h will suffice. Avoid using conical bottom tubes since the tissue may get stuck on the bottom and the fixation will be uneven. Try to minimize the time between dissection and immersion in the fixative. Rinse the blood off the tissue, as presence of blood can decrease fixation efficiency.
6. In order to fix adult mouse brain, liver, or lung, perfusion with the fixative solution is required prior to fixation by immersion. For brain, perfuse the mice through the heart, for lung through the trachea, and for liver through the portal vein. Relatively large tissue samples (>1 cm^3) should be cut into smaller pieces for better fixative penetration. Inadequate fixation may result in uneven staining pattern and/or background on the edges of the tissue ("edge effect").
7. Sucrose is a cryoprotectant and prevents the damage due to formation of intracellular ice crystals during freezing. Most tissues will initially float in 30% sucrose solution. Keep the tissues in sucrose until they sink to the bottom. Lung or tissues

with air spaces may never sink. For these samples, overnight incubation is sufficient.

8. For mouse embryo, 1 h is enough, but for whole adult mouse brain, 3-h incubation is necessary for good penetration.
9. Before transferring the tissue onto OCT, place the tissue on filter paper or tissue paper to make sure all the water in the tissue has been absorbed by the paper. Do not use sharp forceps. Use spatulas to transfer the tissues.
10. Precool the long forceps by dipping it in the chilled isopentane; metal at room temperature can crack the OCT mold and damage the samples within. Some plastic beakers that are not soft and flexible enough may crack in the cold isopentane. Use soft flexible plastic beaker or a metal container. Freezing tissues on dry ice, dry ice/ethanol bath, or directly by submerging them into liquid nitrogen is not considered a good practice and should be avoided.
11. Allow 10–15 min for the blocks to equilibrate to cryostat temperature before attempting to section them. Folds in the sections may cause detection reagents to get trapped and create nonspecific signal. Imperfect sections may be lost during staining or can be prone to uneven staining pattern and/or background on the edges of the tissue section called "edge effect."
12. If the sections are not attaching to the slides very well, try heating the slides on 56°C overnight. Also, coating slides with 0.01% poly-l-lysine may help the tissues adhere better and prevent them from being lifted off during staining. Dilute 0.1% stock solution 10 times in deionized water. Treat the slides for 5 min at room temperature in coplin jars. Air-dry the slides and if necessary, store them at 4°C for several days. 0.01% Poly-l-lysine solution can be stored in 4°C and reused for up to 6 months.
13. While *in situ* hybridization signal tends to be stronger in cryo-sections compared to paraffin-embedded sections, some probes, such as the one for *Shh*, show great results even in paraffin sections. If your tissues are embedded in paraffin, you must first dewax your section as follows: treat with histoclear, a xylene substitute, 3 × 3 min each; absolute ethanol, 2 × 3 min each; 95% ethanol, 1 × 3 min; 70% ethanol, 1 ×3 min; DEPC-water, 1 × 3 min.
14. Proteinase K is a hemolytic serine protease that partially reverses the cross-linking effect of aldehyde fixation, removing the masking proteins that hinder riboprobe binding and permeabilizes the tissue. Too low concentration of Proteinase K may not expose the nucleic acids enough and too high a concentration may overdigest the tissue and the target

is lost. Proteinase K concentration, incubation duration, and temperature should be optimized for each riboprobe and tissue sample. Other proteolytic enzymes such as pepsin, trypsin, and pronase E could also be tested.

15. Formamide decreases the melting temperature of RNA/RNA hybrids, lowering the affinity of RNA probes to nonspecific sequences. The presence of formamide should be taken into consideration when calculating T_m values of the riboprobe. Yeast tRNA and Denhard's solution both saturate nonspecific binding sites.
    - Dextran sulfate provides volume exclusion by sequestering water and thereby effectively increasing the probe concentration without decreasing the volume of hybridization mixture. It increases the rate of hybridization without the concomitant increase in background.
    - NaCl is a neutral salt that provides the ionic strength of the hybridization solution. As the concentration of salt increases, the RNA/RNA hybrids are less stable creating less stringent hybridization conditions and increasing nonspecific binding.
    - EDTA acts as a detergent and is used for permeabilization.
    - Na-phosphate and Tris buffers minimize the changes in pH.
    - pH is an important factor in determining the stringency of hybridization. The higher the pH is, the more stringent the hybridization conditions are.
16. If high nonspecific staining is noticed, incubate the slides with hybridization buffer without the riboprobe for 1–2 h at hybridization temperature. This step is usually called prehybridization blocking.
17. Temperature affects the stability of RNA/RNA hybrids during hybridization. The higher the temperature is, the less stable the RNA/RNA hybrids are. Increasing the temperature during hybridization will result in more stringent hybridization conditions. Decrease in temperature may result in nonspecific binding and false-positive signal. The hybridization temperature should be optimized for each particular riboprobe and/or tissue sample. If a hybridization oven is not available, thoroughly tape the slide boxes shut with black electrical tape and incubate upright in a dry dish inside a 55°C water bath.
18. Do not try to force the cover slips off, otherwise the tissue will tear. The cover slips should float off in about a minute of incubation in prewarmed 5× SSC.
19. You can put the coplin jar back in the hybridization oven and raise the temperature to 65°C, or you can use a water bath.

High-stringency wash with formamide and high salt concentration decreases the melting temperature of the RNA/RNA hybrids, aiming to wash away most of the nonspecific RNA/RNA hybrid formations.

20. RNase A is an endonuclease and digests only single-stranded RNA molecules. It reduces background due to single-stranded riboprobes trapped in the tissue.
21. Endogenous peroxidases/pseudoperoxidases in the tissue will cause TSA reagents to deposit in regions where there is no probe binding giving rise to false-positive signal. High concentration of H_2O_2 will saturate the endogenous enzymes and eliminate this source of background.
22. Blocking with goat serum prevents nonspecific binding of the anti-DIG antibody. Other blocking reagents such as bovine serum albumin, casein, or serum from other species may also be used.
23. The anti-DIG antibody is specifically raised against the DIG incorporated into the riboprobe. Antibody Fab fragments are used as they penetrate tissue easily, and show less nonspecific binding than intact antibodies (IgGs). The Fab fragments are conjugated to POD (horseradish peroxidase-HRP). DIG (Digoxigenin) is a plant-based molecule absent in animal tissues. Background due to endogenous DIG in animal tissue is nonexistent.
24. TSA or tyramide reagent is a substrate for HRP. HRP, in the presence of trace amounts of H_2O_2, will convert the tyramide into very short-lived extremely reactive intermediate that will bind to protein moieties in very close proximity of the HRP and will not be able to bind to moieties away from HRP due to its short life. Tyramide is a powerful amplification reagent and can also amplify the signal as well as the background if present. Use it with caution!
25. It is very important to mount the slides with aqueous mounting media that contains glycerol and antifading reagent. Glycerol is an antifreeze and prevents the media from freezing allowing the slides to be stored at –20 or –80°C freezer for longer storage. Mowiol® is a proprietary blend of polyvinyl alcohols and serves as an antifading reagent and polymerizes to keep the coverslip in place making the use of nail polish unnecessary. Antifade reagents prolong the life of fluorescence molecules and prevent bleaching during imaging. Mowiol® and glycerol are water soluble and coverslips can be easily removed in PBS and slides can be processed with other staining protocols after imaging and/or storage.
26. Aldehyde fixatives like PFA are cross-linking fixatives and may denature or hinder some epitopes so that the antibody will have difficulty recognizing them. Antigen-retrieval process

makes use of high heat in different buffers or proteolitic treatments to reverse the effect of epitope denaturation or hindrance after fixation and tissue processing. 10 mM Citric acid buffer pH = 6.0 treatment in microwave close to boiling temperature is one of the most common procedures for antigen retrieval. Other buffers used for antigen retrieval include: 50 mM EDTA pH = 9.0, Tris–HCl Buffer pH = 7.4, other commercially available buffers with undisclosed composition. Occasionally, detergents such as Trion-X-100, Tween-20, NP-40, saponin, CHAPS, digitonin, urea, and SDS could be added to the buffers to aid the retrieval process. The heat-generating devices include but are not limited to: Microwave, conventional vegetable steamer, pressure cooker, autoclave, water bath, etc. Alternative to heat-induced epitope retrieval is treatment with proteolytic enzymes such as ProteinaseK, Trypsin, Pepsin, Chymotrypsin, and Pronase E. Usually, relatively low concentrations (<0.5%) of the above enzymes are used in appropriate buffers for duration of several minutes at room temperature or 37°C water bath. Please contact the vendor of the antibody or check the literature for suggestions about antigen retrieval.

27. Blocking solution is used to minimize the nonspecific binding of the primary or secondary antibody. In general, serum from the species in which the secondary antibody was raised is used on the blocking solution at concentration from 2 to 20% depending on the level of background. Longer incubation is also suggested, if background is relatively high.
28. If your primary antibody is made in mouse and you are trying to stain mouse tissues, block with Mouse IgG blocking reagent (MOM Basic Kit, Vector Laboratories) with 0.1 M glycine for 30 min.
29. If background staining remains high, blocking treatment can be prolonged up to 2 h in moist chamber at room temperature or overnight at 4°C.
30. If the IgG concentration is not written on the tube or on the specification sheet of the antibody, contact the vendor to inquire about the IgG concentration. Antibodies come in different formulations such as whole serum, crude extract, affinity-purified IgG, ascites fluid, or culture supernatant. Each of the above formulation has completely different IgG concentrations. Each lot of the same antibody may have different IgG concentration. Working with dilutions instead of IgG concentration will be misleading. When using an antibody for the first time, titer the concentration of the antibody. In general, most of the antibodies will perform optimally at IgG concentration between 0.5 and 10 μg/ml. Testing 1, 5, and 10 μg/ml will give you a pretty good estimate of the optimal concentration. Optimal concentration will give strong

signal with minimum nonspecific background staining. If the signal is weak and there is no background, consider increasing the antibody concentration and/or incubation time. If the signal is very strong with relatively high background, consider decreasing the antibody concentration and/or increase blocking.

31. Work with slides one by one. Do not let them become dry at any point during staining.
32. Some antibodies may work better by incubating for 1–3 h at RT or shorter duration at 37°C.
33. Please keep in mind that although tyramide amplification boosts the signal, it may also increase the background. Another alternative is to use a secondary antibody directly conjugated to Fluorophore (completely omitting the use of tyramide in Steps 15–19) and proceed with Steps 21–23.

Acknowledgments

We thank the members of the Molecular Cytology Core Facility for discussions and support and Dr. A.L. Joyner for Shh cDNA.

References

1. Koleva, M., Kappler, R., Vogler, M., Herwig, A., Fulda, S., and Hahn, H. (2004) Pleiotropic effects of sonic hedgehog on muscle satellite cells. *Cell. Mol. Life Sci* **62**, 1863–1870.
2. Kruger, M., Mennerich, D., Fees, S., Schafer, R., Mundlos, S., and Braun, T. (2001) Sonic hedgehog is a survival factor for hypaxial muscles during mouse development. *Development* **128**, 743–783.
3. Duprez, D., Fournier-Thibault, C., Le Douarin, N. (1998) Sonic Hedgehog induces proliferation of committed skeletal muscle cells in the chick limb. *Development* **125**, 495–505.
4. Handrigan, G. R., and Richman, J.M. (2010) Autocrine and paracrine Shh signaling for tooth morphogenesis, but not tooth replacement in snakes and lizards (Squamata). *Dev. Biol.* **337**, 171–186.
5. Seto, M., Ohta, M., Asaoka, Y., Ikenoue, T., Tada, M., Miyabayashi, K., et al. (2009) Regulation of the hedgehog signaling by the mitogen-activated protein kinase cascade in gastric cancer. *Mol. Carcinog.* **48**, 703–712.
6. Elia, D., Madhala, D., Ardon, E., Reshef, R., and Halevy, O. (2007) Sonic hedgehog promotes proliferation and differentiation of adult muscle cells: Involvement of MAPK/ERK and PI3K/Akt pathways. *Biochim. Biophys. Acta* **1773**, 1438–1446.

Chapter 16

Identifying Intracellular Sites of Eicosanoid Lipid Mediator Synthesis with EicosaCell Assays

Christianne Bandeira-Melo, Peter F. Weller, and Patricia T. Bozza

Abstract

Eicosanoids, arachidonic acid-derived signaling lipid mediators, are newly formed and nonstorable molecules that have important roles in physiological and pathological processes. EicosaCell is a microscopic assay that enables the intracellular detection and localization of eicosanoid lipid mediator-synthesizing compartments by means of a strategy to covalently cross-link and immobilize eicosanoids at their sites of synthesis followed by immunofluorescent-based localization of the targeted eicosanoid. EicosaCell is a versatile assay which allows analyses of different types of cell preparations, such as cells isolated from humans or harvested cells from in vivo models of inflammation and adherent or suspension cells stimulated in vitro. EicosaCell assays have been successfully used to identify different intracellular compartments of synthesis of prostaglandins and leukotrienes upon cellular activation. This is of particular interest given that over the past decade intracellular compartmentalization of eicosanoid-synthetic machinery has emerged both as a key component in the regulation of eicosanoid synthesis and in delineating functional intracellular and extracellular actions of eicosanoids. This review covers basics of EicosaCell assay including its selection of reagents, immunodetection design as well as some troubleshooting recommendations.

Key words: Eicosanoids, EicosaCell, Immunofluorescence detection, Bioactive lipid mediators, Heterobifunctional cross-linker

1. Introduction

Among bioactive lipid mediators, eicosanoids – including leukotrienes and prostaglandins – are signaling molecules, derived from the enzymatic oxygenation of arachidonic acid (AA), that control key processes, both intracellularly and extracellularly involving cell–cell communication, such as cell activation, proliferation, apoptosis, metabolism, and migration (1, 2). Therefore, these lipid signals have important roles in physiological and pathological

Alexander E. Kalyuzhny (ed.), *Signal Transduction Immunohistochemistry: Methods and Protocols*, Methods in Molecular Biology, vol. 717, DOI 10.1007/978-1-61779-024-9_16,

conditions such as tissue homeostasis, host defense, inflammation, and cancer and great efforts have been aimed at understanding the biochemical, cellular, and molecular aspects of their biosynthetic pathway.

Over the past decade, intracellular compartmentalization of eicosanoid-synthetic machinery has emerged as a key component in the regulation of eicosanoid synthesis (reviewed in (3–6)). However, the direct evaluation of specific subcellular locales of eicosanoid synthesis has been elusive, as those lipid mediators are newly formed, not stored, and often rapidly released upon cell stimulation. Indeed, in the majority of studies to date, intracellular sites of eicosanoid synthesis have been inferred based solely on the identified immunolocalization of eicosanoid-forming enzymes. Moreover, the productive synthesis of specific eicosanoids is not merely determined by the availabilities of AA and eicosanoid-forming enzymes, but requires sequential interactions between specific biosynthetic proteins acting in cascade, and may involve very unique spatial interactions. Therefore, just by immunolocalizing eicosanoid-forming enzyme proteins within discrete subcellular structures, one cannot assure that those sites are indeed accountable for the efficient and enhanced eicosanoid synthesis observed during inflammatory responses. The immunolocalization of eicosanoid-forming proteins ascertains neither that the localized protein is functional nor that it is activated to synthesize a specific eicosanoid lipid at an intracellular site. We previously developed a method to capture and localize the eicosanoid, prostaglandin (PG) E_2, released extracellularly by a nematode parasite (7). By means of a strategy to covalently cross-link, capture, and localize newly formed eicosanoids at their sites of synthesis, we developed a direct approach to detect the intracellular sites of arachidonic acid (AA)-derived lipid mediator formation in leukocytes and other cell types.

To develop a new strategy for in situ immunolocalization of newly formed eicosanoids to ascertain the intracellular compartmentalization of their synthesis – the EicosaCell assay – modifications of a prior technique was used (7). The EicosaCell rationale relies on the specific features of the heterobifunctional cross-linker 1-ethyl-3-(3-dimethylamino-propyl) carbodiimide ($C_8H_{17}N_3$–HCl; EDAC). EDAC immobilizes newly synthesized eicosanoids intracellularly by cross-linking their eicosanoid carboxyl groups to the amines of adjacent proteins localized at eicosanoid-synthesizing sites. Such EDAC-mediated reaction forms a bond without any spacer length between the two molecules, favoring an accurate positioning of the newly synthesized eicosanoid within the cell. In addition, while other cross-linkers formed bonds that often generate foreign molecules, EDAC-driven eicosanoid-bond is homologous to native eicosanoid that allows immunoassays like EicosaCell. Besides the precise positioned coupling of an immuno-detectable eicosanoid at its sites of formation, EDAC enables: (1) the ending of cell stimulation step; (2) cell fixation; (3) cell permeabilization, allowing the penetration of both anti-eicosanoid antibody (Ab) and the secondary detecting fluorochrome-conjugated Ab into cells; and, importantly, (4) the relative preservation of lipid domains, such as membranes and lipid bodies, which dissipate with air drying or commonly used alcohol fixation.

The EicosaCell technique described herein enables one to directly pinpoint the intracellular locales of eicosanoid synthesis by detecting the newly formed lipids and has been successfully able to confirm the dynamic aspect involved in the localization of eicosanoid synthesis, providing direct evidence of compartmentalization within three distinct sites. As schematized in Fig. 1, the perinuclear envelope (8–11), phagosomes (12), or lipid bodies (10, 11, 13–15) were found to compartmentalize eicosanoid synthesis, in accord with specific cell types and stimulatory conditions studied. So far, EicosaCell assays have been used to identify the production of LTC_4 (10, 12, 13, 16), LTB_4 (14, 17), PGE_2 (9, 11, 15), and PGD_2 (unpublished observations) in different cell types and under different stimulatory conditions. Moreover, it could in principle be adapted to intracellular detection of other lipid mediators as long as specific antibodies are available.

Of note the EicosaCell has high sensitivity enabling the detection of low levels of intracellular generated eicosanoids even when extracellular released eicosanoids could not be detected by

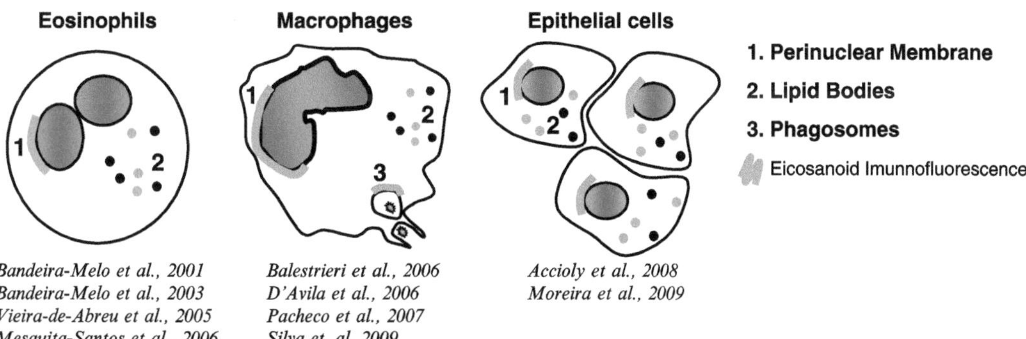

Fig. 1. Schematic summary of the EicosaCell applications. EicosaCell-derived reports have identified three distinct intracellular sites of eicosanoid synthesis, including nuclear envelope, cytoplasmic lipid bodies, and zymozan-driven phagosomes within human or murine eosinophils, murine macrophages as well as rat or human epithelial cells (8–17).

conventional eicosanoid enzyme immune assay (10, 11). Indeed, it has been shown that besides paracrine/autocrine activities, eicosanoids may display intracrine functions (18, 19). For instance, by employing EicosaCell technique, it has been uncovered that a lipid body-derived LTC_4 has intracellular functions in controlling cytokine release from eosinophils (20). Therefore, by identifying compartmentalized levels of eicosanoids, besides providing new insights into the regulation of eicosanoid biosynthesis, EicosaCell assays may contribute to identification of likely intracellular functions of newly synthesized eicosanoids.

2. Materials

2.1. EicosaCell

1. EDAC: 1-ethyl-3-(3-dimethylamino-propyl) carbodiimide hydrochloride ($C_8H_{17}N_3$–HCl) (Sigma-Aldrich Inc., St. Louis, MO; cat no. E7750).
2. HBSS: Hanks buffered salt solution without calcium chloride and magnesium chloride.
3. Water bath at 37°C.
4. Cytocentrifuge.
5. Glass microscope slides and coverslips.
6. Mounting medium: Aqua Poly/Mount (Polysciences Inc., Warrington, PA; cat. no. 18606) or Vectashield (Vector, Inc., Burlingame, CA; cat. no. H 1000).

2.2. Double-Labeling Applications

1. DAPI (4′,6-Diamidino-2-phenylindole dihydrochloride).
2. TO-PRO-3 (642/661) – 1 mM solution in dimethylsulfoxide (DMSO) (Molecular Probes, Eugene, OR).
3. Rat monoclonal IgG2a against lysosome-associated membrane protein (LAMP)-1 (BD Pharmingen, San Jose, CA).
4. BODIPY® 493/503 (4,4-difluoro-1,3,5,7,8-pentamethyl-4-bora-3a,4a-diaza-s-indacene) (Molecular Probes, Eugene, OR; cat no. D-3922). Molecular weight: 262. To prepare BODIPY stock solution, BODIPY should be dissolved in DMSO (1 mM), aliquoted in small Eppendorf tubes (~10 μl/tube) and stored –20°C protected from light. BODIPY working solution should be diluted fresh 1,000× in $HBSS^{-/-}$ and kept from light.
5. Mouse monoclonal antibody (mAb) to adipose differentiation-related protein (ADRP) (Fitzgerald Industries Intl., Concord, MA; cat no. 10R-A117ax) or guinea pig polyclonal Ab to ADRP (reactivity human/mouse/rat/bovine) (Fitzgerald Industries Intl., Concord, MA; cat no. 20R-AP002).

3. Methods

3.1. Preparation of EDAC Solution for EicosaCell

1. Water soluble EDAC hydrochloride should be diluted in HBSS$^{-/-}$. Refer to Note 1 (below) for EDAC solution handling.
2. The working solution should have twice concentration of the final concentration with cells.
3. EDAC final concentration with cells varies according cell type and protocol used (see next subheadings), as exemplified below:

Example 1. With purified human eosinophils stimulated as a cell suspension, EDAC final concentration within eosinophil suspensions should be 0.1% in HBSS$^{-/-}$, therefore the EDAC working solution should be 0.2%.

Example 2. With adherent macrophages stimulated in 6-well plates, EDAC final concentration should be 0.5% in HBSS$^{-/-}$, therefore the EDAC working solution should be 1.0%.

3.2. EicosaCell with Cells in Suspension

1. EicosaCell can be easily performed with various cell types in suspension, such as purified human blood leukocytes, cultured cell lines, as well as peritoneal, pleural, or bronchoalveolar animal cells.
2. After in vivo or in vitro stimulation of these cell populations, incubation with EDAC instantaneously immobilizes eicosanoids at their sites of synthesis within the cell, just before cytospin slides are prepared to allow microscopic analysis.
3. As schematically illustrated in Fig. 2, after preparing a cell suspension from, for instance, a mouse peritoneal cavity, EDAC working solution should be added to cell suspension and incubated for a period of time to ensure cell fixation, eicosanoid immobilization, and cell permeabilization (refer to Note 2 for details).
4. As suggested in the scheme (Fig. 2), slide preparations can be done by cytocentrifuging the cells onto slides. Labeling of newly formed eicosanoids can be done with a various already tested antibodies, as already published elsewhere (10, 14–16).
5. At the end of the staining procedure, cytospun cells should be always extensively washed with HBSS$^{-/-}$ and then mounted in an aqueous medium to be visualized with 100× objective by both phase-contrast and fluorescence microscopy (see Note 3 for details).
6. The specificity of the eicosanoid immunolabeling using the EicosaCell system should be always ascertained by including some mandatory control conditions: (1) nonstimulated EDAC-treated cells labeled with the proper anti-eicosanoid Ab;

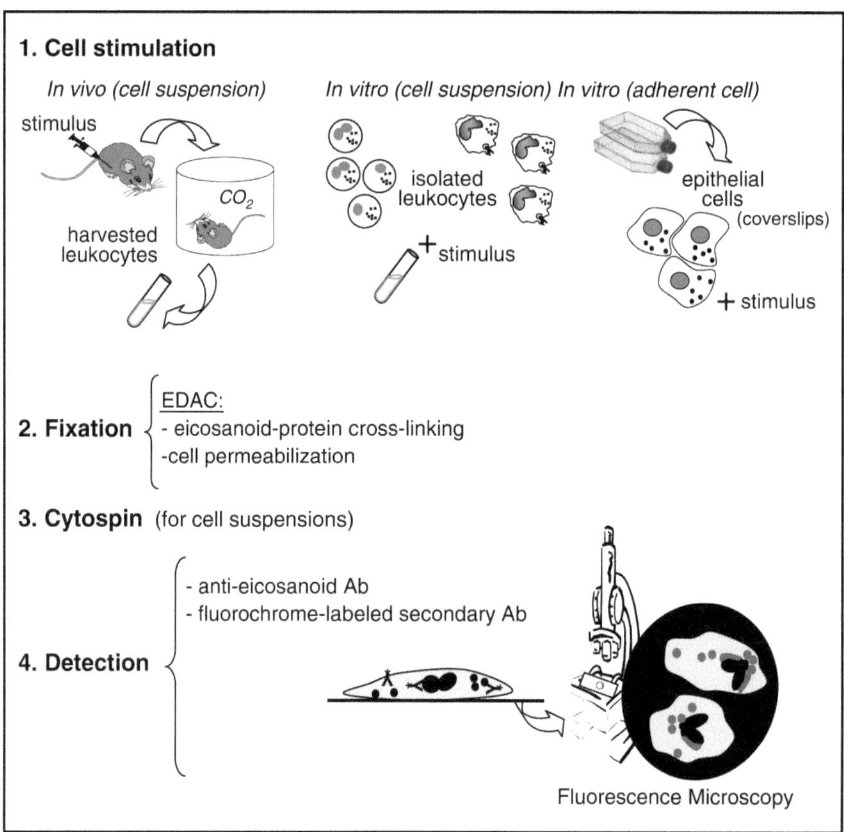

Fig. 2. Schematic illustration of EicosaCell methodology. EicosaCell preparations, which undergo EDAC-dependent capturing and fixation of newly formed eicosanoids at their sites of synthesis, are analyzed by phase-contrast and fluorescent microscopy and can employ cytospun or adherent cells.

(2) the incubation (1 h before EDAC) with eicosanoid synthesis inhibitors, such as cPLA$_2$-α inhibitor (e.g., pyrrolidine-2; 1 μM), COX inhibitor (e.g., indomethacin; 1 μg/ml), FLAP inhibitor (e.g., MK886; 50 μg/animal or 10 μM for in vitro incubations), or 5-LO inhibitor (e.g., zileuton; 50 μg/animal or 10 μM for in vitro incubations) to inhibit eicosanoid synthesis and guarantee the specificity of the observed reaction; and (3) the use of an irrelevant Ab control. Optionally, other suitable controls to check specificity and performance of EicosaCell are: (1) to use, instead of EDAC, paraformadehyde which will not immobilize the newly synthesized eicosanoid within cells; (2) in parallel, to carry out the EicosaCell in a different cell type that lacks the ability to synthesize the targeted eicosanoid (for instance, to use neutrophils to check specificity of LTC$_4$ immunodetection by EicosaCell); or (3) to analyze mixed populations of responsive plus unresponsive cells to a specific stimulus, so you can reassure that the targeted eicosanoid is specifically detected only within stimulated cells.

As shown in Fig. 3, EicosaCell system was successfully employed on macrophages recovered from peritoneal cavities of

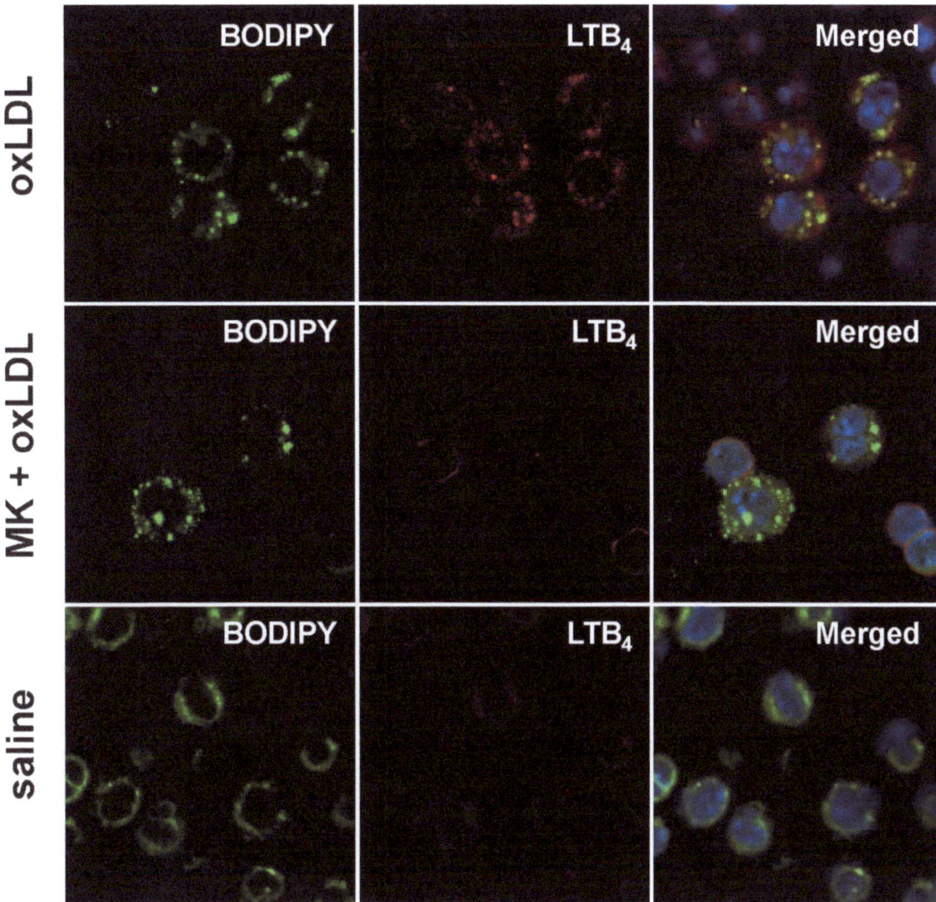

Fig. 3. Leukotriene B_4 detection in leukocytes. Leukocytes were obtained from in vivo oxLDL- or saline-stimulated mice and EicosaCell for LTB_4 was performed as described in (17). Cells were fixed and permeabilized by carbodiimide and stained with bodipy 493/503 (*green*) for lipid body visualization (*left panels*). *Upper panels* show peritoneal macrophages isolated from oxLDL-stimulated animals that were incubated with anti-LTB_4 Ab (*red*) for detection of newly formed LT. *Middle panels*, oxLDL-stimulated animals were treated with MK886 (50 μg/animal) 2 h prior to sacrifice to inhibit FLAP-dependent LT synthesis and labeling. *Lower panel*, show peritoneal leukocytes from nonstimulated animals and incubated with anti-LTB_4 Ab (*red*). TO-PRO-3 nuclei staining are shown in *blue*. Slides were analyzed by laser scanning confocal microscopy.

oxidized LDL-stimulated or control mice (17). Briefly, cells obtained 4 h after stimulation with oxidized LDL and controls were recovered from the pleural cavity with 500 μl of $HBSS^{-/-}$, and immediately mixed with 500 μl of EDAC (1% in $HBSS^{-/-}$). After 30 min incubation at 37°C with EDAC, leukocytes were then washed with $HBSS^{-/-}$, cytospun onto glass slides, and incubated with mouse anti-LTB_4 (Cayman Chemical) in 0.1% normal goat serum. BODIPY 493/503 and TO-PRO-3 were added together with the secondary Ab for 1 h to distinguish cytoplasmic lipid bodies and nuclei, respectively (see Subheading 2.2). Cells were washed twice and incubated with Cy2-labeled anti-rabbit IgG secondary Ab for 1 h. The slides were washed (three times, 10 min each) and mounted with aqueous mounting medium.

Cells were analyzed by confocal laser microscopy. As a control for LTB_4 specificity of detection, one group of oxLDL-stimulated animals was treated with MK886 (50 μg/animal) 2 h before sacrificing the animals for cell recovery, and the treatment with MK886 was maintained throughout the subsequent incubation steps at a concentration of 10 μM for MK886.

The EicosaCell technique is particularly suitable to distinguish intracellular sites of eicosanoid synthesis. As mentioned before, EicosaCell studies have provided direct evidence of compartmentalized eicosanoid synthesis within three distinct sites, including perinuclear envelope (8–11), phagosomes (12), or lipid bodies (10, 11, 13–15). As exemplified in Fig. 4, while, eosinophils stimulated with eotaxin for 1 h showed punctate cytoplasmic immunoreactive LTC_4 colocalized at lipid bodies, LTC_4 biosynthesis after cross-linking of eosinophil membrane-expressed CD9 was found exclusively as a perinuclear immunolabelling (8).

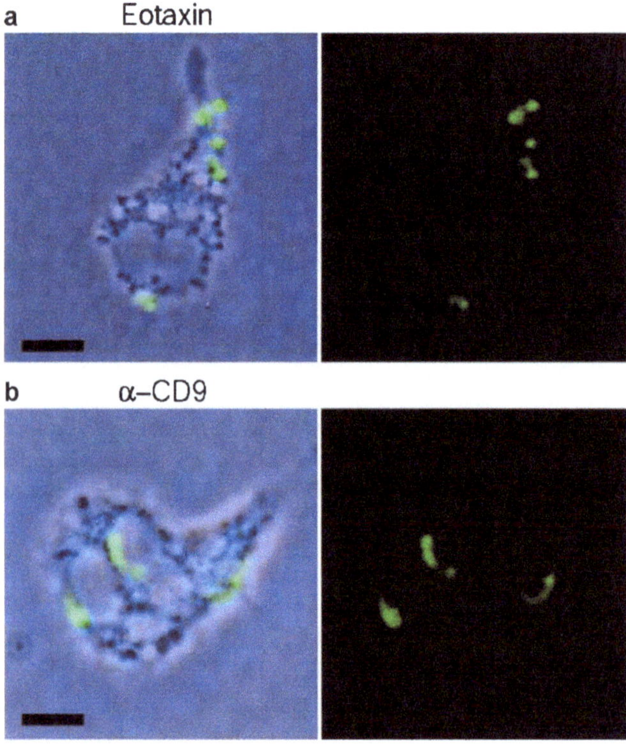

Fig. 4. Intracellular localization of LTC_4 biosynthesis within human eosinophils. Eosinophils were stimulated with 12 nM eotaxin (**a**) or 2.5 μg/ml mAb to CD9 (**b**) for 1 h and EicosaCell for LTC_4 was performed as described in (8). Eosinophils were fixed and permeabilized by carbodiimide and stained with Alexa 488-labeled anti-cysteinyl LT mAb. To facilitate intracellular localization, anti-LTC_4 immunoreactive sites (*green staining*) were overlaid on phase-contrast images. Figure reproduced with permission from (8) (*Bar* 5 μm).

3.3. EicosaCell with Adherent Cells

To study the intracellular compartmentalization of eicosanoid synthesis by EicosaCell in adherent cells, extra care should be taken to ensure the conservation of cell adherence and morphology during the EDAC step. As published, EicosaCell assays have succeeded to immunolocalize PGE_2 within at least three distinct adherent cell types: plated murine macrophages (D'Avila et al., unpublished) and two lineages of intestinal cells (as illustrated in Fig. 1), CACO-2 (a human colon adenocarcinoma cell line) (9), and IEC-6 (a rat epithelial cell line) (11).

1. While adherent CACO-2 cells (on glass coverslips), for instance, should be incubated during 1 h at 37°C with EDAC at 0.5% in $HBSS^{-/-}$ to cross-link PGE_2 carboxyl groups to amines in adjacent proteins without affecting cell morphology, IEC-6 cells must be incubated for at most 30 min (at same concentration; 0.5% in $HBSS^{-/-}$) to retain reasonable cell morphology and PGE_2 immunodetection at synthesizing compartments (refer to the original articles (9, 11) for details of blocking and staining conditions with anti-PGE_2 mAb (Cayman Chemicals) and proper secondary Abs).
2. At the end of the staining procedure, coverslip-adherent cells should be always extensively washed with HBSS and then mounted in an aqueous mounting medium to be visualized with 100× objective by both phase-contrast and fluorescence microscopy. As for cytospun cells, the specificity of the eicosanoid immunolabeling using EicosaCell system should be ascertained by including mandatory controls as listed in Subheading 3.2, step 6.

3.4. Double-Labeling Procedures to Identify Eicosanoid-Synthesizing Intracellular Sites

To better visualize perinuclear eicosanoid synthesis by EicosaCell, a double labeling with DAPI or TO-PRO-3 is recommended.

3.4.1. Nuclear Localization

1. After EDAC and Ab incubation steps, EicosaCell slides preparations should be extensively washed in $HBSS^{-/-}$ and then incubated with DAPI (DAPI working solution 100 ng/ml or 300 nM, see Note 4) for 5 min before aqueous mounting medium application.
2. The morphology of the cells' nuclei is observed using a fluorescence microscope at excitation wavelength 350 nm.

3.4.2. Phagosomal Localization

As performed by Balestrieri et al. (12), phagosome involvement in eicosanoid synthesis can be ascertained by colocalizing the phagosomal protein marker LAMP-1 in EicosaCell preparations.

1. After incubation at 37°C with EDAC, cells should be washed with $HBSS^{-/-}$, cytospun onto glass slides, and incubated with mouse anti-eicosanoid Ab and rat mAb against LAMP-1 (2.5 μg/ml) in blocking buffer (5% normal donkey serum) for 2 h at room temperature.

2. Negative control cells are instead incubated for 2 h with rat IgG (Jackson Immunoresearch).
3. After 2 h, the cells are washed extensively with $HBSS^{-/-}$ and incubated for 1 h at room temperature with fluorescent-labeled secondary Ab to detect the murine anti-eicosanoid Ab and Cy3-conjugated donkey anti-rat IgG (1:200).
4. The cells should be washed five times with $HBSS^{-/-}$ and then mounted in aqueous mounting medium.

3.4.3. Lipid Body Localization

To investigate the roles of lipid bodies in eicosanoid synthesis by EicosaCell assay, two double-labeling strategies can be employed: (1) BODIPY 493/503 lipid probe is an effective dye for staining neutral lipids and, for this reason it is very efficient for lipid body staining, or (2) ADRP immunostaining. ADRP is a structural protein of lipid bodies considered essential for lipid storage and metabolism that is ubiquitously expressed in lipid bodies containing cells (for further information on lipid body labeling refer to (21). Both approaches can be used for adherent, suspension, or agarose-embedded cells.

1. To employ BODIPY 493/503 for lipid body labeling, incubate EicosaCell preparations (coverslips or slides) with 1 μm BODIPY 493/503 (working solution) for 45–60 min at 37°C (water bath) simultaneously with secondary Ab incubation.
2. To remove nonincorporated BODIPY 493/503 after incubation, EicosaCell preparations should be washed at least twice in $HBSS^{-/-}$ before aqueous mounting medium application and coverslip attachment to slides.
3. Alternatively with ADRP to visualize lipid bodies, ADRP immunolabeling may be performed. In this case, slides are incubated for 1 h with guinea pig anti-ADRP Ab together with the primary anti-eicosanoid Ab to localize lipid bodies within leukocytes. The cells are washed with HBSS for 10 min (3×) and incubated with Cy3-labeled anti-guinea pig secondary antibodies for 1 h.

4. Notes

1. EDAC working solution should be prepared fresh, keep protected from light, and discarded after each experiment.
2. Incubation of cells with EDAC can be carried out on either cell in suspension or with the cells already cystopun onto slides by dropping EDAC on top of cells. Even though the latter method is less costly, some differences in preservation of cell morphology, cell permeabilization, and eicosanoid detection may occur and should be analyzed with care.

3. Fluorescent microscopic analyses of EicosaCell preparations should be performed as soon as the slides are mounted, inasmuch as immunofluorescent labeling is usually not stable for a long period and may exhibit bleaching over time. Even though freezing may preserve fluorescence overnight, EDAC-treated cells may display altered cell appearance after freezing-thawing cycle.
4. Prepare DAPI stock solution by dissolving 1 mg/ml of powder in distilled water. Aliquots should be stored at –20°C protected from light.

5. Troubleshooting

Some common problems and nonobvious features found in immunofluorescent-detection of eicosanoids in EDAC preparations by EicosaCell are shown in Table 1 with their possible explanations and potential solutions.

Table 1
Troubleshooting eicosanoid detection

Problems	Common causes	Solutions
Lack of eicosanoid detection (when few or no eicosanoid specific immunostaining is observed, but expected)	*Improper fixation* (when the eicosanoid is not cross-linked by EDAC, but washed-out from cells turning detection impossible)	*Slight increase in EDAC* (by adjusting both EDAC concentration and/or incubation time with cells)
	Inefficient cell stimulation	*Positive control* (mandatory inclusion of a well known agonist)
Eicosanoid detection within nonstimulated cells (eicosanoids are nonstorable in the cell and newly formed upon stimulation, therefore nonstimulated cells should not show any immunostaining for the targeted eicosanoid)	*Improper stimulation* (cell activation during the procedures including cell incubation at 37°C or cell fixation/permeabilization with EDAC can lead to spontaneous, stimulus-independent eicosanoid synthesis)	*Negative control* (nonstimulated cells should always be included as an important negative control. Care is needed to ensure that cells are not mechanically, chemically, or immunologically stimulated)
	Nonspecific detection	(Discussed below)
Poor preservation of cell morphology	*EDAC cell toxicity* (during EDAC incubation step, cell appearance may change from unimportant to severe modification of typical cell morphology)	*Slight decrease in EDAC* (by adjusting both EDAC concentration and incubation time with cells)

(continued)

Table 1 (continued)

Problems	Common causes	Solutions
Non-specific detection (antibodies may non-specifically bind to other cellular structures or to other lipids found within cells)	*Nonspecific binding to cell* (since cross-linking properties of EDAC may favor the tendency for cells to be sticky) *Nonspecific binding to other eicosanoids*	*Irrelevant antibody control* (a proper control using host/isotype-matched antibodies in stimulated cells should always be included) *Eicosanoid synthesis inhibitor control* (stimulated cells treated with a synthesis inhibitor of the targeted eicosanoid is a mandatory control condition that should show no immune-labeling confirming specific detection of targeted eicosanoid)
High non-specific binding (when non-specific staining is too high with >10 % positive cells)	*Improper detecting antibody conditions*	*Change detecting antibody dilution and/or host* *Incubation of detecting antibody with an adsorbing reagent* *Incubation of EDAC-treated cells with a normal serum* (to effectively block out nonspecific sites) *Centrifugation of the detecting antibody* (to eliminate aggregates)
Losing cell adherence with EDAC (the ability of cells to stay adhered to coverslips or other substrates can be affected by EDAC incubation)	*EDAC cell toxicity*	*Slight decrease in EDAC* (previous careful setting of EDAC incubation step is obligatory and should be adjusted for each cell type)

Acknowledgments

The work of the authors is supported by PRONEX-MCT, Conselho Nacional de Desenvolvimento Cientifico e Tecnológico (CNPq, Brazil), PAPES-FIOCRUZ, Fundação de Amparo à Pesquisa do Rio de Janeiro (FAPERJ, Brazil), and NIH grants (AI022571, AI020241, AI051645). The authors are indebted to Dr. Adriana Vieira de Abreu for the contributions to the figures used in the manuscript.

References

1. Yaqoob, P. (2003) Fatty acids as gatekeepers of immune cell regulation. *Trends Immunol.* **24**, 639–645.
2. Wymann, M.P., and Schneiter, R. (2008) Lipid signalling in disease. *Nat. Rev. Mol. Cell Biol.* **9**, 162–176.
3. Peters-Golden, M., and Brock, T.G. (2001) Intracellular compartmentalization of leukotriene synthesis: unexpected nuclear secrets. *FEBS Lett.* **487**, 323–326.
4. Mandal, A.K., Skoch, J., Bacskai, B.J., Hyman, B.T., Christmas, P., Miller, D., et al. (2004) The membrane organization of leukotriene synthesis. *Proc. Natl. Acad. Sci. U S A* **101**, 6587–6592.
5. Bandeira-Melo, C., Bozza, P.T., and Weller, P.F. (2002) The cellular biology of eosinophil eicosanoid formation and function. *J. Allergy Clin. Immunol.* **109**, 393–400.
6. Bozza, P.T., Magalhães, K.G., and Weller, P.F. (2009) Leukocyte lipid bodies – biogenesis and functions in inflammation. *Biochim. Biophys. Acta* doi:10.1016/j.bbalip.2009.01.005.
7. Liu, L.X., Buhlmann, J.E., and Weller, P.F. (1992) Release of prostaglandin E2 by microfilariae of *Wuchereria bancrofti* and *Brugia malayi*. *Am. J. Trop. Med. Hyg.* **46**, 520–523.
8. Tedla, N., Bandeira-Melo, C., Tassinari, P., Sloane, D.E., Samplaski, M., Cosman, D. et al. (2003) Activation of human eosinophils through leukocyte immunoglobulin-like receptor 7. *Proc. Natl. Acad. Sci. U S A* **100**, 1174–1179.
9. Accioly, M.T., Pacheco, P., Maya-Monteiro, C.M., Carrossini, N., Robbsm, B.K., Oliveira, S.S. et al. (2008) Lipid bodies are reservoirs of cyclooxygenase-2 and sites of prostaglandin-E2 synthesis in colon cancer cells. *Cancer Res.* **68**, 1732–40.
10. Bandeira-Melo, C., Phoofolo, M., and Weller, P.F. (2001) Extranuclear lipid bodies, elicited by CCR3-mediated signaling pathways, are the sites of chemokine-enhanced leukotriene C4 production in eosinophils and basophils. *J. Biol. Chem.* **276**, 22779–22787.
11. Moreira, L.S., Piva, B., Gentile, L.B., Mesquita-Santos, F.P., D'Avila, H., Maya-Monteiro, C.M. et al. (2009) Cytosolic phospholipase A2-driven PGE2 synthesis within unsaturated fatty acids-induced lipid bodies of epithelial cells. *Biochim. Biophys. Acta* **1791**, 156–165.
12. Balestrieri, B., Hsu, V.W., Gilbert, H., Leslie, C.C., Han, W.K., Bonventre, J.V. et al. (2006) Group V secretory phospholipase A2 translocates to the phagosome after zymosan stimulation of mouse peritoneal macrophages and regulates phagocytosis. *J. Biol. Chem.* **281**, 6691–6698.
13. Mesquita-Santos, F.P., Vieira-de-Abreu, A., Calheiros, A.S., Figueiredo, I.H., Castro-Faria-Neto, H.C., Weller, P.F. et al. (2006) Cutting edge: prostaglandin D2 enhances leukotriene C4 synthesis by eosinophils during allergic inflammation: synergistic in vivo role of endogenous eotaxin. *J. Immunol.* **176**, 1326–1330.
14. Pacheco, P., Vieira-de-Abreu, A., Gomes, R.N., Barbosa-Lima, G., Wermelinger, L.B., Maya-Monteiro, C.M. et al. (2007) Monocyte chemoattractant protein-1/CC chemokine ligand 2 controls microtubule-driven biogenesis and leukotriene B4-synthesizing function of macrophage lipid bodies elicited by innate immune response. *J. Immunol.* **179**, 8500–8508.
15. D'Avila, H., Melo, R.C., Parreira, G.G., Werneck-Barroso, E., Castro-Faria-Neto, H.C., and Bozza, P.T. (2006) *Mycobacterium bovis* bacillus Calmette-Guerin induces TLR2-mediated formation of lipid bodies: intracellular domains for eicosanoid synthesis in vivo. *J. Immunol.* **176**, 3087–3097.
16. Vieira-de-Abreu, A., Assis, E.F., Gomes, G.S., Castro-Faria-Neto, H.C., Weller, P.F., Bandeira-Melo, C. et al. (2005) Allergic challenge-elicited lipid bodies compartmentalize in vivo leukotriene C4 synthesis within eosinophils. *Am. J. Respir. Cell. Mol. Biol.* **33**, 254–261.
17. Silva, A.R., Pacheco, P., Vieira-de-Abreu, A., Maya-Monteiro, C.M., D'Alegria, B., Magalhães, K.G. et al. (2009) Lipid bodies in oxidized LDL-induced foam cells are leukotriene-synthesizing organelles: a MCP-1/CCL2 regulated phenomenon. *Biochim. Biophys. Acta* **1791**, 1066–1075.
18. Devchand, P.R., Keller, H., Peters, J.M., Vazquez, M., Gonzalez, F.J., and Wahli, W. (1996) The PPARalpha-leukotriene B4 pathway to inflammation control. *Nature* **384**, 39–43.
19. Kliewer, S.A., Lenhard, J.M., Willson, T.M., Patel, I., Morris, D.C., and Lehmann, J,M. (1995) A prostaglandin J2 metabolite binds peroxisome proliferator-activated receptor gamma and promotes adipocyte differentiation. *Cell* **83**, 813–819.
20. Bandeira-Melo, C., Woods, L.J., Phoofolo, M., and Weller, P.F. (2002) Intracrine cysteinyl leukotriene receptor-mediated signaling of eosinophil vesicular transport-mediated interleukin-4 secretion. *J. Exp. Med.* **196**, 841–850.
21. Melo, R.C.N., Bozza, P.T., and Weller, P.F. (2009) Imaging lipid bodies within leukocytes with different light microscopy techniques. *Methods Mol. Biol.* **689**, 149–161.

Chapter 17

Absorption Control in Immunohistochemistry Using Phospho-Peptides Immobilized on Magnetic Beads

Jordan Schoephoerster, Jillian Frisch, Michael Grahek, Chun Wu, Yingwei He, Wei Wang, Jennifer Nguyen, David Schwartz, and Alexander E. Kalyuzhny

Abstract

Although phospho-specific primary antibodies used in immunohistochemistry (IHC) are expected to detect phosphorylated proteins, in some cases these antibodies may also cross-react with nonphosphorylated proteins. Therefore, it is of ultimate importance to employ a control to determine that the staining pattern is specific. One of the frequently used controls in IHC is a so-called absorption control: phospho-specific primary antibodies are first incubated with a phospho-peptide immunogen to block antibody-binding sites, and this mixture is subsequently applied to tissue sections. If the antibody blocked with cognate immunogen does not produce tissue staining, then the antibody is considered specific, but if staining is obtained, the antibody is considered nonspecific. Unfortunately, bound peptide can dissociate from the antibody allowing unblocked antibody to bind to tissue targets, producing unwanted staining. We have developed a simple absorption-control protocol allowing for the efficient neutralization of phospho-specific antibodies with phospho-peptides immobilized on magnetic beads. This technique allows for sequestration of antibody–peptide complex from the incubation solution, minimizing the risk of formation of unblocked antibodies capable of producing tissue staining.

Key words: Fluorescence immunohistochemistry, Absorption control, Phospho-specific antibodies, Phospho-peptides, Magnetic beads, Confocal microscopy, 3T3 cells, NorthernLights™ Guard anti-fade mounting media

1. Introduction

Specificity of primary antibodies is of critical importance for the successful performance of immunohistochemical experiments, and finding a reliable commercial source of antibodies represents a significant challenge (1). In addition to antibodies raised against unmodified tissue proteins, primary antibodies raised against

Alexander E. Kalyuzhny (ed.), *Signal Transduction Immunohistochemistry: Methods and Protocols*, Methods in Molecular Biology, vol. 717, DOI 10.1007/978-1-61779-024-9_17,

phosphorylated proteins (e.g., phospho-specific antibodies) are widely used as IHC tools in signal transduction research (2, 3). Using such antibodies allows for precise tissue localization of various proteins that have been phosphorylated posttranslationally (4–6). However, phospho-specific antibodies can also interact with relevant but nonphosphorylated epitopes as well as nonspecific tissue targets including irrelevant phosphorylated proteins. To validate that tissue staining is specific, a so-called absorption control is required, i.e., antibodies are mixed with corresponding peptide immunogen and then this mixture is added to tissue sections. It is expected that peptide will block antibody-binding sites preventing them from interacting with tissue proteins. If the putative peptide–antibody complexes are still producing tissue labeling, such antibodies are deemed nonspecific and their value becomes questionable. Furthermore, the peptide–antibody complex is reversible, and during the incubation with tissue sections the peptide may dissociate from the antibody. This dissociation results in an unblocked antibody capable of tissue labeling.

To overcome these problems, it would be advantageous to determine the specificity of the antibody by adsorbing it onto peptide-immobilized magnetic beads. The ability to validate the antibody in this manner has been limited by the availability of robust, straightforward, efficient, and cost-effective chemistries to immobilize peptides on beads. This has been overcome by using Solulink's peptide immobilization chemistry. The chemistry is schematically presented in Fig. 1. Here, the HyNic linker is incorporated on a peptide during its solid-phase synthesis and directly added to 4FB-NanoLink beads to effect peptide immobilization. Following washing, the peptide-immobilized magnetic beads are ready for use.

Using this methodology, we describe herein a simple protocol based on the preabsorption of phospho-specific antibodies with phospho-peptide-immobilized magnetic beads and their separation, using a simple magnetic stand.

2. Materials

2.1. Immobilization of Peptides to Magnetic Beads

1. NanoLink 4FB magnetic beads (0.8 nm; 4FB loading: 32.6 nmoL 4FB/mg beads; Solulink Biosciences; www.solulink.com).
2. BOC-HNA for incorporation of HyNic group on peptide during solid-phase peptide synthesis (Solulink Biosciences; www.solulink.com).
3. Conjugation buffer (100 mM Sodium Phosphate, 150 mM NaCl, pH 6.0, 0.01% Tween-20).

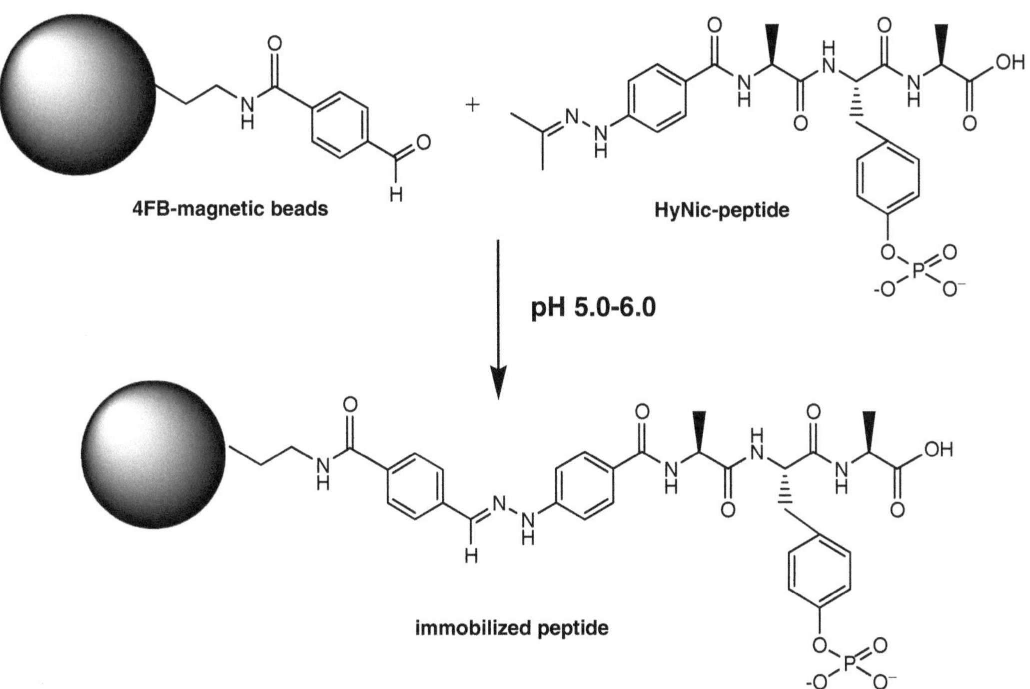

Fig. 1. Schematic representation of the chemical reaction used to immobilize HyNic-peptides on 4FB-magnetic beads. Simple addition of HyNic-peptide to 4FB-magnetic beads and incubation for 1 h followed by washing efficiently yields the immobilized peptide.

4. Phosphate buffered saline (PBS; pH 7.4).
5. Blocker™ Casein in TBS (Cat # 37532; ThermoPierce, Rockford, IL).
6. TurboLink Catalyst (Solulink Biosciences; www.solulink.com).
7. MagCellect magnet (R&D Systems, Inc.).
8. Microcentrifuge.
9. NanoDrop™ 2000 microvolume spectrophotometer (Thermo Scientific).
10. Rotator.
11. N-terminus HyNic-modified Akt phospho-peptide: HyNic-RPHFPQFpSYSASGTA.
12. N-terminus HyNic-modified ERK phospho-peptide: HyNic-HDHTGFLpTEpYVATRW.

2.2. Cell Culture and Stimulation

1. Chemical fume hood.
2. Class II sterile hood.
3. 37°C/CO_2 incubator.
4. Multiwell chamber slides (BD Falcon, VWR Cat# 62405-178).

5. 3T3 cells (ATCC).
6. Dulbecco's Modified Eagle's Medium (DMEM) (Gibco/BRL, Bethesda, MD) supplemented with 10% fetal bovine serum (FBS, HyClone, Ogden, UT) and 2 mM L-Glutamate.
7. Poly-L-lysine solution (0.01%, Sigma Cat# P4832).
8. Laminin (Sigma, Cat#L2020) diluted to 1.16 mg/cm^2 in sterile water, aliquoted for single use and stored at –20°C.
9. Recombinant human PDGF (Cat # 220-BB, R&D Systems).

2.3. Immunohistochemistry

1. Rocking plate for absorption control experiment.
2. Primary antibody: rabbit anti-phospho Akt1 (Cat # AF887, R&D Systems) affinity-purified antibody (see Note 1). Make a 15 μg/mL working solution using antibody diluent. Working solution can be stored 4°C for no longer than 1 week.
3. Secondary antibody: donkey anti-rabbit NL557 (Cat #NL004, R&D Systems) fluorescent antibody. Make 1:200 working solution using antibody diluent. Working solution can be stored at 4°C for no longer than 1 week.
4. Dako Pen: to draw a hydrophobic circular line around tissue sections to prevent leakage of primary and secondary antibodies applied to tissue sections.
5. PBS: Fill a 1 L beaker with 900 mL of deionized water and dissolve 0.23 g of NaH_2PO_4 (anhydrous), 1.15 g Na_2HPO_4 (anhydrous), and 9 g NaCl. Adjust pH to 7.4 using 1 M NaOH and/or 1 M HCl. Adjust volume to 1 L with deionized water.
6. Fixative: 4% formaldehyde in Sorenson's phosphate buffer. Wear mask and gloves and use chemical fume hood when preparing paraformaldehyde fixative. Start by making Sorenson's phosphate buffer by dissolving 8.06 g potassium phosphate and 19.99 g dibasic sodium phosphate in 900 mL deionized water. pH to 7.2, and fill with deionized water to 1 L. Then, make 8% formaldehyde solution by dissolving 10 g of paraformaldehyde powder (Sigma) in 95 mL of deionized water using heating stir plate. Heat this solution during stirring. Turn the heat off after temperature reaches 56–58°C and add 1–2 drops of 1 M NaOH to clear the solution. Continue stirring for another 20–30 min and then filter this solution using regular filter paper (for example Whatman #1). Add the Paraformaldehyde solution to 125 mL Sorenson's phosphate buffer, and fill with deionized water to 250 mL (see Note 2). Working fixing solution is made of equal volumes of 4% PFA and culture media (i.e., 400 μL PFA is added to 400 μL culture media).
7. Antibody diluent: PBS containing 1% bovine serum albumin, 1% normal donkey serum, 0.3% Triton X-100 (v/v), and 0.01% sodium azide. Store at –20°C.

8. Anti-fade mounting medium: NorthernLights™ Guard (Catalog # NL996; R&D Systems).
9. Coverslips for histological slides: 24×50 mm, thickness No.1.
10. Nuclear counterstain: 300 nM DAPI.
11. Microscopy: fluorescence microscope Provis equipped with cooled DP71 color digital camera (Olympus, Melville, NY) and fluorescence filter set to visualize NL557 fluorescent tag (557 nm excitation and 574 nm emission), and DAPI (345–360 nm excitation and 456–460 nm emission).

3. Methods

3.1. Conjugation of Peptides to Magnetic Beads

Unless otherwise stated, all procedures are performed at room temperature. If the protocol calls for incubation at room temperature, reagents stored at 4°C should be warmed to room temperature before use. Wear gloves to avoid contact of reagents with skin. It is recommended that each experiment is performed in duplicate to ensure consistency.

1. Bead Preparation: NanoLink 4FB Magnetic Beads are exchanged into Blocker™ Casein and incubated for 1 h. The blocked beads are exchanged into the Conjugation buffer by magnetizing the beads, removing the supernatant, and adding Conjugation buffer. This step is repeated three times.
2. HyNic-Peptide Preparation: To ~1 mg of peptide, add water to make a 10 mg/mL solution (see Note 3). HyNic-peptides are prepared by using BOC-HNA for incorporation of the HyNic group on the peptides during solid-phase peptide synthesis (see Note 4).
3. Peptide Immobilization (Fig. 1): To blocked, buffer-exchanged NanoLink 4FB magnetic beads, add peptide (326 nmol, 5× equiv./4FB equiv.) followed by the addition of conjugation buffer to have 450 μL of total volume and 50 μL TurboLink catalyst. The reaction tubes are placed on a rotator for 2 h and then exchanged into PBS employing the magnetic stand by washing 3 times. These are suspended in 1 mL PBS.

3.2. Culture and Stimulation of 3T3 Cell In Vitro

1. In a sterile hood, add 100 μL of poly-L-lysine solution into each well of a chamber slide, close the lid, and incubate for 30 min.
2. Discard poly-L-lysine solution and let it dry in the sterile hood for 1 h.
3. Slowly defrost frozen Laminin at 4°C and add 100 μL of the thawed Laminin solution into each well and incubate at 37°C/5%CO_2 for 1–2 h.

4. Dilute 3T3 cells with DMEM culture medium to make 1×10^6 cells/mL cell suspension.
5. Discard Laminin solution from the chamber slide and add 400 μL of 3T3 cells in DMEM culture medium.
6. Incubate cells at 37°C/5%CO_2 for 3 h.
7. Transfer chamber slides with cells from the 37°C/CO_2 humidified incubator into the sterile hood. Wait for approximately 20 min to allow the temperature of the culture medium in the chamber slide to decrease from 37°C to ambient room temperature.
8. Stimulate 3T3 cells by adding PDGF so that its final concentration in the cell suspension is 10 ng/mL, and incubate for 10 min.
9. Fix the cells by adding 400 μL 4% formaldehyde and incubating for 15 min (this brings the formaldehyde solution to 2%).
10. Gently remove the culture media with formaldehyde from each well by positioning the tip of the pipette into the corner of the well to avoid disturbing cells.
11. Wash 3×5 min with PBS.

3.3. Immunohistochemical Staining

1. Make two identical working solutions of primary anti-Akt1 phospho-specific antibodies: one will be used for an absorption-control experiment and the other will be used for IHC as usual.
2. Prepare primary antibody solution for absorption-control experiment as follows:
   (a) Mix primary anti-Akt1 antibody with peptide-conjugated magnetic beads and incubate the mixture on a rocking plate for 2 h (see Note 5).
   (b) Place vial with the above mixture into a MagCellect magnet for 15 min.
   (c) Collect supernatant into a plastic tube and label it "preabsorbed antibody."
3. Make a chamber slide layout map designating wells for (i) regular IHC, (ii) absorption control, (iii) no-primary antibody control, and (iv) no-primary/no-secondary antibody control.
4. Discard PBS from the chamber slide and add (i) unprocessed primary antibody (regular IHC group), (ii) preabsorbed primary antibody (absorption control group), (iii) antibody diluent (no-primary antibody control group), and (iv) antibody diluent (no-primary/no-secondary antibody control group) into designated wells (see Note 6).

5. Incubate overnight at 4°C.
6. Discard incubation solutions and wash cells 3×15 min in PBS.
7. Add fluorescent secondary antibodies to groups i–iii and only antibody diluent to group iv, and incubate for 30 min.
8. Repeat Step 6.
9. Using a Slide Separator included with chamber slide, pry up the chamber compartment detaching it from the slide.
10. To prevent fading of fluorescent probes, mount stained tissue sections under coverslips using anti-fade mounting media (see Note 7 on mixing mounting media with fluorescent nuclear counterstain DAPI). Wipe off excess mounting media and examine cells (Fig. 2) under the fluorescence microscope (see Note 8).

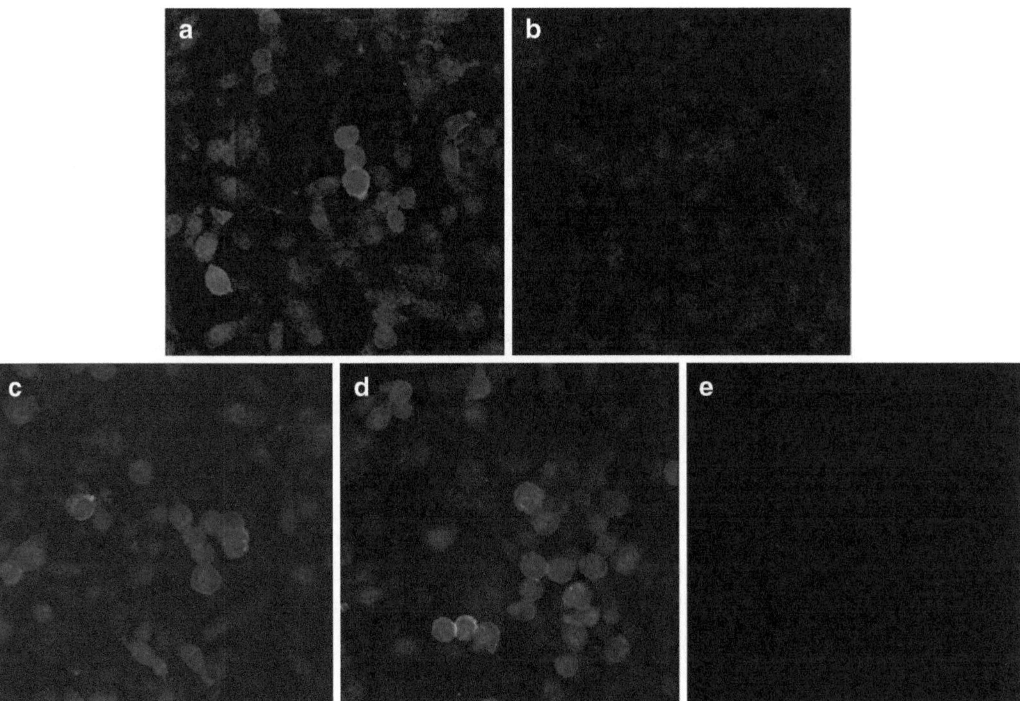

Fig. 2. Immunofluorescence detection of phosphorylated Akt1 in 3T3 cells stimulated with recombinant human PDGF protein. (**a**) Cells were incubated with untreated phospho-specific Akt1 antibodies; (**b**) cells were incubated with phospho-specific Akt1 antibodies absorbed by magnetic beads conjugated with Akt1 phospho-peptide (absorption control); (**c**) cells were incubated with phospho-specific Akt1 antibodies absorbed by magnetic beads conjugated with nonphosphorylated Akt1 peptide (phospho-peptide absorption control); (**d**) cells were incubated with phospho-specific Akt1 antibodies absorbed by magnetic beads conjugated with Erk2 phospho-peptide (irrelevant phospho-peptide absorption control); (**e**) cells were incubated with secondary antibodies only (primary antibody control).

4. Notes

1. Akt1 pS473 antibodies are raised against the peptide sequence RPHFPQFpSYSASGTA, which is found in Akt1 (this sequence is also very similar to the ones found in Akt1, Akt2, and Akt3, and thus anti-Akt1 antibody is expected to cross-react with all three isoforms).
2. The solution may need to be carefully heated (use a stirring hot-plate in a fume hood) to dissolve paraformaldehyde and then cooled to room temperature before use. Avoid overheating formaldehyde solution by monitoring the temperature with a thermometer. If the solution is accidentally heated above 58°C, discard it and make a fresh one. Avoid using this fixative for more than 3 weeks. Instead of adding formaldehyde directly to cell-culture media, culture media can be removed and the cells rinsed. To make a 2% formaldehyde solution, the 4% formaldehyde solution can be diluted 1:1 in PBS.
3. If the peptide is not completely soluble, an equal volume of DMF can be added to make 5 mg/mL peptide solution.
4. HyNic-peptides can be synthesized by any peptide company or core lab, using the Boc-HNA reagent (Solulink Biosciences; www.solulink.com). HyNic-peptides used in our study were prepared by Abgent Inc. (www.abgent.com).
5. An important factor to consider is the amount (i.e., nmol) of peptide-conjugated beads needed to neutralize primary antibodies. It is accepted in IHC that for an absorption control, the mole ratio of antigen/antibody in an absorption control mixture should be 10:1 for primary antibodies taken at their working dilution concentration (e.g., 15 μg/mL). Given that the peptide/beads conjugation ratio is close to 1:1, the amount of antigenic peptide needed to be mixed with antibodies can be calculated based on the amount of beads, which is known. In addition to using specific phospho-peptides, it is recommended to run an absorption control using an irrelevant phospho-peptide to make sure that the adsorption is specific and is not simply due to the presence of a phosphate group on amino acids composing the peptide. For example, in our study we also used magnetic beads conjugated to an irrelevant ERK2 phosphopeptide (HDHTGFLpTEpYVATRW) to determine if the latter can neutralize phospho-specific Akt1 antibodies (refer to Fig. 2). To save time, absorption control supernatants can be made while the cells are incubating in the chamber slide. One can increase the concentration

of the beads and/or decrease the concentration of the primary antibody, followed by removal and remagnetization of the supernatant for 15 min if boosting absorption efficiency is needed.

6. Do not allow cell and tissue samples to dry out during the incubation and washing steps. Cell or/and tissue samples that were found dry should be excluded from the experiment. Also, watch for partial drying of tissue section margins: this may result in strong and not necessarily specific cell and tissue labeling. Since partial drying may be overlooked (i.e., when staining a large number of slides) during the staining procedure, it is recommended that the researcher interprets labeling on the edges of the sample cautiously. It appears that labeling in the central part of the tissue will be more specific than that one on the tissue edges. When liquids are added to, or removed from, chamber slides with cells, great care should be taken to avoid disrupting the cells. Each well of an 8-well chamber slide holds 1 mL of liquid, so 500 μL of cells and 500 μL of formaldehyde fixative can be added, but this can result in overflow. Therefore, a mixture 400 μL cells and 400 μL of formaldehyde fixative might be appropriate. Less total volume can be used, as long as an equal volume of fixative is mixed with the cell-culture media. During the incubation with antibodies and washing steps, some cells will be lost from the chambers. To compensate for losses and for better final cell density, the researcher might increase the original cell concentration. PDGF needs to be added to each well individually in a very small volume: to minimize a pipetting error, make a 1:100 dilution of PDGF in culture media, and use it as a stock solution for further serial dilutions.
7. If needed, fluorescent nuclear counterstain DAPI can be mixed with anti-fade mounting media before applying it to cells and tissue sections.
8. When manipulating digital images, adjust brightness and contrast simultaneously on "control" and "experimental" samples to avoid bias and inaccurate interpretation of cell and tissue staining.

Acknowledgments

We thank Ernesto Resnik at R&D Systems, Inc. for his assistance with magnetic beads and MagCellect protocol.

References

1. Kalyuzhny, A.E. (2009) The dark side of the immunohistochemical moon: Industry. *J. Histochem. Cytochem.* **57**, 1099–1101.
2. Mandell, J.W. (2003) Phosphorylation state-specific antibodies: applications in investigative and diagnostic pathology. *Am. J. Pathol.* **163**, 1687–1698.
3. Mandell, J.W. (2008) Immunohistochemical assessment of protein phosphorylation state: the dream and the reality. *Histochem. Cell Biol.* **130**, 465–471.
4. Chen, H.F., Xie, L.D., and Xu, C.S. (2010) The signal transduction pathways of heat shock protein 27 phosphorylation in vascular smooth muscle cells. *Mol. Cell. Biochem.* **333**, 49–56.
5. Kaminska, B. (2009) Molecular characterization of inflammation-induced JNK/c-Jun signaling pathway in connection with tumorigenesis. *Methods Mol. Biol.* **512**, 249–264.
6. Baba, H.A., Stypmann. J., Grabellus, F., Kirchhof, P., Sokoll, A., Schäfers, M., et al. (2003) Dynamic regulation of MEK/Erks and Akt/GSK-3beta in human end-stage heart failure after left ventricular mechanical support: myocardial mechanotransduction-sensitivity as a possible molecular mechanism. *Cardiovasc. Res.* **59**, 390–399.

Index

Alexander E. Kalyuzhny (ed.), *Signal Transduction Immunohistochemistry: Methods and Protocols*, Methods in Molecular Biology, vol. 717, DOI 10.1007/978-1-61779-024-9,

MIX
Papier aus verantwortungsvollen Quellen
Paper from responsible sources
FSC® C105338

If you have any concerns about our products,
you can contact us on
ProductSafety@springernature.com

In case Publisher is established outside the EU,
the EU authorized representative is:
Springer Nature Customer Service Center GmbH
Europaplatz 3, 69115 Heidelberg, Germany

Printed by Libri Plureos GmbH
in Hamburg, Germany